HEINEMANN MODULAR MATHEMATICS *for* LONDON AS AND A-LEVEL

Pure Mathematics 2

Geoff Mannall Michael Kenwood

Heinemann Educational Publishers,
a division of Heinemann Publishers (Oxford) Ltd,
Halley Court, Jordan Hill, Oxford, OX2 8EJ

OXFORD LONDON EDINBURGH MADRID ATHENS BOLOGNA
PARIS MELBOURNE SYDNEY AUCKLAND SINGAPORE
TOKYO IBADAN NAIROBI HARARE GABORONE
PORTSMOUTH NH (USA)

First published 1995

95 96

10 9 8 7 6 5 4

ISBN 0 435 51808 9

Original design by Geoffrey Wadsley: additional design work by Jim Turner

Typeset and illustrated by TecSet Limited, Wallington, Surrey

Printed in Great Britain by the Bath Press, Avon

Acknowledgements:

The publisher's and authors' thanks are due to the University of London Examinations and Assessment Council (ULEAC) for permission to reproduce questions from past examination papers. These are marked with an [L].

The answers have been provided by the authors and are not the responsibility of the examining board.

About this book

This book is designed to provide you with the best preparation possible for your London Modular Mathematics P2 examination. The series authors are examiners and exam moderators themselves and have a good understanding of the exam board's requirements.

Finding your way around

To help to find your way around when you are studying and revising use the:

- **edge marks** (shown on the front page) – these help you to get to the right chapter quickly;
- **contents list** – this lists the headings that identify key syllabus ideas covered in the book so you can turn straight to them;
- **index** – if you need to find a topic the **bold** number shows where to find the main entry on a topic.

Remembering key ideas

We have provided clear explanations of the key ideas and techniques you need throughout the book. Key ideas you need to remember are listed in a **summary of key points** at the end of each chapter and marked like this in the chapters:

■ $$\log_a b = \frac{\log_c b}{\log_c a}$$

Exercises and exam questions

In this book questions are carefully graded so they increase in difficulty and gradually bring you up to exam standard.

- **past exam questions** are marked with an L;
- **review exercises** on pages 108, 250 and 379 help you practise answering questions from several areas of mathematics at once, as in the real exam;
- **exam style practice paper** – this is designed to help you prepare for the exam itself;
- **answers** are included at the end of the book – use them to check your work.

Contents

Algebra I

1

In Book P1 you were shown how to use some basic algebraic processing skills. In this chapter further processing skills will be explained and you will find out how to process algebraic fractions, how to find the remainder when one algebraic expression is divided by another, how to find a factor or factors of a polynomial by using the factor theorem and how to put an expression into partial fractions.

1.1 Identities

In chapter 1 of Book P1 the difference between an equation and an identity was explained. An equation is true for a limited number of values of the variable; for example, $x^2 = 9$ is true only for $x = +3$ and $x = -3$. But an identity is true for *all* values of the variable; for example, $x^2 + 2x + 1 \equiv (x + 1)^2$ is true for $x = 0$ because $0 + 0 + 1 = (0 + 1)^2$, for $x = 1$ because $1 + 2 + 1 = (1 + 1)^2$, for $x = 2$ because $4 + 4 + 1 = (2 + 1)^2$, and so on.

In order to proceed further with the study of algebra you must be able to deal confidently with identities. There are two common methods used in the processing of identities.

The first method is based on using the coefficients of the terms in x, x^2, x^3, etc., on the left and right-hand sides of the identity. This relies on the fact that the terms on the left and right-hand sides of an identity are the same. So that when like terms (the terms in x^0, terms in x^1, terms in x^2, etc.) have been collected together on the left-hand side and like terms have been collected together on the right-hand side of the identity, the coefficients of x^0 are equal, the coefficients of x^1 are equal, the coefficients of x^2 are equal, and so on.

That is, if

$$a_nx^n + a_{n-1}x^{n-1} + \ldots + a_2x^2 + a_1x + a_0 \equiv b_nx^n + b_{n-1}x^{n-1} + \ldots + b_2x^2 + b_1x + b_0$$

then: $$a_0 = b_0,\ a_1 = b_1,\ a_2 = b_2,\ \ldots,\ a_n = b_n$$

Example 1

Find the constants A and B such that $A(x+2)+B(x+1) \equiv x$.

The left-hand side can be written as

$$\begin{aligned} &Ax + 2A + Bx + B \\ &= Ax + Bx + 2A + B \\ &= (A+B)x + (2A+B) \end{aligned}$$

So: $$(A+B)x + (2A+B) \equiv x$$

Equate coefficents of x:

$$A + B = 1 \qquad (1)$$

Equate the constant terms:

$$2A + B = 0 \qquad (2)$$

There are now two equations in A and B which can be solved simultaneously as shown in chapter 2 of Book P1.

Subtract (1) from (2):

$$A = -1$$

Substitute the value of A in (1):

$$\begin{aligned} -1 + B &= 1 \\ B &= 2 \end{aligned}$$

Example 2

Find the values of A and B for which

$$A(4-x) + B(x-7) \equiv 30 - 6x$$

The left-hand side can be rewritten:

$$\begin{aligned} &4A - Ax + Bx - 7B \\ &= (-A+B)x + (4A-7B) \end{aligned}$$

So we have: $$(-A+B)x + (4A-7B) \equiv -6x + 30$$

Equate coefficients of x:

$$-A + B = -6 \qquad (1)$$

Equate the constant terms:

$$4A - 7B = 30 \qquad (2)$$

Multiply (1) by 4:

$$-4A + 4B = -24 \qquad (3)$$

Add (3) and (2):

$$-3B = 6$$
$$B = -2$$

Substitute the value of B into (1):

$$-A - 2 = -6$$
$$A = 4$$

Example 3

Find the values of A, B and C such that

$$A(x^2 + 4) + (x - 2)(Bx + C) \equiv 7x^2 - x + 14$$

Rewrite the left-hand side (LHS) as a polynomial in descending powers of x:

$$\text{LHS} = Ax^2 + 4A + Bx^2 - 2Bx + Cx - 2C$$
$$= (A + B)x^2 + (-2B + C)x + (4A - 2C)$$

So: $\quad (A + B)x^2 + (-2B + C)x + (4A - 2C) \equiv 7x^2 - x + 14$

Equate coefficients of x^2:

$$A + B = 7 \qquad (1)$$

Equate coefficients of x:

$$-2B + C = -1 \qquad (2)$$

Equate the constant terms:

$$4A - 2C = 14 \qquad (3)$$

You have three equations in three unknowns to be solved simultaneously. Substitute the expression for A given in (1) into equation (3) so that you have two equations in B and C which you can solve in the usual way.

From (1): $\quad A = 7 - B$

Substitute in (3):

$$4(7 - B) - 2C = 14$$
$$28 - 4B - 2C = 14$$
$$-4B - 2C = -14$$
$$4B + 2C = 14$$
$$2B + C = 7 \qquad (4)$$

Add (2) and (4): $\quad 2C = 6$

$$C = 3$$

Substitute in (2): $-2B + 3 = -1$

$$-2B = -4$$

$$B = 2$$

Substitute in (1): $A + 2 = 7$

$$A = 5$$

Example 4

Find the constants A, B and C such that

$$A(x+2)(x+3) + B(x+1)(x+3) + C(x+1)(x+2) \equiv 4x + 6$$

Rearrange the left-hand side as a polynomial in descending powers of x:

$$A(x^2 + 5x + 6) + B(x^2 + 4x + 3) + C(x^2 + 3x + 2)$$
$$= (A + B + C)x^2 + (5A + 4B + 3C)x + (6A + 3B + 2C)$$

So: $(A + B + C)x^2 + (5A + 4B + 3C)x + (6A + 3B + 2C) \equiv 4x + 6$

Equate coefficients of x^2:

$$A + B + C = 0 \qquad (1)$$

Equate coefficients of x:

$$5A + 4B + 3C = 4 \qquad (2)$$

Equate constant terms:

$$6A + 3B + 2C = 6 \qquad (3)$$

From (1): $A = -B - C$

Substitute into (2) and (3):

$$5(-B - C) + 4B + 3C = 4$$
$$-5B - 5C + 4B + 3C = 4$$
$$-B - 2C = 4 \qquad (4)$$

$$6(-B - C) + 3B + 2C = 6$$
$$-6B - 6C + 3B + 2C = 6$$
$$-3B - 4C = 6 \qquad (5)$$

Multiply (4) by 2: $-2B - 4C = 8 \qquad (6)$

Subtract (6) from (5):

$$-B = -2$$
$$B = 2$$

Substitute for B in (6):

$$-4 - 4C = 8$$
$$4C = -12$$
$$C = -3$$

Substitute for B and C in (1):

$$A = -2 + 3$$
$$A = 1$$

The second method used in the processing of identities is substituting particular values of the variable into the identity. This is allowed, since an identity is true for *all* values of the variable. You must choose values to substitute that will make some of A, B and C disappear from the identity. This will make it easier to find the others.

Example 5

Find the constants A and B such that $A(x+2) + B(x+1) \equiv x$.

Substitute $x = -1$ since this makes the bracket $(x+1)$ zero.

$$A(-1+2) + B(-1+1) = -1$$
$$A = -1$$

Substitute $x = -2$ and the bracket $(x+2)$ becomes zero:

$$A(-2+2) + B(-2+1) = -2$$

$$-B = -2$$
$$B = 2 \text{ as in example 1}$$

Example 6

Find the values of A and B for which

$$A(4-x) + B(x-7) \equiv 30 - 6x$$

Substitute $x = 4$:

$$A(4-4) + B(4-7) = 30 - 24$$
$$-3B = 6$$
$$B = -2$$

Substitute $x = 7$:

$$A(4-7) + B(7-7) = 30 - 42$$
$$-3A = -12$$
$$A = 4 \text{ as in example 2}$$

Example 7

Find the values of A, B and C such that

$$A(x+2)(x+3)+B(x+1)(x+3)+C(x+1)(x+2) \equiv 4x+6$$

Substitute $x=-1$:

$$A(-1+2)(-1+3)+B(-1+1)(-1+3)+C(-1+1)(-1+2)$$
$$= -4+6$$
$$A(1)(2)+B(0)(2)+C(0)(1) = 2$$
$$2A = 2$$
$$A = 1$$

Substitute $x=-2$:

$$A(-2+2)(-2+3)+B(-2+1)(-2+3)+C(-2+1)(-2+2)$$
$$= -8+6$$
$$A(0)(1)+B(-1)(1)+C(-1)(0) = -2$$
$$-B = -2$$
$$B = 2$$

Substitute $x=-3$:

$$A(-3+2)(-3+3)+B(-3+1)(-3+3)+C(-3+1)(-3+2)$$
$$= -12+6$$
$$A(-1)(0)+B(-2)(0)+C(-2)(-1) = -6$$
$$2C = -6$$
$$C = -3 \text{ as in example 4}$$

You will gather from the above examples that, in general, it is much easier to solve problems like this by substitution than it is by equating coefficients. However, sometimes the method of substitution will not work completely. For example, if one of the brackets in the identity is, say, (x^2+9) there is no real value of x that will make this bracket zero. For if $x^2+9=0$ then $x^2=-9$ and a negative number has no real square roots. Under these circumstances it is usually easier to use the method by substitution as far as you can and then finish off the problem by equating coefficients, that is, to use a combination of the two methods.

Example 8

Find the values of A, B and C such that

$$A(x^2+4)+(x-2)(Bx+C) \equiv 7x^2 - x + 14$$

Substitute $x = 2$:

$$A(4+4)+(2-2)(2B+C) = 28 - 2 + 14$$
$$8A = 40$$
$$A = 5$$

Since there is no real value of x that will make (x^2+4) equate to zero, revert to equating coefficients:

$$\begin{aligned} &A(x^2+4)+(x-2)(Bx+C) \\ &= Ax^2 + 4A + Bx^2 - 2Bx + Cx - 2C \\ &= (A+B)x^2 + (-2B+C)x + (4A-2C) \end{aligned}$$

But since $A = 5$ the LHS of the identity is:

$$(5+B)x^2 + (-2B+C)x + (20-2C)$$

So: $\quad (5+B)x^2 + (-2B+C)x + (20-2C) \equiv 7x^2 - x + 14$

Equate coefficients of x^2:

$$5 + B = 7$$
$$B = 2$$

Equate the constant terms:

$$20 - 2C = 14$$
$$C = 3$$

Another approach would be to take two other values of x and form equations that can be solved to find B and C.

For $x = 0$, $\quad 4A - 2C = 14$

So: $\quad 20 - 2C = 14$

and: $\quad C = 3$

For $x = 1$, $\quad 5A - B - C = 20$

So: $\quad 25 - B - 3 = 20$

and: $\quad B = 2$

You should appreciate that the two approaches are essentially equivalent.

Example 9

Find the values of A, B and C for which

$$A(2x^2+1)+(2x-3)(Bx+C)\equiv 11x$$

Substitute $x = 1\frac{1}{2}$:

$$A(\tfrac{18}{4}+1)+(3-3)(1\tfrac{1}{2}B+C)=16\tfrac{1}{2}$$
$$5\tfrac{1}{2}A=16\tfrac{1}{2}$$
$$A=3$$

Since there is no real number that will make $(2x^2+1)$ equate to zero, revert to the method of comparing coefficients.

$$A(2x^2+1)+(2x-3)(Bx+C)$$
$$=2Ax^2+A+2Bx^2-3Bx+2Cx-3C$$
$$=(2A+2B)x^2+(-3B+2C)x+(A-3C)$$

Since $A = 3$, this can be written:

$$(6+2B)x^2+(-3B+2C)x+(3-3C)$$

So: $$(6+2B)x^2+(-3B+2C)x+(3-3C)\equiv 11x$$

Equate coefficients of x^2:

$$6+2B=0$$
$$B=-3$$

Equate the constant terms:

$$3-3C=0$$
$$C=1$$

Exercise 1A

1 Each of the following is either

(a) true for all values of x, or

(b) true for some values of x, or

(c) true for no values of x.

Distinguish which is which and explain why.

(i) $(3x-1)^2=9x^2-1$

(ii) $(3x-1)^2=25$

(iii) $(3x-1)^2=9x^2-6x+1$

(iv) $(2x+3)^2=x^2-4$

(v) $x^2+(x+2)^2=(x+1)^2$

Find the values of the constants A, B, and C:

2 $A(x+3)+B(x+2) \equiv 4x+9$
3 $A(x+5)+B(x+2) \equiv x+8$
4 $A(x+2)+B(x-1) \equiv 6x+3$
5 $A(x+4)+B(x+2) \equiv x+12$
6 $A(x+1)+B(x-3) \equiv 8x+16$
7 $A(x+3)(x+4)+B(x+2)(x+4)+C(x+2)(x+3) \equiv 6x^2+34x+46$
8 $A(x+1)(x+3)+B(x-1)(x+1)+C(x-1)(x+3) \equiv 6x+2$
9 $A(x+2)(x+3)+B(x+1)(x+3)+C(x+1)(x+2) \equiv 4x+6$
10 $A(x+2)(x-1)+B(2x+1)(x-1)+C(2x+1)(x+2) \equiv 4x^2-17x-14$
11 $A(x-1)(x+4)+B(2x+1)(x+4)+C(2x+1)(x-1) \equiv 12x^2+59x-26$
12 $(Ax+B)(x+1)+C(x^2+3) \equiv x-3$
13 $(Ax+B)(x^2+3)+Cx^2(5x-2) \equiv -13x^3+18x^2+6x+36$
14 $A(x^2+x+3)+(Bx+C)(2x+1) \equiv x^2-x+2$
15 $A(x-2)^2+B(x+1)(x-2)+C(x+1) \equiv 3$
16 $A(x^2+4)+(x-2)(Bx+C) \equiv 4x^2-7x+22$
17 $A(x^2+x+2)+(Bx+C)x \equiv -x^2-2x+4$
18 $A(x^2+x+1)+(Bx+C)(2x-5) \equiv 7x^2-7x+3$
19 $A(x^2+1)+(Bx+C)(2x+3) \equiv 8x^2+4x+1$
20 $A(x^2+5)+(Bx+C)(x-3) \equiv 3x^2+5x+28$
21 $(Ax+B)(3x-4)+C(x^2+x+1) \equiv x^2-12x+6$
22 Find the values of A, B and C for which

$$4x^2-12x+25 \equiv A(x+B)^2+C$$

Hence find the minimum value of $4x^2-12x+25$.

23 Find the values of A, B and C for which

$$3x^2+18x-5 \equiv A(x+B)^2+C$$

Hence find the minimum value of $3x^2+18x-5$.

24 Find the values of A, B and C for which

$$16+4x-x^2 \equiv A-(B-Cx)^2$$

Hence find the maximum value of $16+4x-x^2$.

25 Find the values of A and B for which

$$9x^2+30x+A \equiv (3x+B)^2$$

1.2 Long division

Not many years ago questions such as 'divide 12 603 by 14' were commonplace in mathematics lessons, not only in secondary schools but also in primary schools. The method that was used was called long division. Students used to practise questions using long division for weeks on end!

Today most students have an electronic calculator that produces an answer to such questions in a fraction of a second and so the method of long division has become largely redundant. However, most calculators today are still unable to divide one algebraic expression by another and so the method of long division has to be used in these circumstances. It is a technique with which you therefore need to become familiar.

The method of long division is best explained by using an example.

Example 10
Divide $2x^4 - 9x^3 + 13x^2 - 17x + 15$ by $x - 3$.

You first need to ask, 'If I divide the first term of $2x^4 - 9x^3 + 13x^2 - 17x + 15$ by the first term of $x - 3$, what will be the answer?' In this case, $2x^4$ divided by x gives $2x^3$. (You learned how to manipulate indices in Book P1). Now multiply this answer of $2x^3$ by $x - 3$ to give $2x^4 - 6x^3$.

The solution thus far is written:

$$\begin{array}{r@{}l} & 2x^3 \\ x-3\,\big) & \overline{2x^4 - 9x^3 + 13x^2 - 17x + 15} \\ & 2x^4 - 6x^3 \end{array}$$

Notice that $2x^3$, the term you obtain when you divide $2x^4$ by x, is written immediately above $2x^4$. Notice also that the terms of $2x^4 - 6x^3$ are written immediately underneath the corresponding terms of $2x^4 - 9x^3 + 13x^2 - 17x + 15$, i.e. $2x^4$ is written underneath $2x^4$ and $-6x^3$ is written underneath $-9x^3$.

You now subtract $2x^4 - 6x^3$ from $2x^4 - 9x^3$.

So:

$$\begin{aligned} & 2x^4 - 9x^3 - (2x^4 - 6x^3) \\ &= 2x^4 - 9x^3 - 2x^4 + 6x^3 \\ &= -3x^3 \end{aligned}$$

The solution now looks like this:

$$
\begin{array}{r|l}
 & 2x^3 \\
\cline{2-2}
x - 3 & 2x^4 - 9x^3 + 13x^2 - 17x + 15 \\
 & \underline{2x^4 - 6x^3} \\
 & \quad\; - 3x^3
\end{array}
$$

At this stage copy the next term from $2x^4 - 9x^3 + 13x^2 - 17x + 15$ that does not have anything written directly underneath it, next to $-3x^3$. In this case the $13x^2$ needs to be copied:

$$
\begin{array}{r|l}
 & 2x^3 \\
\cline{2-2}
x - 3 & 2x^4 - 9x^3 + 13x^2 - 17x + 15 \\
 & \underline{2x^4 - 6x^3} \\
 & \quad\; - 3x^3 + 13x^2
\end{array}
$$

Now repeat the whole process from the beginning. So ask, 'If I divide the first term of $-3x^3 + 13x^2$ by the first term of $x - 3$, what will be the answer?' This time the answer is $-3x^2$, which you write next to $2x^3$. You then multiply $-3x^2$ by $x - 3$ to give $-3x^3 + 9x^2$.

Write these two terms underneath the corresponding terms of $-3x^3 + 13x^2$ and subtract:

$$
\begin{aligned}
& -3x^3 + 13x^2 - (-3x^3 + 9x^2) \\
&= -3x^3 + 13x^2 + 3x^3 - 9x^2 \\
&= 4x^2
\end{aligned}
$$

The solution now looks like this:

$$
\begin{array}{r|l}
 & 2x^3 - 3x^2 \\
\cline{2-2}
x - 3 & 2x^4 - 9x^3 + 13x^2 - 17x + 15 \\
 & \underline{2x^4 - 6x^3} \\
 & \quad\; -3x^3 + 13x^2 \\
 & \quad\; \underline{-3x^3 + 9x^2} \\
 & \qquad\qquad\; 4x^2
\end{array}
$$

Copy the next term from $2x^4 - 9x^3 + 13x^2 - 17x + 15$ that has nothing written directly underneath it next to $4x^2$, and repeat the process.

$$\begin{array}{r|l}
 & 2x^3 - 3x^2 + 4x \\
\hline
x - 3 & 2x^4 - 9x^3 + 13x^2 - 17x + 15 \\
 & \underline{2x^4 - 6x^3} \\
 & \quad -3x^3 + 13x^2 \\
 & \quad \underline{-3x^3 + 9x^2} \\
 & \qquad\quad 4x^2 - 17x \\
 & \qquad\quad \underline{4x^2 - 12x} \\
 & \qquad\qquad\quad -5x
\end{array}$$

Repeat the process once more:

$$\begin{array}{r|l}
 & 2x^3 - 3x^2 + 4x - 5 \\
\hline
x - 3 & 2x^4 - 9x^3 + 13x^2 - 17x + 15 \\
 & \underline{2x^4 - 6x^3} \\
 & \quad -3x^3 + 13x^2 \\
 & \quad \underline{-3x^3 + 9x^2} \\
 & \qquad\quad 4x^2 - 17x \\
 & \qquad\quad \underline{4x^2 - 12x} \\
 & \qquad\qquad\quad -5x + 15 \\
 & \qquad\qquad\quad \underline{-5x + 15}
\end{array}$$

There are no more terms to copy so the solution is finished. Thus $2x^4 - 9x^3 + 13x^2 - 17x + 15$ divided by $x - 3$ gives a result of $2x^3 - 3x^2 + 4x - 5$. This can easily be checked by multiplication.

Example 11

Divide $6x^3 + x^2 + 13x + 7$ by $2x + 1$.

$$\begin{array}{l r|l}
 & & 3x^2 - x + 7 \\
\cline{3-3}
2x \text{ multiplied by } 3x^2 = 6x^3: & 2x + 1 & 6x^3 + x^2 + 13x + 7 \\
\text{Multiply } 2x + 1 \text{ by } 3x^2: & & \underline{6x^3 + 3x^2} \\
\text{Subtract and copy } 13x: & & \quad -2x^2 + 13x \\
\text{Multiply } 2x + 1 \text{ by } -x: & & \quad \underline{-2x^2 - x} \\
\text{Subtract and copy } 7: & & \qquad\qquad 14x + 7 \\
\text{Multiply } 2x + 1 \text{ by } 7: & & \qquad\qquad \underline{14x + 7}
\end{array}$$

The answer is $3x^2 - x + 7$.

Sometimes the polynomial that you are given to divide has one or more terms in x missing. For example, $3x^5 + 2x^3 + 7x - 6$ has the terms in x^4 and x^2 missing. In questions involving long division these must be added and a zero placed in front of them. So the polynomial above would be written $3x^5 + 0x^4 + 2x^3 + 0x^2 + 7x - 6$.

Example 12

Divide $-4x^4 - 5x^2 + 5x + 4$ by $2x + 1$.

Write $-4x^4 - 5x^2 + 5x + 4$ as

$$-4x^4 + 0x^3 - 5x^2 + 5x + 4$$

$$\begin{array}{rl} & \quad -2x^3 + x^2 - 3x + 4 \\ \text{$2x$ multiplied by $-2x^3 = -4x^4$:} \qquad 2x+1 & \overline{)\,-4x^4 + 0x^3 - 5x^2 + 5x + 4} \\ \text{Multiply $2x+1$ by $-2x^3$:} & \underline{-4x^4 - 2x^3} \\ \text{Subtract and copy $-5x^2$:} & \qquad\quad 2x^3 - 5x^2 \\ \text{Multiply $2x+1$ by x^2:} & \qquad\quad \underline{2x^3 + x^2} \\ \text{Subtract and copy $5x$:} & \qquad\qquad\quad -6x^2 + 5x \\ \text{Multiply $2x+1$ by $-3x$:} & \qquad\qquad\quad \underline{-6x^2 - 3x} \\ \text{Subtract and copy 4:} & \qquad\qquad\qquad\quad 8x + 4 \\ \text{Multiply $2x+1$ by 4:} & \qquad\qquad\qquad\quad \underline{8x + 4} \end{array}$$

The answer is $-2x^3 + x^2 - 3x + 4$.

Example 13

Divide $-3x^4 + 8x^3 - 10x^2 + 18x - 14$ by $-x + 2$.

$$\begin{array}{rl} & \quad 3x^3 - 2x^2 + 6x - 6 \\ \text{$-x$ multiplied by $3x^3 = -3x^4$:} \qquad -x+2 & \overline{)\,-3x^4 + 8x^3 - 10x^2 + 18x - 14} \\ \text{Multiply $-x+2$ by $3x^3$:} & \underline{-3x^4 + 6x^3} \\ \text{Subtract and copy $-10x^2$:} & \qquad\quad 2x^3 - 10x^2 \\ \text{Multiply $-x+2$ by $-2x^2$:} & \qquad\quad \underline{2x^3 - 4x^2} \\ \text{Subtract and copy $18x$:} & \qquad\qquad\quad -6x^2 + 18x \\ \text{Multiply $-x+2$ by $6x$:} & \qquad\qquad\quad \underline{-6x^2 + 12x} \\ \text{Subtract and copy -14:} & \qquad\qquad\qquad\qquad\quad 6x - 14 \\ \text{Multiply $-x+2$ by -6:} & \qquad\qquad\qquad\qquad\quad \underline{6x - 12} \\ \text{Subtract:} & \qquad\qquad\qquad\qquad\qquad\quad -2 \end{array}$$

Since there is no other term to copy next to -2 you cannot go any further. So $-x+2$ divided into $-3x^4+8x^3-10x^2+18x-14$ gives an answer of $3x^3-2x^2+6x-6$ (called the **quotient**) and leaves a **remainder** of -2. (This is a similar situation to dividing 39 by 5. The number 5 divided into 39 gives a quotient of 7 and leaves a remainder of 4. That is, $39 = (7 \times 5) + 4$.) This result is written as

$$-3x^4+8x^3-10x^2+18x-14 = (3x^3-2x^2+6x-6)(-x+2)-2$$

Example 14

Divide $3x^5-8x^4+8x^3-11x^2+15x-7$ by x^2-2x+1.

	$3x^3-2x^2+x-7$
x^2 multiplied by $3x^3 = 3x^5$:	$x^2-2x+1\ \overline{)3x^5-8x^4+8x^3-11x^2+15x-7}$
Multiply x^2-2x+1 by $3x^3$:	$\underline{3x^5-6x^4+3x^3}$
Subtract and copy $-11x^2$:	$-2x^4+5x^3-11x^2$
Multiply x^2-2x+1 by $-2x^2$:	$\underline{-2x^4+4x^3-2x^2}$
Subtract and copy $15x$:	x^3-9x^2+15x
Multiply x^2-2x+1 by x:	$\underline{x^3-2x^2+x}$
Subtract and copy -7:	$-7x^2+14x-7$
Multiply x^2-2x+1 by -7:	$\underline{-7x^2+14x-7}$

Example 15

Divide $-3x^4+28x^3-43x^2+16x+21$ by $-x^2+7x+4$.

	$3x^2-7x+6$
$-x^2$ multiplied by $3x^2 = -3x^4$:	$-x^2+7x+4\ \overline{)-3x^4+28x^3-43x^2+16x+21}$
Multiply $-x^2+7x+4$ by $3x^2$:	$\underline{-3x^4+21x^3+12x^2}$
Subtract and copy $16x$:	$7x^3-55x^2+16x$
Multiply $-x^2+7x+4$ by $-7x$:	$\underline{7x^3-49x^2-28x}$
Subtract and copy 21:	$-6x^2+44x+21$
Multiply $-x^2+7x+4$ by 6:	$\underline{-6x^2+42x+24}$
Subtract:	$2x-3$

So the answer when $-3x^4+28x^3-43x^2+16x+21$ is divided by $-x^2+7x+4$ is $3x^2-7x+6$ with a remainder of $2x-3$. This can be expressed as

$$-3x^4+28x^3-43x^2+16x+21 \equiv (-x^2+7x+4)(3x^2-7x+6)+(2x-3)$$

where $3x^2-7x+6$ is the quotient and $2x-3$ is the remainder.

Exercise 1B

Divide:

1 $2x^3 - 3x^2 - 3x + 2$ by $x - 2$
2 $2x^3 + 3x^2 - 1$ by $2x - 1$
3 $2x^3 - 11x^2 + 12x - 35$ by $x - 5$
4 $-3x^4 + 11x^3 - 13x^2 + 26x - 15$ by $x - 3$
5 $-18x^3 + 33x^2 - 29x + 10$ by $-3x + 2$
6 $2x^4 + 12x^3 + 14x^2 - 8x$ by $x + 4$
7 $4x^3 + 4x^2 - x + 1$ by $2x + 1$
8 $15x^4 - x^3 + 7x^2 + 5$ by $3x + 1$
9 $3x^4 + 19x^3 - 25x^2 - 57x + 130$ by $-x - 7$
10 $-8x^4 + 24x^3 - 12x^2 + 17x - 26$ by $-2x + 5$
11 $3x^5 + 4x^4 + x^2 + 3x + 1$ by $x^2 + 2x + 1$
12 $-2x^4 + 13x^3 - 3x^2 + 37x + 35$ by $2x^2 - x + 7$
13 $-2x^4 + 3x^3 - 4x^2 - 24x + 7$ by $x^2 - 3x + 7$
14 $2x^5 + 6x^4 + 11x^3 - 11x^2 + 5x - 7$ by $2x^2 + 1$
15 $-6x^5 - 12x^4 + 25x^3 - 10x^2 - 14x + 12$ by $-3x^2 + 2$
16 $3x^4 - 4x^3 + 12x^2 - 6x + 11$ by $x^2 + 2$
17 $x^5 + x^4 + 10x^2 + 10x + 7$ by $x^2 + 2x$
18 $-6x^5 + 3x^4 - 4x^3 + 20x^2 - 3x + 10$ by $3x^2 + 2$
19 $4x^4 - 4x^3 - 4x^2 + 6x - 7$ by $2x^2 + 2x - 1$
20 $-3x^5 + 3x^4 - 19x^3 - 3x^2 + 17x - 9$ by $-x^2 + x - 7$

1.3 Algebraic fractions

You should have already studied fractions as part of your GCSE course.

The number at the top of a fraction is called the **numerator** and the number at the bottom is called the **denominator**. Thus the numerator in the fraction $\frac{5}{17}$ is 5 and 17 is the denominator. The same is true for an algebraic fraction. The numerator of the fraction $\frac{3x^2}{x^3-1}$ is $3x^2$ and the denominator is $x^3 - 1$.

Multiplying fractions

To multiply two numerical fractions, you first 'cancel' each factor in either numerator with the same factor, if it exists, in either denominator. Then you multiply the numbers that are left in the numerators to give the numerator in the answer, and you multiply the numbers that are left in the denominators to give the denominator in the answer.

For example:

$$\begin{aligned}
&\tfrac{5}{81} \times \tfrac{33}{40} \\
&= \tfrac{1}{81} \times \tfrac{33}{8} \qquad \text{(cancel by 5)} \\
&= \tfrac{1}{27} \times \tfrac{11}{8} \qquad \text{(cancel by 3)} \\
&= \tfrac{1 \times 11}{27 \times 8} \\
&= \tfrac{11}{216}
\end{aligned}$$

The same methods apply when you multiply algebraic fractions. You first cancel any factor common to both a numerator and a denominator and continue doing this for as many times as you can. You then multiply together the terms remaining in the numerators and multiply together the terms remaining in the denominators.

Example 16

Simplify $\dfrac{14}{x^2 - 1} \times \dfrac{x - 1}{21}$.

$$\begin{aligned}
&\frac{14}{x^2 - 1} \times \frac{x - 1}{21} \\
&= \frac{14}{(x - 1)(x + 1)} \times \frac{x - 1}{21} \qquad \text{(factorise } x^2 - 1\text{)} \\
&= \frac{2}{(x - 1)(x + 1)} \times \frac{x - 1}{3} \qquad \text{(cancel by 7)} \\
&= \frac{2}{x + 1} \times \frac{1}{3} \qquad \text{(cancel by } x - 1\text{)} \\
&= \frac{2 \times 1}{3(x + 1)} \\
&= \frac{2}{3(x + 1)}
\end{aligned}$$

Example 17

Simplify $$\frac{x}{2x^2 - x - 6} \times \frac{x - 2}{3x^2}$$

$$\frac{x}{2x^2 - x - 6} \times \frac{x-2}{3x^2}$$

$$= \frac{x}{(2x+3)(x-2)} \times \frac{x-2}{3x^2} \qquad \text{(factorise } 2x^2 - x - 6\text{)}$$

$$= \frac{1}{(2x+3)(x-2)} \times \frac{x-2}{3x} \qquad \text{(cancel by } x\text{)}$$

$$= \frac{1}{2x+3} \times \frac{1}{3x} \qquad \text{(cancel by } x-2\text{)}$$

$$= \frac{1 \times 1}{(2x+3)3x}$$

$$= \frac{1}{6x^2+9x} = \frac{1}{3x(2x+3)}$$

Dividing fractions

As you should know from your GCSE studies, if you are asked to evaluate $\frac{9}{35} \div \frac{27}{25}$ you write $\frac{9}{35} \times \frac{25}{27}$ and then evaluate this. In the same way, to divide one algebraic fraction by another, you take the fraction which comes after the division symbol and turn it upside down. You then change the division symbol into a multiplication symbol and continue as above.

Example 18

Simplify $\dfrac{5x^2}{x^2-9} \div \dfrac{3x^3}{2(x+3)}$.

$$\frac{5x^2}{x^2-9} \div \frac{3x^3}{2(x+3)}$$

$$= \frac{5x^2}{x^2-9} \times \frac{2(x+3)}{3x^3}$$

$$= \frac{5x^2}{(x-3)(x+3)} \times \frac{2(x+3)}{3x^3} \qquad \text{(factorise } x^2 - 9\text{)}$$

$$= \frac{5}{(x-3)(x+3)} \times \frac{2(x+3)}{3x} \qquad \text{(cancel by } x^2\text{)}$$

$$= \frac{5}{x-3} \times \frac{2}{3x} \qquad \text{(cancel by } x+3\text{)}$$

$$= \frac{5 \times 2}{(x-3)3x}$$

$$= \frac{10}{3x^2-9x} = \frac{10}{3x(x-3)}$$

Example 19

Simplify $\dfrac{15(x-7)}{6x^2+10x-4} \div \dfrac{10(2x+3)}{3x^2+2x-1}$.

$$\begin{aligned}
&\frac{15(x-7)}{6x^2+10x-4} \div \frac{10(2x+3)}{3x^2+2x-1} \\
&= \frac{15(x-7)}{6x^2+10x-4} \times \frac{3x^2+2x-1}{10(2x+3)} \\
&= \frac{15(x-7)}{(2x+4)(3x-1)} \times \frac{(3x-1)(x+1)}{10(2x+3)} \qquad \text{(factorise)} \\
&= \frac{3(x-7)}{(2x+4)(3x-1)} \times \frac{(3x-1)(x+1)}{2(2x+3)} \qquad \text{(cancel by 5)} \\
&= \frac{3(x-7)}{2x+4} \times \frac{x+1}{2(2x+3)} \qquad \text{(cancel by } 3x-1\text{)} \\
&= \frac{3(x-7)(x+1)}{2(2x+4)(2x+3)} \\
&= \frac{3x^2-18x-21}{8x^2+28x+24}
\end{aligned}$$

Adding fractions

Consider the sum $\frac{2}{7}+\frac{3}{5}$. In order to add these fractions you must first write them as **equivalent fractions** that have the same denominator, often called a **common denominator**. In this case a common denominator is 35, because 35 is a multiple of both 7 and 5.

So:

$$\begin{aligned}
\frac{2}{7}+\frac{3}{5} &= \frac{2\times 5}{7\times 5}+\frac{3\times 7}{5\times 7} \\
&= \frac{10}{35}+\frac{21}{35} \\
&= \frac{10+21}{35} \\
&= \frac{31}{35}
\end{aligned}$$

You should know that the common denominator which you use when adding fractions is not *always* the product of the two original denominators. For example, consider the sum $\frac{5}{12}+\frac{7}{18}$. A common denominator of $12 \times 18 = 216$ would work, but it is much simpler to use the **lowest common multiple** (LCM) of 12 and 18.

The lowest common multiple of two integers is the smallest integer that can be divided exactly by each of them.

Since: $12 = 2 \times 2 \times 3$

and: $18 = 2 \times 3 \times 3$

the LCM is $2 \times 2 \times 3 \times 3 = 36$.

So:
$$\frac{5}{12} + \frac{7}{18}$$
$$= \frac{5 \times 3}{12 \times 3} + \frac{7 \times 2}{18 \times 2}$$
$$= \frac{15}{36} + \frac{14}{36}$$
$$= \frac{29}{36}$$

The same rules apply for the addition of algebraic fractions.

Example 20

Express as a single fraction:

$$\frac{3}{2x+5} + \frac{x-7}{4x^2+10x}$$

$$\frac{3}{2x+5} + \frac{x-7}{4x^2+10x}$$
$$= \frac{3}{2x+5} + \frac{x-7}{2x(2x+5)}$$
$$= \frac{2x \times 3}{2x(2x+5)} + \frac{x-7}{2x(2x+5)} \quad \text{(common denominator } 2x(2x+5)\text{)}$$
$$= \frac{6x}{2x(2x+5)} + \frac{x-7}{2x(2x+5)}$$
$$= \frac{6x+x-7}{2x(2x+5)}$$
$$= \frac{7x-7}{2x(2x+5)}$$
$$= \frac{7(x-1)}{2x(2x+5)}$$

Example 21

Express as a single fraction:

$$\frac{2x+1}{2x^2+14x+20}+\frac{x-3}{3x^2-75}$$

$$\begin{aligned}
&\frac{2x+1}{2x^2+14x+20}+\frac{x-3}{3x^2-75}\\
&=\frac{2x+1}{2(x+2)(x+5)}+\frac{x-3}{3(x+5)(x-5)}\\
&=\frac{3(x-5)(2x+1)}{2\times 3(x+2)(x+5)(x-5)}+\frac{2(x+2)(x-3)}{2\times 3(x+2)(x+5)(x-5)}\\
&=\frac{6x^2-27x-15}{6(x+2)(x+5)(x-5)}+\frac{2x^2-2x-12}{6(x+2)(x+5)(x-5)}\\
&=\frac{6x^2-27x-15+2x^2-2x-12}{6(x+2)(x+5)(x-5)}\\
&=\frac{8x^2-29x-27}{6(x+2)(x+5)(x-5)}
\end{aligned}$$

Subtracting fractions

The method used to subtract algebraic fractions is similar to that for adding fractions.

Example 22

Express as a single fraction:

$$\frac{5x-2}{2(x+3)^2}-\frac{x-4}{6(6-x-x^2)}$$

$$\begin{aligned}
&\frac{5x-2}{2(x+3)^2}-\frac{x-4}{6(6-x-x^2)}\\
&=\frac{5x-2}{2(x+3)^2}-\frac{x-4}{2\times 3(x+3)(2-x)}\\
&=\frac{3(2-x)(5x-2)}{2\times 3(x+3)^2(2-x)}-\frac{(x+3)(x-4)}{2\times 3(x+3)^2(2-x)}\\
&=\frac{3(2-x)(5x-2)-(x+3)(x-4)}{6(x+3)^2(2-x)}\\
&=\frac{-15x^2+36x-12-x^2+x+12}{6(x+3)^2(2-x)}
\end{aligned}$$

$$= \frac{-16x^2 + 37x}{6(x+3)^2(2-x)}$$

$$= \frac{x(-16x+37)}{6(x+3)^2(2-x)}$$

Exercise 1C

Express as a single fraction:

1 $\dfrac{3x^2}{(2+x)(3+2x)} \times \dfrac{4(3+2x)}{9x^5}$

2 $\dfrac{2-x}{x(x-4)} \times \dfrac{6x}{(2-x)(1+x)}$

3 $\dfrac{1-x^2}{4x(7+x)} \times \dfrac{6x^2}{3+x-2x^2}$

4 $\dfrac{x^2+6x+9}{3(x+2)(2x-1)} \times \dfrac{12x^2-42x+18}{4x^2-36}$

5 $\dfrac{x^4-25x^2}{2x-6} \times \dfrac{4x^2-20x-24}{2x^2+11x+5}$

6 $\dfrac{4(x+7)}{9(x^2-25)} \div \dfrac{2(x+7)(3x-2)}{15x(x+5)}$

7 $\dfrac{7x^2(2-x)}{27(2x-1)(x+4)} \div \dfrac{14x(2x-3)}{9(x+2)(2x-1)}$

8 $\dfrac{15x^2+30x}{11x^2-11x-22} \div \dfrac{21(3x^2+4x-4)}{44(x^2-2x)}$

9 $\dfrac{2(x+3)^2}{x^2(x-1)} \div \dfrac{10(x^2-9)}{2x^2+x-3}$

10 $\dfrac{x^2+3x+2}{x^2-3x+2} \div \dfrac{x^2+4x+3}{2x^2-3x+1}$

11 $\dfrac{1}{4} + \dfrac{1}{x-2}$

12 $\dfrac{1}{2x+1} + \dfrac{2}{x+3}$

13 $\dfrac{3}{4x-5} - \dfrac{2}{6-2x}$

14 $\dfrac{5}{4x+3} + \dfrac{3}{2x-1} - \dfrac{1}{2}$

15 $\dfrac{6}{2x-3} - \dfrac{3}{3-2x}$

16 $\dfrac{2}{(x+3)(x+1)} + \dfrac{3}{2x-1}$

17 $\dfrac{x}{(x-1)(2x+1)} - \dfrac{2}{x+4}$

18 $\dfrac{6x}{(5-2x)(6+x)} - \dfrac{7x+2}{(5-2x)(1-x)}$

19 $\dfrac{2x-7}{5(x^2-4)} + \dfrac{x+4}{2(x^2+x-6)}$

20 $\dfrac{5}{3-x} + \dfrac{2+x}{21-4x-x^2} - \dfrac{2x-3}{2x^2+13x-7}$

1.4 Partial fractions

Now that you have learned how to add and subtract fractions to produce a single fraction, you are going to learn how to reverse the process. You know that:

$$\frac{2}{x+5} + \frac{3}{x+4}$$
$$= \frac{2x+8+3x+15}{(x+5)(x+4)}$$
$$= \frac{5x+23}{(x+5)(x+4)}$$

Now you need to be able to start with

$$\frac{5x+23}{(x+5)(x+4)}$$

and split it into

$$\frac{2}{x+5} + \frac{3}{x+4}$$

This process of taking a single fraction and breaking it up into the sum (or difference) of two or more fractions is known as **splitting an expression into its partial fractions.**

Linear factors in the denominator

When you need to split an expression into its partial fractions, the first thing you must do is to factorise the denominator of the given expression. With, for example, the fraction $\dfrac{5x+1}{x^2+x-2}$ you first rewrite it as

$$\frac{5x+1}{(x-1)(x+2)}$$

- **For a fraction with only linear factors in the denominator, and where the degree of the denominator exceeds that of the numerator, e.g. $\frac{5x+1}{(x-1)(x+2)}$, the partial fractions are of the form**

$$\frac{A}{x-1}+\frac{B}{x+2}$$

where A and B are constants.

So: $$\frac{5x+1}{(x-1)(x+2)} \equiv \frac{A}{x-1}+\frac{B}{x+2}$$

Note that this is an identity, because when you have found the values of A and B, $\frac{5x+1}{(x-1)(x+2)}$ will be equal to $\frac{A}{x-1}+\frac{B}{x+2}$ for *all* values of x, and not just for *some* values of x.

Now: $$\frac{A}{x-1}+\frac{B}{x+2} \equiv \frac{A(x+2)+B(x-1)}{(x-1)(x+2)}$$

So: $$\frac{5x+1}{(x-1)(x+2)} \equiv \frac{A(x+2)+B(x-1)}{(x-1)(x+2)}$$

However, if these two fractions are identical, and also the denominators of these two fractions are identical, then the numerators of the two fractions must also be identical. That is:

$$5x+1 \equiv A(x+2)+B(x-1)$$

The values of A and B can now be found using the methods shown in section 1.1.

$x=1 \Rightarrow$ $$6=3A$$ $$A=2$$

$x=-2 \Rightarrow$ $$-9=-3B$$ $$B=3$$

So: $$\frac{5x+1}{(x-1)(x+2)} \equiv \frac{2}{x-1}+\frac{3}{x+2}$$

and this can easily be checked by adding $\frac{2}{x-1}$ and $\frac{3}{x+2}$.

Example 23

Express $\dfrac{4x-2}{x^3-x}$ in partial fractions.

$$\frac{4x-2}{x^3-x}$$

$$= \frac{4x-2}{x(x^2-1)} \qquad \text{(take out the common factor } x\text{)}$$

$$= \frac{4x-2}{x(x-1)(x+1)} \qquad \text{(factorise the difference of two squares)}$$

Since each of the three factors in the denominator is linear:

$$\frac{4x-2}{x(x-1)(x+1)} \equiv \frac{A}{x} + \frac{B}{x-1} + \frac{C}{x+1}$$

Now: $$\frac{A}{x} + \frac{B}{x-1} + \frac{C}{x+1} \equiv \frac{A(x-1)(x+1) + Bx(x+1) + Cx(x-1)}{x(x-1)(x+1)}$$

So: $$4x-2 \equiv A(x-1)(x+1) + Bx(x+1) + Cx(x-1)$$

$x = 0 \Rightarrow$ $$-2 = -A$$
$$A = 2$$

$x = 1 \Rightarrow$ $$2 = 2B$$
$$B = 1$$

$x = -1 \Rightarrow$ $$-6 = 2C$$
$$C = -3$$

Thus: $$\frac{4x-2}{x^3-x} \equiv \frac{2}{x} + \frac{1}{x-1} - \frac{3}{x+1}$$

Example 24

Express $\dfrac{3x^2+7x-8}{(x+1)(x+2)(x-3)}$ in partial fractions.

Since the denominator has factors that are each linear, and the degree of the numerator is 2 which is less than the degree of the denominator, the form of the partial fractions is:

$$\frac{A}{x+1} + \frac{B}{x+2} + \frac{C}{x-3} \equiv \frac{A(x+2)(x-3) + B(x+1)(x-3) + C(x+1)(x+2)}{(x+1)(x+2)(x-3)}$$

Therefore, equating numerators:

$$3x^2 + 7x - 8 \equiv A(x+2)(x-3) + B(x+1)(x-3) + C(x+1)(x+2)$$

$x = -1 \Rightarrow$
$$3 - 7 - 8 = -4A$$
$$-12 = -4A$$
$$A = 3$$

$x = -2 \Rightarrow$
$$12 - 14 - 8 = 5B$$
$$-10 = 5B$$
$$B = -2$$

$x = 3 \Rightarrow$
$$27 + 21 - 8 = 20C$$
$$40 = 20C$$
$$C = 2$$

Therefore:
$$\frac{3x^2 + 7x - 8}{(x+1)(x+2)(x-3)} \equiv \frac{3}{x+1} - \frac{2}{x+2} + \frac{2}{x-3}$$

Quadratic factor in the denominator

If any of the factors in the denominator is *not* linear then the partial fractions cannot take the form shown in the previous section.

- **For a fraction that has a non-reducible quadratic factor in the denominator, and where the degree of the denominator exceeds that of the numerator, e.g.**

$$\frac{x^2 - 5x + 1}{(x^2 + 1)(x - 2)}$$

the partial fractions are of the form

$$\frac{Ax + B}{x^2 + 1} + \frac{C}{x - 2}$$

where A, B and C are constants.

It is *essential* to ensure that each partial fraction with a quadratic denominator has a numerator of the form $Ax + B$. Any numerator of the form A will not, in general, work if the denominator is quadratic.

Example 25

Express $\dfrac{5x^2 + 4x + 4}{(x+2)(x^2+4)}$ in partial fractions.

Since one of the factors in the denominator is quadratic the partial fractions are of the form

$$\frac{A}{x+2} + \frac{Bx + C}{x^2 + 4} \equiv \frac{A(x^2 + 4) + (x+2)(Bx + C)}{(x+2)(x^2+4)}$$

So, equating numerators:

$$5x^2 + 4x + 4 \equiv A(x^2 + 4) + (x + 2)(Bx + C)$$

$x = -2 \Rightarrow$

$$20 - 8 + 4 = 8A$$
$$16 = 8A$$
$$A = 2$$

Equating coefficients of x^2:

$$5 = A + B$$
$$5 = 2 + B$$
$$B = 3$$

Equating coefficients of x:

$$4 = 2B + C$$
$$4 = 6 + C$$
$$C = -2$$

Thus:

$$\frac{5x^2 + 4x + 4}{(x + 2)(x^2 + 4)} \equiv \frac{2}{x + 2} + \frac{3x - 2}{x^2 + 4}$$

You must make sure that you have factorised the denominator of the given fraction *completely*, before you try to express it in partial fractions. If the denominator is $(x^2 - 3x + 2)(x^2 - x + 3)$ then the partial fractions are *not* of the form

$$\frac{Ax + B}{x^2 - 3x + 2} + \frac{Cx + D}{x^2 - x + 3}$$

because $(x^2 - 3x + 2)(x^2 - x + 3)$ is not completely factorised. It can be factorised further into $(x - 1)(x - 2)(x^2 - x + 3)$. So the partial fractions are of the form

$$\frac{A}{x - 1} + \frac{B}{x - 2} + \frac{Cx + D}{x^2 - x + 3}$$

Example 26

Express

$$\frac{-2x - 1}{(x^2 - 3x + 2)(x^2 - x + 3)}$$

in partial fractions.

Since the denominator can be factorised to
$(x - 1)(x - 2)\ (x^2 - x + 3)$ the partial fractions are of the form

$$\frac{A}{x - 1} + \frac{B}{x - 2} + \frac{Cx + D}{x^2 - x + 3}$$

$$\equiv \frac{A(x - 2)(x^2 - x + 3) + B(x - 1)(x^2 - x + 3) + (Cx + D)(x - 1)(x - 2)}{(x - 1)(x - 2)(x^2 - x + 3)}$$

Equating numerators:

$$-2x - 1 \equiv A(x-2)(x^2 - x + 3) + B(x-1)(x^2 - x + 3) + (Cx + D)(x-1)(x-2)$$

$x = 1 \Rightarrow$

$$\begin{aligned} -3 &= A(-1)(1 - 1 + 3) \\ -3 &= -3A \\ A &= 1 \end{aligned}$$

$x = 2 \Rightarrow$

$$\begin{aligned} -5 &= B(1)(4 - 2 + 3) \\ -5 &= 5B \\ B &= -1 \end{aligned}$$

Equating coefficients of x^3:

$$\begin{aligned} 0 &= A + B + C \\ 0 &= 1 - 1 + C \\ 0 &= C \end{aligned}$$

Equating constant terms:

$$\begin{aligned} -1 &= -6A - 3B + 2D \\ -1 &= -6 + 3 + 2D \\ 2D &= 2 \\ D &= 1 \end{aligned}$$

So: $$\frac{-2x - 1}{(x^2 - 3x + 2)(x^2 - x + 3)} \equiv \frac{1}{x-1} - \frac{1}{x-2} + \frac{1}{x^2 - x + 3}$$

Repeated factor in the denominator

If the factors in the denominator include one that is repeated, e.g. $(x + 4)^2$, then the partial fractions take yet another form, which you must remember.

- **For a fraction that has a repeated factor in the denominator, and where the degree of the denominator exceeds that of the numerator, e.g.**

$$\frac{2(x^2 - 2x - 1)}{(x+1)(x-1)^2}$$

the partial fractions are of the form

$$\frac{A}{x+1} + \frac{B}{x-1} + \frac{C}{(x-1)^2}$$

where A, B and C are constants.

Example 27

Express $\dfrac{2(x^2 - 2x - 1)}{(x+1)(x-1)^2}$ in partial fractions.

The partial fractions are of the form

$$\frac{A}{x+1} + \frac{B}{x-1} + \frac{C}{(x-1)^2}$$
$$\equiv \frac{A(x-1)^2 + B(x+1)(x-1) + C(x+1)}{(x+1)(x-1)^2}$$

Equating numerators gives

$$2(x^2 - 2x - 1) \equiv A(x-1)^2 + B(x+1)(x-1) + C(x+1)$$

$x = -1 \Rightarrow$

$$2(1 + 2 - 1) = 4A$$
$$4 = 4A$$
$$A = 1$$

$x = 1 \Rightarrow$

$$2(1 - 2 - 1) = 2C$$
$$-4 = 2C$$
$$C = -2$$

Equating constant terms:

$$-2 = A - B + C$$
$$-2 = 1 - B - 2$$
$$B = 1$$

So:

$$\frac{2(x^2 - 2x - 1)}{(x+1)(x-1)^2} \equiv \frac{1}{x+1} + \frac{1}{x-1} - \frac{2}{(x-1)^2}$$

Example 28

Express $\dfrac{x^2 + 3x + 4}{(x+2)^3}$ in partial fractions.

The partial fractions are of the form:

$$\frac{A}{x+2} + \frac{B}{(x+2)^2} + \frac{C}{(x+2)^3}$$
$$\equiv \frac{A(x+2)^2 + B(x+2) + C}{(x+2)^3}$$

Equating numerators gives

$$x^2 + 3x + 4 \equiv A(x+2)^2 + B(x+2) + C$$

$x = -2 \Rightarrow$ $\quad 4 - 6 + 4 = C$

$$C = 2$$

Equating coefficients of x^2: $\quad 1 = A$

Equating coefficients of x: $\quad 3 = 4A + B$

$$3 = 4 + B$$

$$B = -1$$

So: $$\frac{x^2 + 3x + 4}{(x+2)^3} \equiv \frac{1}{x+2} - \frac{1}{(x+2)^2} + \frac{2}{(x+2)^3}$$

Improper algebraic fractions

The polynomial $3x + 2$ is of degree 1 and the polynomial $x^2 - 5x + 2$ is of degree 2. The degree of a polynomial in x is the same as the degree of the highest power of x in the polynomial.

A fraction where the degree of the numerator is less than the degree of the denominator is called a **proper fraction**. A fraction where the degree of the numerator is either equal to the degree of the denominator or higher than the degree of the denominator is called an **improper fraction**.

So

$$\frac{2x + 3}{3x^2 - 2x + 1}$$

is a proper fraction because the numerator has degree 1 and the denominator has degree 2. But

$$\frac{3x^2 + 2x + 6}{5x^2 - 7}$$

is an improper fraction because the numerator has degree 2, which is equal to the degree of the denominator, for this is also of degree 2. Similarly,

$$\frac{2x^3 + 3x^2 - 7}{5x - 3}$$

is an improper fraction because the numerator is of degree 3, which is greater than the degree of the denominator, for this is of degree 1.

When you try to split a fraction into its component partial fractions you must make sure that it is a proper fraction. If it is improper, you must first divide the denominator into the numerator before you try to split up the fraction into its partial fractions.

Example 29

Express $\frac{2x^2+8x+7}{(x+2)(x+3)}$ in partial fractions.

Since the degree of both the numerator and the denominator is 2, the fraction is improper. So the denominator must be divided into the numerator.

First multiply out the denominator:

$$\frac{2x^2+8x+7}{(x+2)(x+3)} \equiv \frac{2x^2+8x+7}{x^2+5x+6}$$

$$\begin{array}{r} 2 \\ x^2+5x+6\overline{)2x^2+\;\;8x+\;\;7} \\ \underline{2x^2+10x+12} \\ -\;2x-\;\;5 \end{array}$$

So x^2+5x+6 divides into $2x^2+8x+7$ twice and leaves a remainder $-2x-5$.

Now when 33 is divided by 7, the quotient is 4 with a remainder 5. This answer can be written $4\frac{5}{7}$, that is $4+\frac{5}{7}$. In the same way,

$$\frac{2x^2+8x+7}{x^2+5x+6}$$

can be written:

$$2+\frac{-2x-5}{x^2+5x+6}$$

$$\equiv 2+\frac{-(2x+5)}{x^2+5x+6}$$

$$\equiv 2-\frac{2x+5}{x^2+5x+6}$$

The fraction

$$\frac{2x+5}{x^2+5x+6}$$

is proper since the degree of the numerator (1) is less than the degree of the denominator (2). It can therefore now be split into partial fractions.

$$\frac{2x+5}{x^2+5x+6} \equiv \frac{2x+5}{(x+2)(x+3)}$$

The partial fractions are of the form

$$\frac{A}{x+2}+\frac{B}{x+3} \equiv \frac{A(x+3)+B(x+2)}{(x+2)(x+3)}$$

Equating numerators gives:

$$2x + 5 \equiv A(x + 3) + B(x + 2)$$

$x = -2 \Rightarrow$ $\quad 1 = A$

$x = -3 \Rightarrow$ $\quad -1 = -B$

$$B = 1$$

So:
$$\frac{2x^2 + 8x + 7}{(x + 2)(x + 3)} \equiv 2 - \frac{2x + 5}{x^2 + 5x + 6}$$

$$\equiv 2 - \left[\frac{1}{x + 2} + \frac{1}{x + 3}\right]$$

$$\equiv 2 - \frac{1}{x + 2} - \frac{1}{x + 3}$$

Example 30

Express $\frac{x^3 + 3x^2 + 10}{(x + 1)(x + 4)}$ in partial fractions.

The degree of the numerator (3) is higher than the degree of the denominator (2), so the fraction is improper.

$$\frac{x^3 + 3x^2 + 10}{(x + 1)(x + 4)} \equiv \frac{x^3 + 3x^2 + 10}{x^2 + 5x + 4}$$

So by division:

$$\begin{array}{r} x - 2 \\ x^2 + 5x + 4 \,\overline{\smash{)}\, x^3 + 3x^2 + 0x + 10} \\ \underline{x^3 + 5x^2 + 4x \quad\;\;} \\ -2x^2 - 4x + 10 \\ \underline{-2x^2 - 10x - 8} \\ 6x + 18 \end{array}$$

That is, $\quad x^3 + 3x^2 + 10 \equiv (x^2 + 5x + 4)(x - 2) + 6x + 18$

So:
$$\frac{x^3+3x^2+10}{x^2+5x+4} \equiv x-2+\frac{6x+18}{x^2+5x+4}$$

$$\frac{x^3+3x^2+10}{(x+1)(x+4)} \equiv x-2+\frac{6x+18}{(x+1)(x+4)}$$

Now $\dfrac{6x+18}{(x+1)(x+4)}$ is proper, since the degree of the numerator (1) is less than the degree of the denominator (2). We can therefore split it into partial fractions.

$$\frac{6x+18}{(x+1)(x+4)} \equiv \frac{A}{x+1}+\frac{B}{x+4}$$
$$\equiv \frac{A(x+4)+B(x+1)}{(x+1)(x+4)}$$

Equating numerators gives

$$6x+18 \equiv A(x+4)+B(x+1)$$

$x=-1 \Rightarrow$
$$12 = 3A$$
$$A = 4$$

$x=-4 \Rightarrow$
$$-6 = -3B$$
$$B = 2$$

So:
$$\frac{x^3+3x^2+10}{(x+1)(x+4)} \equiv x-2+\frac{4}{x+1}+\frac{2}{x+4}$$

Exercise 1D

Express as partial fractions:

1 $\dfrac{2x+5}{(x+2)(x+3)}$

2 $\dfrac{2x+2}{(x-1)(x+3)}$

3 $\dfrac{x+1}{(x+3)(x+4)}$

4 $\dfrac{x+7}{x^2+5x+6}$

5 $\dfrac{2x^2+12x-10}{(x-1)(2x-1)(x+3)}$

6 $\dfrac{3x^2-x+6}{(x^2+4)(x-2)}$

7 $\dfrac{x^2-2x+9}{(x^2+3)(x-3)}$

8 $\dfrac{-2x^2+4x-4}{(x^2+5)(2x+3)}$

9 $\dfrac{-6x^2+x-12}{(5+2x^2)(x+3)}$

10 $\dfrac{-2x^2+13}{(2x+1)(x^2+2x+7)}$

11 $\dfrac{2x-7}{(x-5)^2}$

12 $\dfrac{x^2+4x+7}{(x+3)^3}$

13 $\dfrac{-3x^2+10x+5}{(x+2)(x-1)^2}$

14 $\dfrac{-5x^2+8x+9}{(x+2)(x-1)^2}$

15 $\dfrac{10x+9}{(2x+1)(2x+3)^2}$

16 $\dfrac{x}{x-1}$

17 $\dfrac{x^2}{x-1}$

18 $\dfrac{x^2+1}{x^2-1}$

19 $\dfrac{x^2+2}{x(x-1)}$

20 $\dfrac{x^3}{x^2-1}$

21 $\dfrac{9-2x-2x^2}{(1+x)(2-x)}$

22 $\dfrac{4x^2-3x+2}{2x^2-x-1}$

23 $\dfrac{2x^3+10x^2+12x+1}{(x+2)(x+3)}$

24 $\dfrac{x^3+x^2-2x+4}{x^2-4}$

25 $\dfrac{-x^4-x^3+2x^2-x-2}{x^2(x+1)}$

26 $\dfrac{13}{(2x-3)(3x+2)}$

27 $\dfrac{4x^2+5x+9}{(2x-1)(x+2)^2}$

28 $\dfrac{x^3+4x^2+3x+4}{(x^2+1)(x+1)^2}$

29 $\dfrac{4x^4+6x^3+4x^2+x-3}{x^2(2x+3)}$

30 $\dfrac{4x+3}{(2x-1)(3x+1)}$

31 $\dfrac{x^3+3x^2-2x-5}{(x-1)^2(x^2+2)}$

32 $\dfrac{3x^2+12x+8}{(2x+3)(x^2-4)}$

33 $\dfrac{x^3-x^2-1}{x(x^2+x+1)}$

34 $\dfrac{x^2+2x+3}{x^2(x+1)}$

35 $\dfrac{3}{x^3+1}$

1.5 The remainder theorem and the factor theorem

When a polynomial is divided by a linear expression there is usually a remainder. For example,

$$(3x^4 - 7x^3 - 10x^2 + 19x - 19) \div (x - 3)$$

$$\begin{array}{r}
3x^3 + 2x^2 - 4x + 7 \\
x - 3\,\overline{)\,3x^4 - 7x^3 - 10x^2 + 19x - 19} \\
\underline{3x^4 - 9x^3} \\
2x^3 - 10x^2 \\
\underline{2x^3 - 6x^2} \\
-\ 4x^2 + 19x \\
\underline{-\ 4x^2 + 12x} \\
7x - 19 \\
\underline{7x - 21} \\
2
\end{array}$$

So: $\qquad (3x^4 - 7x^3 - 10x^2 + 19x - 19) \div (x - 3)$

gives a quotient of $3x^3 + 2x^2 - 4x + 7$ and a remainder of 2. You can write this sentence mathematically as:

$$\frac{3x^4 - 7x^3 - 10x^2 + 19x - 19}{x - 3} \equiv 3x^3 + 2x^2 - 4x + 7 + \frac{2}{x - 3}$$

Now if both sides of the identity are multiplied by $(x - 3)$ then

$$\frac{3x^4 - 7x^3 - 10x^2 + 19x - 19}{\cancel{x - 3}} \times (\cancel{x - 3})$$

$$\equiv (x - 3)\left(3x^3 + 2x^2 - 4x + 7 + \frac{2}{x - 3}\right)$$

$$3x^4 - 7x^3 - 10x^2 + 19x - 19 \equiv (x - 3)(3x^3 + 2x^2 - 4x + 7) + (\cancel{x - 3})\left(\frac{2}{\cancel{x - 3}}\right)$$

That is:

$$3x^4 - 7x^3 - 10x^2 + 19x - 19 \equiv (x - 3)(3x^3 + 2x^2 - 4x + 7) + 2$$

But this is the definition of a quotient and remainder. In other words,

$$\mathrm{f}(x) \equiv (x - a)\mathrm{g}(x) + R$$

'original polynomial $\equiv$ (divisor) $\times$ (quotient) + remainder'

The 'divisor' is the linear expression that the polynomial was divided by.

Let's try another example.

Divide $6x^5 + 2x^3 - 5x^2 + 2x - 1$ by $x - 1$.

$$
\begin{array}{r|l}
 & 6x^4 + 6x^3 + 8x^2 + 3x + 5 \\
\hline
x - 1 & 6x^5 + 0x^4 + 2x^3 - 5x^2 + 2x - 1 \\
 & \underline{6x^5 - 6x^4} \\
 & \quad 6x^4 + 2x^3 \\
 & \quad \underline{6x^4 - 6x^3} \\
 & \qquad 8x^3 - 5x^2 \\
 & \qquad \underline{8x^3 - 8x^2} \\
 & \qquad\quad 3x^2 + 2x \\
 & \qquad\quad \underline{3x^2 - 3x} \\
 & \qquad\qquad 5x - 1 \\
 & \qquad\qquad \underline{5x - 5} \\
 & \qquad\qquad\qquad 4
\end{array}
$$

So: $\dfrac{6x^5 + 2x^3 - 5x^2 + 2x - 1}{x - 1} \equiv 6x^4 + 6x^3 + 8x^2 + 3x + 5 + \dfrac{4}{x - 1}$

Multiply by $x - 1$:

$$\frac{6x^5 + 2x^3 - 5x^2 + 2x - 1}{\cancel{x - 1}} \times (\cancel{x - 1}) \equiv (x - 1)\left(6x^4 + 6x^3 + 8x^2 + 3x + 5 + \frac{4}{x - 1}\right)$$

$$6x^5 + 2x^3 - 5x^2 + 2x - 1 \equiv (x - 1)(6x^4 + 6x^3 + 8x^2 + 3x + 5) + (\cancel{x - 1})\left(\frac{4}{\cancel{x - 1}}\right)$$

$$6x^5 + 2x^3 - 5x^2 + 2x - 1 \equiv (x - 1)(6x^4 + 6x^3 + 8x^2 + 3x + 5) + 4$$

Once again:

original polynomial $\equiv$ (divisor) $\times$ (quotient) $+$ remainder

In the first example the divisor was $x - 3$. Look at the left-hand side of the identity:

$$3x^4 - 7x^3 - 10x^2 + 19x - 19 \equiv (x - 3)(3x^3 + 2x^2 - 4x + 7) + 2$$

and put $x = 3$.

$$\begin{aligned}\text{LHS} &= (3 \times 3^4) - (7 \times 3^3) - (10 \times 3^2) + (19 \times 3) - 19 \\ &= 243 - 189 - 90 + 57 - 19 \\ &= 300 - 298 \\ &= 2\end{aligned}$$

which is the remainder when the polynomial is divided by $x - 3$.

You will see the reason for this if you consider the right-hand side of the identity and put $x = 3$:

$$\begin{aligned}\text{RHS} &= (x - 3)(3x^3 + 2x^2 - 4x + 7) + 2 \\ &= (3 - 3)(81 + 18 - 12 + 7) + 2 \\ &= (0 \times 94) + 2 \\ &= 0 + 2 \\ &= 2\end{aligned}$$

That is, as soon as you put $x = 3$, the first bracket becomes zero, so the whole of the first expression becomes zero since anything multiplied by zero gives an answer of zero. So the right-hand side of the identity reduces to 'zero + the remainder of 2'.

You can do the same with the second example. Put $x = 1$ in the right-hand side of the identity:

$$6x^5 + 2x^3 - 5x^2 + 2x - 1 \equiv (x - 1)(6x^4 + 6x^3 + 8x^2 + 3x + 5) + 4$$

$$\text{LHS} = 6 + 2 - 5 + 2 - 1 = 4$$

which, once again, is the remainder. Again, you can see why this is so by considering the right-hand side of the identity with $x = 1$:

$$\begin{aligned}\text{RHS} &= (1 - 1)(6 + 6 + 8 + 3 + 5) + 4 \\ &= (0 \times 28) + 4 \\ &= 4\end{aligned}$$

Let us now consider the general situation. Suppose that when the polynomial $\mathrm{f}(x)$, of degree n, where

$$\mathrm{f}(x) \equiv a_n x^n + a_{n-1}x^{n-1} + \ldots + a_2x^2 + a_1x + a_0$$

is divided by $(x - \alpha)$, the quotient is a polynomial $\mathrm{g}(x)$ (of degree $n - 1$) and there is a remainder R.

Then: $$\frac{\mathrm{f}(x)}{x - \alpha} \equiv \mathrm{g}(x) + \frac{R}{x - \alpha}$$

or $$\frac{\mathrm{f}(x)}{\cancel{x - \alpha}} \times (\cancel{x - \alpha}) \equiv (x - \alpha) \times \mathrm{g}(x) + (\cancel{x - \alpha}) \times \frac{R}{\cancel{x - \alpha}}$$

$$\mathrm{f}(x) \equiv (x - \alpha)\mathrm{g}(x) + R$$

Consider the LHS and RHS of this identity with $x = \alpha$.

$$\text{LHS} = f(\alpha)$$

$$\begin{aligned}\text{RHS} &= (\alpha - \alpha) \times g(\alpha) + R \\ &= 0 \times g(\alpha) + R \\ &= R\end{aligned}$$

So: $$f(\alpha) = R$$

That is, if you substitute $x = \alpha$ into the polynomial you obtain the remainder that you would get if you divided $f(x)$ by $(x - \alpha)$. This the **remainder theorem**. It can be stated more formally as:

- **If a polynomial $f(x)$ is divided by $x - \alpha$, the remainder is $f(\alpha)$.**

Example 31

Find the remainder when $3x^3 + 7x^2 + 2x + 1$ is divided by $x - 2$.

Let $f(x) \equiv 3x^3 + 7x^2 + 2x + 1$.

The remainder is

$$\begin{aligned}f(2) &= (3 \times 8) + (7 \times 4) + (2 \times 2) + 1 \\ &= 24 + 28 + 4 + 1 \\ &= 57\end{aligned}$$

Example 32

Find the remainder when $3x^4 - 2x^2 + 6x + 6$ is divided by $x + 1$.

Let $f(x) \equiv 3x^4 - 2x^2 + 6x + 6$.

The remainder is

$$\begin{aligned}f(-1) &= 3 - 2 - 6 + 6 \\ &= 1\end{aligned}$$

A more general case of the remainder theorem can be found by considering the remainder when the polynomial $f(x)$ is divided by $\alpha x - \beta$. Suppose that the answer is, again, another polynomial $g(x)$ and the remainder is R.

So: $$\frac{f(x)}{\alpha x - \beta} \equiv g(x) + \frac{R}{\alpha x - \beta}$$

Multiply by $\alpha x - \beta$:

$$\frac{f(x)}{\cancel{\alpha x - \beta}} \times (\cancel{\alpha x - \beta}) \equiv (\alpha x - \beta) \times g(x) + (\cancel{\alpha x - \beta}) \times \frac{R}{\cancel{\alpha x - \beta}}$$

$$f(x) \equiv (\alpha x - \beta) \times g(x) + R$$

Now put $x = \frac{\beta}{\alpha}$:

$$f\left(\frac{\beta}{\alpha}\right) = \left(\alpha\frac{\beta}{\alpha} - \beta\right) \times g\left(\frac{\beta}{\alpha}\right) + R$$

$$f\left(\frac{\beta}{\alpha}\right) = 0 \times g\left(\frac{\beta}{\alpha}\right) + R$$

That is: $$f\left(\frac{\beta}{\alpha}\right) = R$$

This is a more general form of the remainder theorem.

- **If a polynomial f(x) is divided by $\alpha x - \beta$, the remainder is $f\left(\frac{\beta}{\alpha}\right)$.**

Example 33

Find the remainder when $2x^3 + 4x^2 - 6x + 1$ is divided by $2x - 1$.

Let $f(x) \equiv 2x^3 + 4x^2 - 6x + 1$.

Put $x = \frac{1}{2}$.

The remainder is:

$$\begin{aligned} f(\tfrac{1}{2}) &= (2 \times \tfrac{1}{8}) + (4 \times \tfrac{1}{4}) - (6 \times \tfrac{1}{2}) + 1 \\ &= \tfrac{1}{4} + 1 - 3 + 1 \\ &= -\tfrac{3}{4} \end{aligned}$$

Example 34

Find the remainder when $2x^2 + 3x - 1$ is divided by $3x + 2$.

Let $f(x) \equiv 2x^2 + 3x - 1$.

The remainder is:

$$\begin{aligned} f(-\tfrac{2}{3}) &= (2 \times \tfrac{4}{9}) + (3 \times -\tfrac{2}{3}) - 1 \\ &= \tfrac{8}{9} - 2 - 1 \\ &= -2\tfrac{1}{9} \end{aligned}$$

Example 35

When $2x^3 + ax^2 + x + 1$ is divided by $x + 2$ the remainder is -29. Find the value of the constant a.

Let $f(x) \equiv 2x^3 + ax^2 + x + 1$.

The remainder is:

$$\begin{aligned} f(-2) = -16 + 4a - 2 + 1 &= -29 \\ 4a - 17 &= -29 \\ 4a &= -12 \\ a &= -3 \end{aligned}$$

If $x - \alpha$ is a factor of a polynomial $f(x)$ then $x - \alpha$ divides exactly into $f(x)$ and leaves no remainder. So:

$$\frac{f(x)}{x - \alpha} \equiv g(x)$$

That is: $$f(x) \equiv (x - \alpha) \times g(x)$$

Put $x = \alpha$:
$$\begin{aligned} f(\alpha) &= (\alpha - \alpha) \times g(\alpha) \\ &= 0 \times g(\alpha) \\ &= 0 \end{aligned}$$

This is the **factor theorem**.

- **If $x - \alpha$ is a factor of the polynomial $f(x)$, then $f(\alpha) = 0$.**

Example 36

Factorise completely $x^3 - 6x^2 + 11x - 6$.

Let $f(x) \equiv x^3 - 6x^2 + 11x - 6$.

Try $x = 1$: $$f(1) = 1 - 6 + 11 - 6 = 0$$

So $x - 1$ is a factor of $f(x)$.

Try $x = 2$: $$f(2) = 8 - 24 + 22 - 6 = 0$$

So $x - 2$ is a factor.

Try $x = 3$: $$f(3) = 27 - 54 + 33 - 6 = 0$$

So $x - 3$ is a factor.

Thus: $$x^3 - 6x^2 + 11x - 6 \equiv (x - 1)(x - 2)(x - 3)$$

Example 37

The polynomial $ax^3 - x^2 + bx + 6$ has a factor of $x + 2$, and when it is divided by $x + 1$ there is a remainder of 10. Find the values of the constants a and b. Find the values of x for which the polynomial is zero.

Let $f(x) \equiv ax^3 - x^2 + bx + 6$.

Since $(x + 2)$ is a factor of $f(x)$,

$$f(-2) = -8a - 4 - 2b + 6 = 0 \qquad (1)$$

Since there is a remainder of 10 when $f(x)$ is divided by $(x + 1)$,

$$f(-1) = -a - 1 - b + 6 = 10 \qquad (2)$$

From (1): $$-8a - 2b = -2$$

From (2): $$-a - b = 5$$

Solving these simultaneously:

$$4a + b = 1 \quad (1)$$

$$a + b = -5 \quad (2)$$

(1) – (2): $3a = 6$

$$a = 2$$

Substitute in (2): $2 + b = -5$

$$b = -7$$

So: $\mathrm{f}(x) \equiv 2x^3 - x^2 - 7x + 6$

You know that $(x + 2)$ is a factor of $\mathrm{f}(x)$.

Divide $\mathrm{f}(x)$ by $(x + 2)$:

$$\begin{array}{r} 2x^2 - 5x + 3 \\ x + 2\ \overline{)\,2x^3 - x^2 - 7x + 6} \\ \underline{2x^3 + 4x^2} \qquad\qquad \\ -5x^2 - 7x \qquad \\ \underline{-5x^2 - 10x} \qquad \\ 3x + 6 \\ \underline{3x + 6} \end{array}$$

So : $2x^3 - x^2 - 7x + 6 = (x + 2)(2x^2 - 5x + 3)$
$= (x + 2)(2x - 3)(x - 1)$

When $2x^3 - x^2 - 7x + 6 = 0$,

$$(x + 2)(2x - 3)(x - 1) = 0$$

That is, $x = -2$, $\frac{3}{2}$ and 1.

Exercise 1E

Factorise:

1 $x^3 - 4x + 3x^2 - 12$

2 $x^3 + x^2 - 10x + 8$

3 $2x^3 + x^2 - 13x + 6$

4 $6x^3 - x^2 - 32x + 20$

5 $3x^3 + 17x^2 - 27x + 7$

Find the remainder when $f(x)$ is divided by $g(x)$:

6 $f(x) \equiv 3x^3 + 2x^2 - 6x + 1, \quad g(x) \equiv x - 2$

7 $f(x) \equiv 5x^4 - 2x^3 + 3x - 2, \quad g(x) \equiv x + 3$

8 $f(x) \equiv 2x^3 + 3x^2 - 7x - 14, \quad g(x) \equiv x + 5$

9 $f(x) \equiv 4x^4 + 2x^2 - x - 7, \quad g(x) \equiv 2x - 1$

10 $f(x) \equiv 4x^3 + 6x^2 + 3x + 2, \quad g(x) \equiv 2x + 3$

11 When divided by $x + 1$, the polynomial $ax^3 - x^2 - x + 6$ leaves a remainder of 4. Find the value of the constant a.

12 When divided by $x - 2$, the polynomial $x^3 - ax^2 + 7x + 2$ leaves a remainder of -4. Find the value of the constant a.

13 When divided by $x + 3$, the polynomial $2x^3 + x^2 + ax + 1$ leaves a remainder of -53. Find the value of the constant a.

14 When divided by $3x - 1$, the polynomial $9x^3 + 9x^2 + ax + 2$ leaves a remainder of $4\frac{1}{3}$. Find the value of the constant a.

15 When divided by $2x - 3$, the polynomial $4x^3 - ax^2 - 2x + 7$ leaves a remainder of 13. Find the value of the constant a.

16 When divided by $x - 1$, the polynomial $ax^3 + x^2 + bx - 4$ leaves a remainder of -6. Given that $x - 2$ is a factor of the polynomial, find the values of the constants a and b.

17 When divided by $x + 1$, the polynomial $ax^3 + bx^2 - 13x + 6$ leaves a remainder of 18. Given that $2x - 1$ is a factor of the polynomial, find the values of the constants a and b.

18 When divided by $x - 1$, the polynomial $3x^3 + ax^2 - 5x + 2$ leaves a remainder of -4. Find the value of the constant a and hence factorise the polynomial.

19 Given that $2x^3 + ax^2 + x - 12$ leaves a remainder of 6 when divided by $x + 2$, find the value of the constant a. Hence solve the equation

$$2x^3 + ax^2 + x - 12 = 0$$

20 When divided by $x - 1$ the polynomial $ax^3 - 3x^2 + bx + 6$ leaves a remainder of -6. When divided by $x + 2$ it leaves a remainder of zero. Find the values of the constants a and b and hence solve the equation

$$ax^3 - 3x^2 + bx + 6 = 0$$

21 $$P(x) \equiv x^4 + x^3 - 5x^2 + ax + b$$

Given that $(x - 2)$ and $(x + 3)$ are factors of $P(x)$, find the constants a and b.

Factorise $P(x)$ completely. [L]

22 $$f(x) \equiv 2x^3 + px^2 + qx + 6$$

where p and q are constants.

When $f(x)$ is divided by $(x + 1)$, the remainder is 12. When $f(x)$ is divided by $(x - 1)$, the remainder is -6.

(a) Find the value of p and the value of q.

(b) Show that $f(\frac{1}{2}) = 0$ and hence write $f(x)$ as the product of three linear factors. [L]

23 Find the value of the constant k so that the polynomial $P(x)$, where

$$P(x) \equiv x^2 + kx + 11$$

has a remainder 3 when it is divided by $(x - 2)$.

Show that, with this value of k, $P(x)$ is positive for all real x. [L]

24 The polynomial $f(x)$, where

$$f(x) \equiv 2x^3 + Ax^2 + Bx - 3$$

is exactly divisible by $(x - 1)$ and has remainder $+9$ when divided by $(x + 2)$. Find the values of the constants A and B.

Hence solve the equation $f(x) = 0$. [L]

SUMMARY OF KEY POINTS

1 Identities can be processed by

(a) comparing coefficients of LHS and RHS of the identity,

(b) substituting particular values of the variable into the identity,

(c) a combination of (a) and (b).

2 When using the method of long division, if the polynomial into which you are dividing is of degree n, then it must be written down with $n + 1$ terms even if some of these have zero coefficients, e.g. $x^3 + 3$ must be written as $x^3 + 0x^2 + 0x + 3$.

3 To multiply two algebraic fractions, you can cancel each common factor in either numerator with the same common factor in either denominator. You then multiply what remains of each numerator to give the numerator of the answer and multiply what remains of each denominator to give the denominator of the answer.

4 To divide one fraction by another, you turn the divisor upside down and multiply.

5 To add or subtract two fractions you change each fraction to an equivalent fraction with a denominator equal to the lowest common multiple of the two original denominators, and then add or subtract.

6 A proper fraction with linear factors in the denominator such as

$$\frac{\mathrm{f}(x)}{(x-1)(x-2)(x-3)},$$

has partial fractions of the form

$$\frac{A}{x-1}+\frac{B}{x-2}+\frac{C}{x-3}$$

7 A proper fraction with a non-reducible quadratic factor in the denominator, such as

$$\frac{\mathrm{f}(x)}{(x+2)(x^2+2x+3)}$$

has partial fractions of the form

$$\frac{A}{x+2}+\frac{Bx+C}{x^2+2x+3}$$

8 A proper fraction with a repeated factor in the denominator, such

as $\frac{\mathrm{f}(x)}{(x+2)^3}$ has partial fractions of the form

$$\frac{A}{x+2}+\frac{B}{(x+2)^2}+\frac{C}{(x+2)^3}$$

9 An improper fraction is one where the degree of the numerator is greater than or equal to the degree of the denominator

10 With an improper fraction the denominator must first be divided into the numerator before attempting to split it into partial fractions.

11 The remainder theorem states that if a polynomial $f(x)$ is divided by $\alpha x - \beta$, the remainder is $f\left(\frac{\beta}{\alpha}\right)$.

12 The factor theorem states that if $\alpha x - \beta$ is a factor of $f(x)$ then $f\left(\frac{\beta}{\alpha}\right) = 0$.

The mathematics of uncertainty

2

In many areas of your studies you may already have found it necessary to collect, classify and interpret data. You may have done this during your work in Science, Business Studies, Geography and Mathematics. You may also have used sets of data that have been collected by someone else. Every day the media present you with data on matters of concern and interest together with their interpretation of the data. This interpretation may or may not appear to be fair, depending on the point that is being made and the way in which the data are being used.

Statistics is the branch of mathematics concerned with methods of collecting and interpreting data. It can help us to make decisions in the face of uncertainty. For example, any claims made by a manufacturer about the effectiveness of a new fertiliser on various crops would require careful analysis and investigation before they could be accepted.

2.1 Different types of data

One dictionary defines data as 'a set of observations, facts or measurements'. The something you observe, measure and record is known as a **variable** or **variate**. Here are some examples of variates:

- the colours of snooker balls;
- the number of children in each house on an estate;
- the heights of conifer trees in a plantation.

These are all variates since a set of measurements or observations can be defined and obtained for each of them.

Here are the three variates again, shown with typical measurements or observations:

Variate	*Measurement or observation*
Colour of snooker balls	red, white, yellow, green, blue, pink, black, brown
Number of children in each house on an estate	0, 1, 2, 3, 4, 5, 6, . . .
Height of conifer trees	20 m, 22.65 m, 35.60 m

By considering these variates you can see that there are important differences between them.

The colour of a snooker ball is called a **qualitative** variate because numerical values cannot be assigned to it in the way they can to the other variates. With a qualitative variate it is necessary to assign **non-numerical descriptors**. For the qualitative variate *colour* we use the descriptors red, white, blue, and so on. For the qualitative variate *nationality* we would use descriptors such as English, French, and German.

The number of children on an estate and the heights of conifer trees are both called **quantitative** variates because they can be counted or measured numerically using numbers such as 4 children in a house or a tree of height 12.73 m.

The number of children in each house on an estate is always a whole number. That is, the variate can only take specific values such as 0,1,2,3 . . . and it increases in steps. Variates like these are called **discrete**, meaning the values of the variate are separate or discrete from one another. Shoe size is a typical discrete variate where the sizes go up in halves: $4\frac{1}{2}$, 5, $5\frac{1}{2}$, 6 . . . and so on.

The heights of conifer trees in a plantation can, in theory, take any value in the range between the shortest and the tallest. A variate such as height that can take any value in a range is called a **continuous** variate.

You need to be able to distinguish between the different types of variates. To summarise:

- A variate may be either qualitative or quantitative.
- A quantitative variate may be either discrete or continuous.

2.2 Populations and samples

You can use a statistical **survey** to investigate some property or characteristic of a set of people or items that have something in common. When you collect data about every member of the set or

population of people or items in the survey it is called a **census**. The government conducts a census of everyone living in the United Kingdom every 10 years. This is very costly in terms of both time and money.

A less costly way of conducting a survey is to consider only a **sample** (a limited number of members) of the population. Sample surveys are extremely popular. They are used by the government, by the marketing departments of large industrial and commercial firms to help them develop their products, and by medical researchers to give just a few examples.

Any individual member of a population is sometimes called a **sampling unit**. The collection of all sampling units makes up the population and this is called the **sampling frame**. You could take the electoral roll of a town as a sampling frame for collecting data about the people living in the town who were on the electoral roll when it was last printed.

In any survey you want to obtain data that is as representative of the population as possible and not biased in any way. For example, if you wanted to survey people's sweet-eating habits your sample would be extremely biased if it only included college students.

The best way of obtaining representative data is to use a process known as **random sampling**. A random sample is one in which every member of the population has an equal chance of being selected in the sample. The idea of a random sample is easy to understand in principle, but it is not easy in practice to obtain one from a large population. If the population is reasonably small then each member can be assigned a different numbered card, and cards can be selected at random as required until a sample of the required size is obtained. If the population is large so that the numbered card approach is unrealistic, then a different number can be assigned to each member of the population and a table of random sampling numbers can be used to choose the members of the sample.

Example 1

Comment critically on the following methods of obtaining a representative sample of a population.

(a) A survey is to be conducted about the sporting interests of young adult males in Wolverhampton. One suggestion for collecting data is to interview a sample of those leaving a home match at the Wolves Football Ground.

(b) A survey is to be conducted about the employment of young females living on a housing estate. One suggestion for collecting data is to call at every fifth house on the estate during a Thursday morning and, for those where no answer is obtained, try the next house and the next until someone is found to be at home.

(a) This method of collecting data would certainly produce a sample biased towards those interested in football in particular and sport in general. Also, 'young adult males' is not a clear enough definition for members of the proposed sample.

(b) This method would certainly introduce serious bias towards women who are unemployed, with most of those having employment being absent from the estate at the time of the survey. Also, 'young females' is not a sufficiently clear description of the members wanted in the survey.

Exercise 2A

1 State whether the following variates are (a) qualitative (b) quantitative. When a variate is quantitative, say whether it is (c) discrete (d) continuous.

(i) the number of days in each month
(ii) the distance, in km, travelled by a headteacher daily
(iii) the makes of all the teachers' cars in a school car park
(iv) the number of potatoes obtained from each plant in a row
(v) the time taken by each member of your class to complete a statistics project.

2 A safety committee is interested in estimating the percentage of commercial vehicles with illegal tyres. Suggest possible ways in which a random sample of commercial vehicles could be obtained so that the tyres can be checked.

3 A questionnaire is posted to every home in a county to collect data about the personal preferences of housewives for different kinds of detergent. Explain briefly why the sample of data collected is biased.

4 (a) Explain the terms 'sampling unit' and 'sampling frame'.
(b) A representative sample of the television viewing habits of the adult population in Avon is required. A sample of 200 people is collected by conducting a verbal questionnaire with:

(i) the first 100 men and the first 100 women coming out of the mainline station in Bristol at 9 a.m. on a weekday

(ii) 100 men and 100 women in a busy shopping precinct at Bath on a Saturday afternoon.

Comment critically on these methods of obtaining a representative sample and suggest a way in which you might select a sample of your own.

5 A time-and-motion study team at a large factory wishes to obtain a random sample of:

(a) electric light bulbs coming off the production line

(b) the lunch-time eating habits of their production line workers

(c) the absentee rate for their workers on different days of the week.

Suggest a possible method of collecting data for each sample.

2.3 Presenting data in a frequency distribution

In the examples and exercises in the rest of this chapter it is assumed that any sample data used are representative of the population being considered and that the data have been collected using a random sampling procedure.

When data have just been collected they are sometimes called **raw data**. That is, the data have not been sorted into any order or put in a form which is easy to analyse. Remember, you will want to use the data to make observations or draw conclusions about the population from which the sample was collected. A **frequency distribution** is often used to summarise raw data and to make it easier to analyse. The following example shows you how to do this.

Example 2

The temperatures in °C at noon for the first 14 days of July in Penzance, Cornwall were recorded to the nearest °C as:

22, 27, 19, 23, 19, 18, 27, 27, 25, 23, 26, 26, 27, 28

Summarise these data using a frequency distribution.

First notice that the temperatures range from 18 °C to 28 °C. Make a table of values from 18 to 28 and record the number of times or frequency with which each temperature occurred. The table then shows the data as a frequency distribution:

Temperature (°C)	18	19	20	21	22	23	24	25	26	27	28
Frequency (number of days)	1	2	0	0	1	2	0	1	2	4	1

In this example the frequency distribution does not show any obvious pattern in the data.

2.4 Using histograms to present data

In the previous example temperature data were recorded to the nearest whole degree and frequencies were collected for each individual temperature between 18 and 28 degrees. The data were not grouped in any way.

From your GCSE work you will be familiar with the idea of grouping data and using a histogram to present them. For example, here are some test data from an experiment given as a grouped frequency distribution.

Temperature T °C	$0 \leqslant T < 2$	$2 \leqslant T < 4$	$4 \leqslant T < 6$	$6 \leqslant T < 8$	$8 \leqslant T < 10$	$10 \leqslant T < 12$
Frequency	2	7	3	9	5	6

The temperatures have been grouped into classes, with $0 \leqslant T < 2$ meaning any temperature greater than or equal to 0 °C and less than 2 °C.

These data can be displayed as a **histogram**:

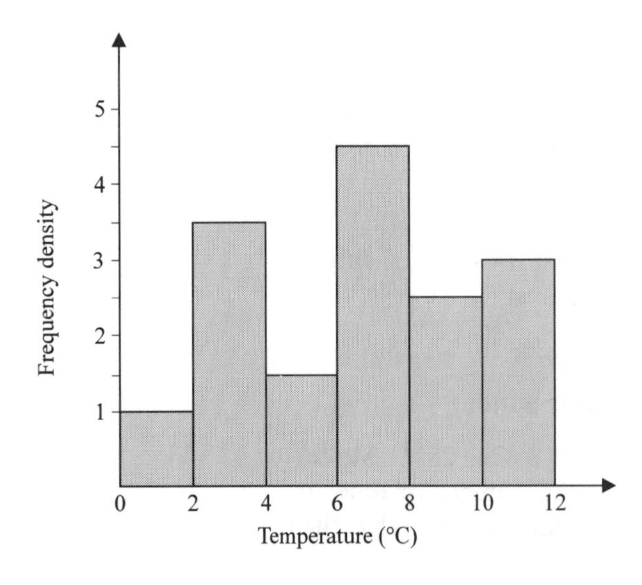

This histogram looks very similar to a bar chart because every rectangle is of equal width. This is because the temperature data have been grouped into equal class widths of two degrees at a time, from 0 to 2 degrees, from 2 to 4 degrees and so on. But there is no reason why data need always be grouped in equal class widths, especially if there are particular ranges in which no data fall at all.

In a bar chart all the bars are the same width and the *length* of each bar is proportional to the frequency it represents. By contrast, the width of a rectangle on a histogram depends on the class width it represents and:

- **The area of a rectangle on a histogram is proportional to the frequency it represents.**

Notice that the vertical axis of a histogram is labelled 'frequency density'. The reason for this will become clear shortly.

The difference between a histogram and a bar chart can be seen by looking at histograms of the same set of data grouped in two different ways. For example, here are two different ways of grouping another set of temperature data:

Temperature T °C	$0 \leqslant T < 2$	$2 \leqslant T < 4$	$4 \leqslant T < 6$	$6 \leqslant T < 8$	$8 \leqslant T < 10$	$10 \leqslant T < 12$
Frequency	2	0	1	0	5	6

The same data can be grouped differently. The three class widths $2 \leqslant T < 4$, $4 \leqslant T < 6$ and $6 \leqslant T < 8$ could be replaced by one class width $2 \leqslant T < 8$ like this:

Temperature T °C	$0 \leqslant T < 2$	$2 \leqslant T < 8$	$8 \leqslant T < 10$	$10 \leqslant T < 12$
Frequency	2	1	5	6

Here are the histograms for the two different groupings of the same data:

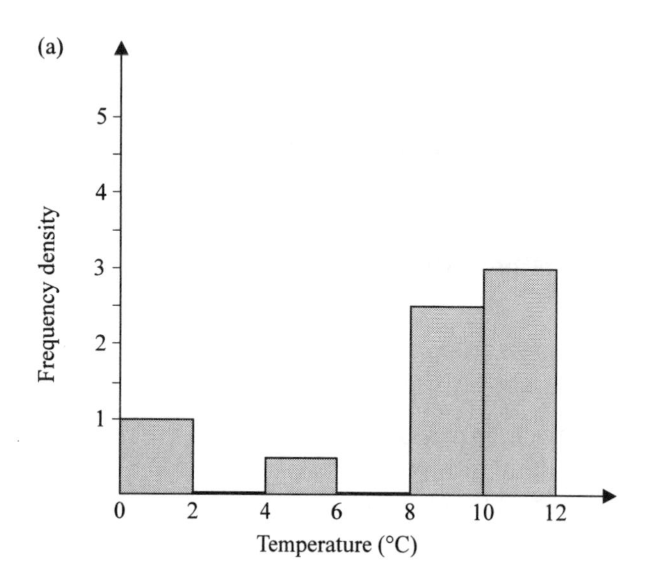

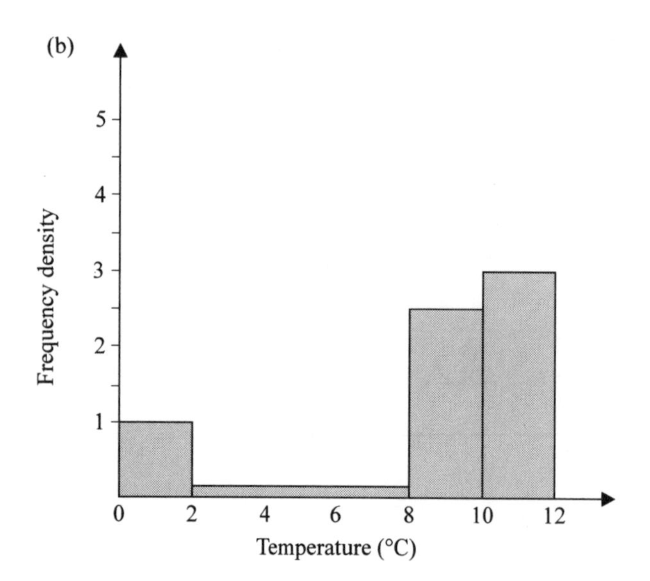

Notice that there are no gaps, as such, between adjacent rectangles of a histogram. (In (a) above, the rectangles are of zero height in the 2-4 and 6-8 intervals.) This is because the variate is continuous for the range of the data. The area of each rectangle is proportional to the frequency it represents. The total area of the histogram is the sum of the areas of all the rectangles in the histogram.

Frequency density

Now let's look at how the height of a histogram rectangle is calculated. In a frequency distribution of a continuous variate the class widths are found by calculating the difference between the upper class boundary and the lower class boundary for each class.

The height of each rectangle in a histogram is found by using the fact that the area of each rectangle is proportional to the frequency it represents. That is:

$$\text{Area} = \text{frequency} \times \text{constant of proportionality}$$

This constant can take any value but for convenience it is usually taken to be 1, giving:

$$\text{Area of rectangle} = \text{frequency}$$

This allows for easy reading and interpretation of histograms.

Now:

Area of rectangle = Class width of rectangle × Height of rectangle

So: $$\text{Height of rectangle} = \frac{\text{Frequency}}{\text{Class width}}$$

This calculation gives the **frequency density per unit of the variate** or, more simply the **frequency density**.

The horizontal axis of the histogram is labelled with the name of the variate. The vertical axis is labelled 'frequency per unit of the variate' or just 'frequency density'.

- **Remember that:** $$\textbf{Frequency density} = \frac{\textbf{Frequency of class}}{\textbf{Class width}}$$

Example 3

The lengths in cm of the stems of 50 daisy flowers are shown in this frequency table. Draw a histogram to display these data.

Length (x cm)	$0 \leqslant x < 20$	$20 \leqslant x < 30$	$30 \leqslant x < 35$	$35 \leqslant x < 40$	$40 \leqslant x < 50$	$50 \leqslant x < 60$
Frequency	4	16	12	10	6	2

First find the class widths. Then use the relationship:

$$\text{Height of rectangle} = \frac{\text{Frequency}}{\text{Class width}}$$

to find the height of each rectangle: the frequency density.

Length (x cm)	$0 \leqslant x < 20$	$20 \leqslant x < 30$	$30 \leqslant x < 35$	$35 \leqslant x < 40$	$40 \leqslant x < 50$	$50 \leqslant x < 60$
Class width (width of rectangle)	20	10	5	5	10	10
Frequency (area of rectangle)	4	16	12	10	6	2
Frequency density (height of rectangle)	0.2	1.6	2.4	2.0	0.6	0.2

Now that you know both the width and the height of each rectangle the histogram can be drawn:

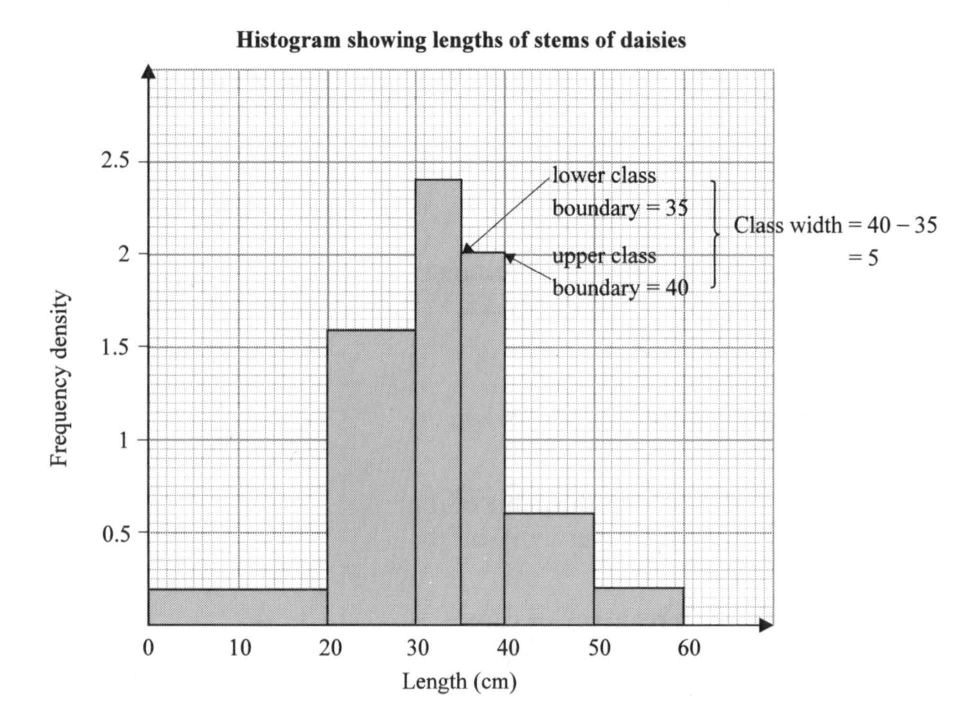

2.5 Using cumulative frequency polygons to present data

The data in a grouped frequency distribution can also be presented as a **cumulative frequency distribution**. The cumulative frequency at any point in the distribution is obtained by adding together all the frequencies up to that point.

Example 4

Here is a grouped frequency distribution showing the volumes of concentrated juice in a sample of 250 bottles that have been filled by an automatic pump. Each volume is measured to the nearest millilitre.

Volume (mℓ)	< 993.5	$993.5 \leqslant V < 995.5$	$995.5 \leqslant V < 997.5$	$997.5 \leqslant V < 999.5$	$999.5 \leqslant V < 1001.5$
No. of bottles	0	10	27	52	60

Volume (mℓ)	$1001.5 \leqslant V < 1003.5$	$1003.5 \leqslant V < 1005.5$	$1005.5 \leqslant V < 1007.5$
No. of bottles	49	39	13

There are no bottles containing a volume of juice less than 993.5 mℓ.

There are 10 bottles containing a volume of juice less than 995.5 mℓ.

There are $(10 + 27)$ bottles containing a volume of juice less than 997.5 mℓ.

You continue this process of **cumulation** until you have considered each class interval. The cumulative frequency distribution can now be summarised in a table:

Volume (mℓ)	< 993.5	< 995.5	< 997.5	< 999.5	< 1001.5	< 1003.5	< 1005.5	< 1007.5
Cumulative frequency	0	10	37	89	149	198	237	250

These data are used to draw a cumulative frequency polygon by plotting the *cumulative frequencies* against the *upper class boundaries* of the corresponding classes. In this case, these data are plotted at (993.5,0), (995.5,10) and so on. Here is the cumulative frequency polygon with the points joined by straight lines:

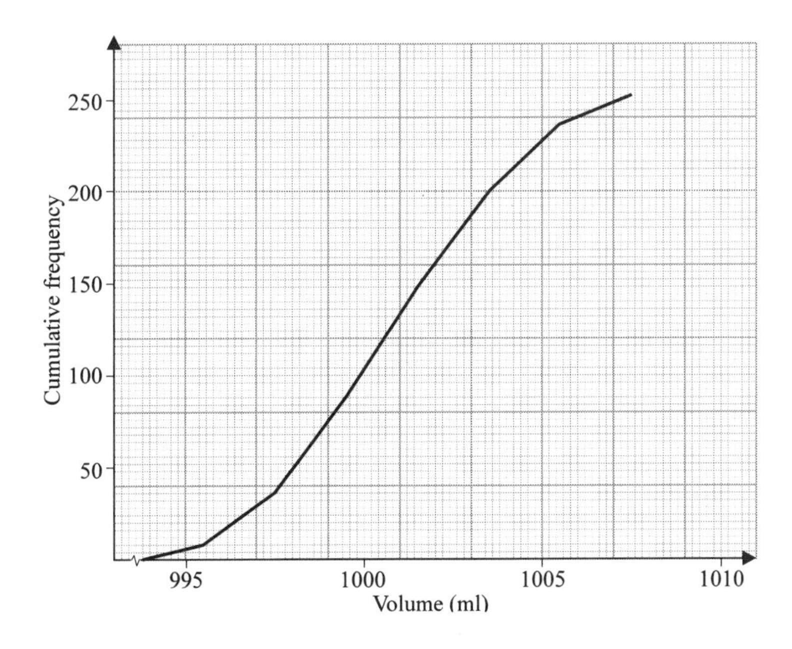

Exercise 2B

1 The masses of strawberries, to the nearest lb, picked by a group of students on a particular day in July were recorded and are shown in the following table.

Mass (lb)	0–19	20–29	30–39	40–49
No. of students	16	26	36	22

Show this information in a carefully marked histogram.

2 In a survey about the mileage covered by a brand of radial tyre, a manufacturer recorded the following:

Mileage (000 miles)	10–12	13–16	17–20	21–25
Number of tyres	24	36	48	60

In this frequency distribution, the mileage covered by each tyre was rounded to the nearest 1000 miles before entry into the table.

Draw an accurate histogram to show this information.

3 The % marks of 200 recruits in an initative test for the Marines are given in the following cumulative frequency table.

Marks	$\leqslant 20$	$\leqslant 30$	$\leqslant 40$	$\leqslant 60$	$\leqslant 100$
Cumulative frequency	12	32	75	149	200

(a) Draw the cumulative frequency polygon for these data.

(b) Draw a histogram to show these data.

4 The Neversofar Ladies Athletic Club have an open 1500 m race every August. Last year 40 runners took part with the following results timed to the nearest second:

Time (s)	240–259	260–269	270–274	275–290
No. of runners	8	10	15	7

Draw a histogram to show this information.

5 Draw a cumulative frequency polygon for the following continuous data where x has been rounded to the nearest whole number.

x	1–5	6–10	11–15	16–20	21–25	26–30	31–35	36–40
Frequency	2	8	17	32	25	12	9	5

6 Reclassify the data in question 5 into the four intervals 1–15, 16–20, 21–30 and 31–40 and draw the corresponding histogram.

7 The lengths, to the nearest second, of the first 100 calls to be answered by the operators at the complaints section of a large hotel chain were recorded and classified in the frequency distribution shown.

Time (s)	–20	–40	–50	–60	–80	–100	–120
Frequency	3	7	12	22	35	16	5

Display this information in a cumulative frequency polygon.

8 Display the information given in question 7 in a histogram.

9 The frequency distribution of a continuous variate is illustrated by the histogram shown.

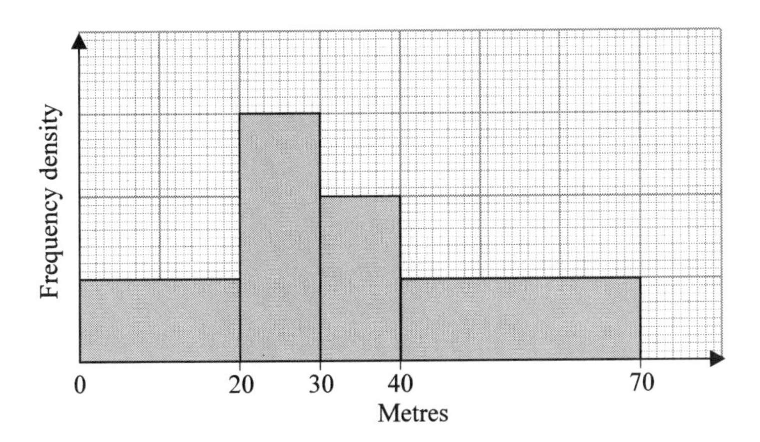

Given that the total frequency is 20, write down the class intervals and the corresponding frequencies.

Draw a cumulative frequency polygon for these data.

10 The cumulative frequency polygon of a continuous variate is shown. Deduce the class boundaries and the corresponding frequencies. Draw a histogram to show these data.

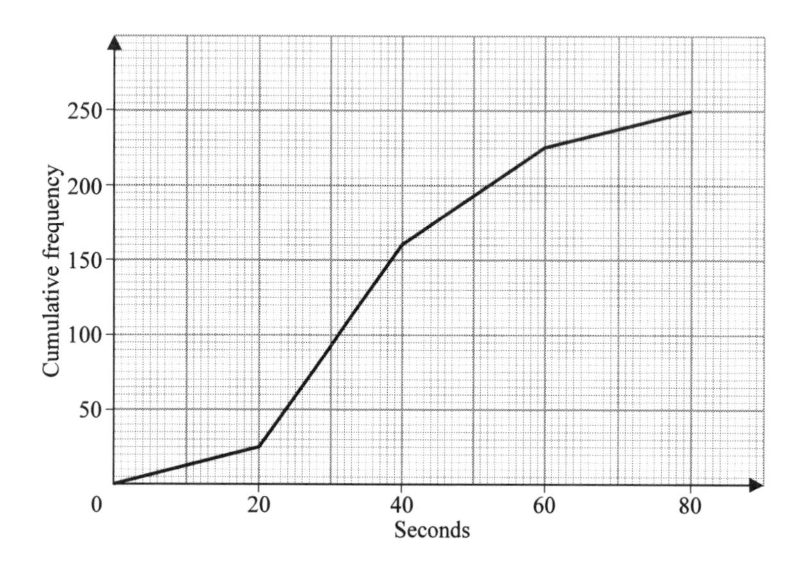

2.6 Measures of location

You may have learned methods for finding the **mode**, the **median** and the **mean** of a set of data in your GCSE course. These measures are known collectively as **measures of location** because they act as a focus for the data and can be used as single values to represent the data. Here is a definition of each one:

The mode

For a list of items or numbers, the **mode** is the one that occurs most often. Sometimes there are two or more items in the list that occur with equal frequency and more often than any other item. So there may be two or more modes in a distribution.

The median

The **median** is the middle value of an ordered set of data. It is obtained by first rewriting the list in order of size. If this ordered list has an odd number of items, find the middle member and this is the median. If the list has an even number of items, find the two central members in the ordered list and the median is halfway between them.

The mean

The **mean** is obtained by dividing the total sum of all the items in the list, or frequency distribution, by the number of items in the list.

Example 5

A young professional golfer made the following scores on the first 18 holes in a tournament:

4, 4, 7, 8, 5, 10, 6, 6, 5, 4, 7, 5, 10, 7, 4, 6, 5, 4

Find her (a) modal score (b) median score (c) mean score.

First rearrange these data in a frequency table:

Score (x)	4	5	6	7	8	9	10
Frequency (f)	5	4	3	3	1	0	2

(a) From the table it is clear that she made a score of 4 at five holes and this was the score that occurred most frequently. So the mode is 4 because this score occurred most often.

(b) When all the scores are arranged in order of size:

4, 4, 4, 4, 4, 5, 5, 5, 5, 6, 6, 6, 7, 7, 7, 8, 10, 10

you can see that the ninth and tenth scores, that is the two middle scores, are 5 and 6 respectively. So the median score is halfway between 5 and 6, which is 5.5.

(c) There are 18 scores in the distribution, so the mean score is obtained by summing all the scores and then dividing the sum by 18:

$$\begin{aligned}\text{Sum of all scores} &= (4 \times 5) + (5 \times 4) + (6 \times 3) + (7 \times 3) + (8 \times 1)\\ &\quad +(9 \times 0) + (10 \times 2)\\ &= 20 + 20 + 18 + 21 + 8 + 0 + 20\\ &= 107\end{aligned}$$

So the mean is: $$\text{Mean} = \frac{107}{18} = 5.94$$

Σ formula for the mean

A formula for the mean can be written using the sigma (Σ) notation (see page 96 in Book P1). In example 5 the scores can be written as $x_1, x_2, x_3, \ldots$ occurring with frequencies $f_1, f_2, f_3, \ldots$ respectively.

The mean of a frequency distribution is often denoted by $\bar{x}$ and from the definition of the mean:

$$\bar{x} = \frac{f_1x_1 + f_2x_2 + f_3x_3 + \ldots + f_nx_n}{f_1 + f_2 + f_3 + \ldots + f_n}$$

So:

$$\bar{x} = \frac{\Sigma fx}{\Sigma f}$$

In example 5:

$$\begin{aligned}\Sigma fx &= (4 \times 5) + (5 \times 4) + (6 \times 3) + (7 \times 3) + (8 \times 1) + (9 \times 0) + (10 \times 2) \\ &= 20 + 20 + 18 + 21 + 8 + 20 = 107\end{aligned}$$

and: $\Sigma f = 18$

So $\bar{x} = \dfrac{107}{18} = 5.94$, as found before.

Finding the modal class, median and mean of a grouped frequency distribution

Often, sample data are collected and summarised in a grouped frequency distribution in which the data are grouped in class intervals. You will often find that you are given just the frequency distribution to work with, particularly when the original sample data are unavailable.

The following examples show how to find estimates of the modal class, the mean, and the median of a grouped frequency distribution. To make these estimates we assume that the members in any class interval are uniformly distributed throughout the interval.

Examples 6, 7 and 8 use the data on the distribution of the lengths of stems of daisy flowers from example 3:

Length (x cm)	$0 \leqslant x < 20$	$20 \leqslant x < 30$	$30 \leqslant x < 35$	$35 \leqslant x < 40$	$40 \leqslant x < 50$	$50 \leqslant x < 60$
Frequency	4	16	12	10	6	2

Example 6
Estimate the modal class of this distribution.

Looking at the histogram you can see that the class interval with the greatest frequency density is $30 \leqslant x < 35$. This is the **modal class**.

Example 7
Estimate the median of the distribution.

Here are two methods of estimating the median:

Method 1

Plot the histogram of the distribution:

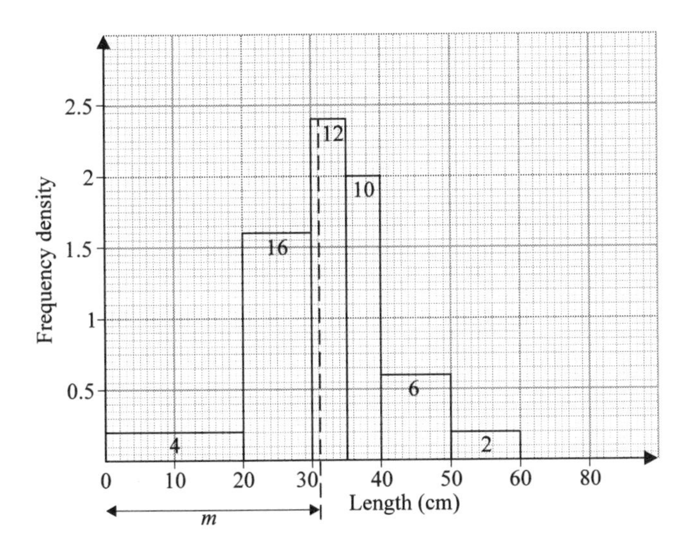

The median is the middle member of the population when all members are arranged in size order. With a frequency distribution this is equivalent to saying that the median occurs at the value of x shown by the dotted line on the histogram *where the area of the whole histogram is divided into two equal parts.*

By looking at the histogram you can see that the dotted line should be drawn in the interval $30 \leqslant x < 35$. Suppose it is at the point on the x-axis where $x = m$, then we have:

Area to left of $x = m$ is: $4 + 16 + 2.4(m - 30) = 2.4m - 52$

Area to right of $x = m$ is: $2.4(35 - m) + 10 + 6 + 2 = 102 - 2.4m$

If the line $x = m$ divides the histogram into two equal areas then:

$$2.4m - 52 = 102 - 2.4m$$

$$4.8m = 154 \text{ and } m = 32.1$$

So the median of the distribution is estimated to be 32.1.

Method 2

This method uses the cumulative frequency polygon. First form the cumulative frequency table:

Length of stem x (cm)	0	<20	<30	<35	<40	<50	<60
Cumulative frequency	0	4	20	32	42	48	50

Plot the points (0,0), (20,4), . . . and so on on graph paper and join successive points with straight lines to form the cumulative frequency polygon:

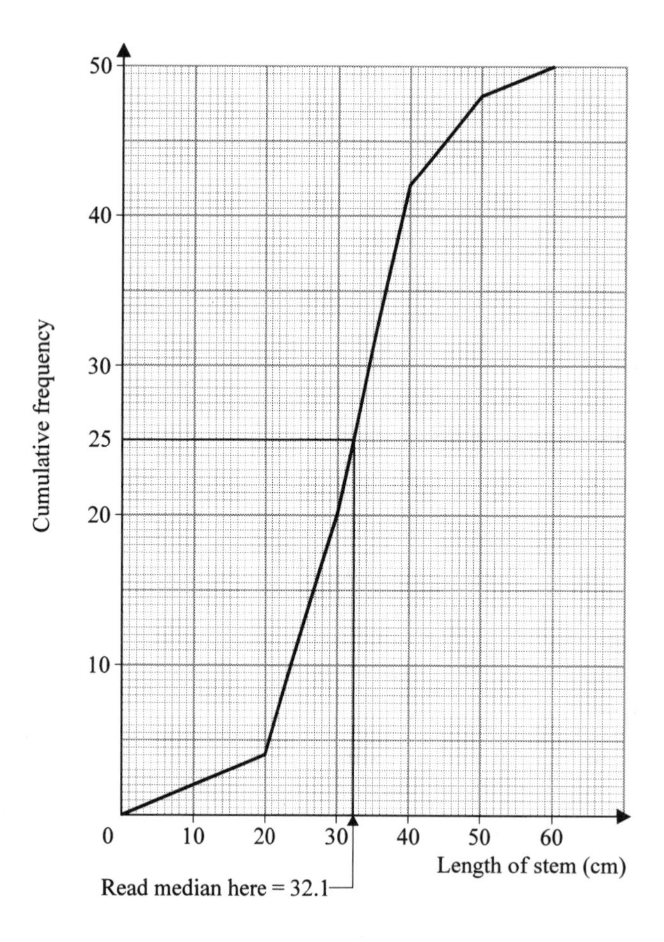

The median estimate is obtained by reading off on the x-axis the point that is equivalent to 25 on the cumulative frequency scale.

Once again the estimate of the median is 32.1.

Example 8

Estimate the mean of the distribution.

As a first step when estimating the mean, tabulate the data:

Length (L cm)	Mid-interval value (M cm)	Frequency f	fM
$0 \leqslant L < 20$	10	4	40
$20 \leqslant L < 30$	25	16	400
$30 \leqslant L < 35$	32.5	12	390
$35 \leqslant L < 40$	37.5	10	375
$40 \leqslant L < 50$	45	6	270
$50 \leqslant L < 60$	55	2	110
	Totals	$\Sigma f = 50$	$\Sigma fM = 1585$

Using the totals in the f and fM columns:

$$\text{Estimate of mean} = \frac{\Sigma fM}{\Sigma f} = \frac{1585}{50} = 31.7$$

Important notes

You will have noticed that in example 7, method 2, the upper class boundaries of the class intervals were used to draw the *cumulative frequency polygon* and to estimate the *median*. But when estimating the *mean* from a frequency distribution, **you must always use the mid-interval values**, as shown in example 8.

Calculations of the mean can be shortened by using coding: see example 14, page 70.

2.7 Measures of dispersion

Measures of location are important statistics when you are considering sample data. The other statistic required is some measure that tells you how much the members of the distribution are spread out. In these two frequency distributions distribution A is much more spread out than distribution B.

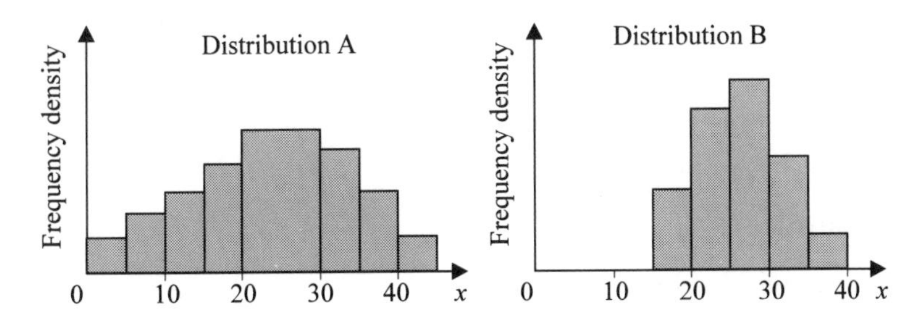

There are several ways of measuring the spread or **dispersion** of distributions like these.

The range

A rather crude but quick way of measuring dispersion is to subtract the smallest value from the largest value in the distribution. This gives a value called the **range**.

One real snag about using the range in this way is that large or small isolated members tend to distort the value and thus give a false impression.

The interquartile range

To offset the snag about using the range, the cumulative frequency polygon or the histogram of the distribution can be used to find more reliable measures of dispersion.

First the data must be divided into a number of equal parts. **Percentiles** divide the data into 100 equal parts and **quartiles** divide the data into four equal parts. The quartiles of an ordered set of data are such that 25% of the data are less than or equal to the first or **lower quartile**, 50% are less than or equal to the second quartile (this is the **median**), and 75% are less than or equal to the third or **upper quartile**.

The median or second quartile of the distribution is often called the **50th percentile** because 50% of the total frequency lies either side of the median. The lower quartile is also called the **25th percentile** and the upper quartile is also called the **75th percentile**.

On the cumulative frequency polygon for a distribution with frequency N, estimates of the lower quartile, the median and the upper quartile can be read off on the horizontal axis of the cumulative frequency polygon at the points with these cumulative frequencies:

$$\frac{N}{4}, \quad \frac{N}{2}, \quad \frac{3N}{4}$$

If N is large, then the points corresponding to $\frac{N}{4}$, $\frac{N}{2}$ and $\frac{3N}{4}$ give good approximations to the lower quartile, the median and the upper quartile. A more detailed description of the method of estimating percentiles is given in chapter 4 of Book T1.

The difference between the upper quartile and the lower quartile is called the **interquartile range** or **IQR**.

- **IQR = Upper quartile – Lower quartile**

The middle 50% of the distribution falls within the IQR.

Note

It is possible to define other measures of spread such as the intersextile range which includes the middle 60% of the distribution. The IQR, however, remains the measure of this kind that is most often used.

Example 9

This frequency distribution shows the number of minutes, to the nearest minute, taken by an ambulance to reach an accident from its station on 50 separate calls. Estimate the IQR.

Time (min)	3	4	5	6	7	8	9	10	11	12
Frequency	2	5	7	12	9	7	4	3	0	1

First rewrite the data as a cumulative frequency distribution:

Time (min)	$\leqslant 2.5$	$\leqslant 3.5$	$\leqslant 4.5$	$\leqslant 5.5$	$\leqslant 6.5$	$\leqslant 7.5$	$\leqslant 8.5$	$\leqslant 9.5$	$\leqslant 10.5$	$\leqslant 11.5$	$\leqslant 12.5$
Cumulative frequency	0	2	7	14	26	35	42	46	49	49	50

Then draw the cumulative frequency polygon:

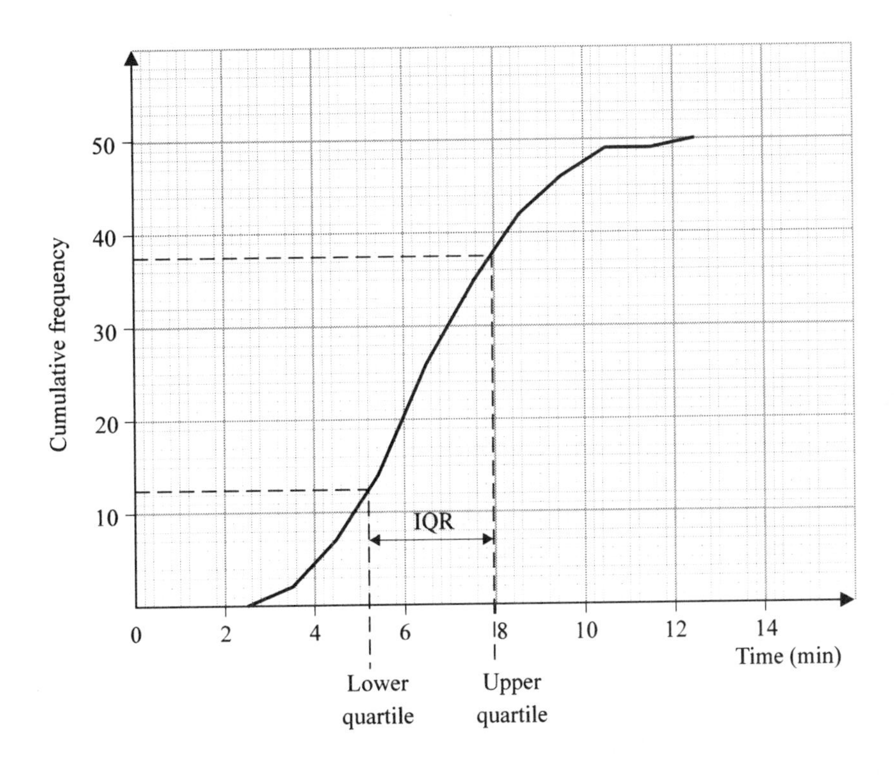

Reading from the cumulative frequency polygon the lower quartile is at:

$$\left(\frac{50}{4}\right) = 12.5\text{th call} = 5.2 \text{ minutes}$$

The upper quartile is at:

$$3 \times \left(\frac{50}{4}\right) = 37.5\text{th call} = 7.9 \text{ minutes}$$

So the interquartile range is:

$$\text{IQR} = (7.9 - 5.2) \text{ minutes} = 2.7 \text{ minutes}$$

Example 10

Estimate the median and the interquartile range of the distribution given in example 4 on page 54.

As $N = 250$, which is large, an estimate of the median can be obtained directly from the cumulative frequency polygon on page 54 by reading off on the horizontal axis the value which corresponds to a cumulative frequency of $\frac{250}{2} = 125$. This estimate of the median is 1000.7 $\text{m}\ell$.

The lower and upper quartiles correspond to cumulative frequencies of $\frac{250}{4} \approx 63$ and $\frac{750}{4} \approx 188$ and are read off as:

$$\text{Lower quartile} = 998.7 \text{ m}\ell$$

$$\text{Upper quartile} = 1003.1 \text{ m}\ell$$

So the interquartile range is:

$$\text{IQR} = \text{Upper quartile} - \text{Lower quartile} = 4.4 \text{ m}\ell$$

To ensure you understand this method, draw your own accurate diagram for these data and use your graph to check the results given.

The variance and the standard deviation σ

The most important measures of dispersion are the **variance** and the **standard deviation** of a set of numbers or of a frequency distribution. The reason for their primary importance is that, together with the mean, they form the basis on which decision making depends in higher statistical work.

Suppose that you have a population of n observations $x_1, x_2, x_3, \ldots x_n$. The mean μ, of this population is found by summing all the observations and dividing the sum by n. (Notice that the symbol μ is used for the mean of a population of numbers instead of $\bar{x}$.) This can be summarised by the formula:

$$\mu = \frac{1}{n}\sum_{r=1}^{n} x_r$$

The variance of the n observations is found by subtracting the mean μ from each observation to give n terms such as $(x_r - \mu)$.

If we tried to sum these terms to get an idea of the overall spread of the data from the mean (μ), we would get:

$$(x_1 - \mu) + (x_2 - \mu) + \ldots + (x_n - \mu)$$
$$= (x_1 + x_2 + \ldots x_n) - n\mu$$

But $$\mu = \frac{(x_1 + x_2 + \ldots x_n)}{n}$$

So this becomes: $$(x_1 + x_2 + \ldots x_n) - n\,\frac{(x_1 + x_2 + \ldots + x_n)}{n}$$
$$= 0$$

This is not a very useful measure! To overcome the difficulty, square each of the $(x_r - \mu)$ terms. This gets rid of the negative terms that cancel to give zero. Then divide the sum of these terms by n to give a measure of the average difference from the mean. This is called the **variance** and it can be written:

- **Variance** $\sigma^2 = \dfrac{\Sigma(x_r - \mu)^2}{n}$

As the units of original data have been squared the units of the variance will be the square of the units of the original data. To return to the original units of the data, take the positive square root of the variance. This gives another measure of dispersion called the **standard deviation**:

- **Standard deviation** $\sigma = \sqrt{\left[\dfrac{\Sigma(x_r - \mu)^2}{n}\right]}$

The units of the standard deviation are the same as those of the data set.

Example 11

Find the mean and the standard deviation of the population of eight numbers 3, 4, 4, 5, 6, 7, 9, 10.

The mean is:

$$\frac{3 + 4 + 4 + 5 + 6 + 7 + 9 + 10}{8} = \frac{48}{8} = 6$$

The variance is:

$$\frac{\Sigma(x-\mu)^2}{n} = \frac{9+4+4+1+0+1+9+16}{8} = 5.5$$

The standard deviation σ is $\sqrt{5.5} = 2.3$ (to 1 decimal place).

Example 12

Find the mean and the standard deviation of the eight numbers

(a) 13, 14, 14, 15, 16, 17, 19, 20
(b) 27, 36, 36, 45, 54, 63, 81, 90
(c) 14, 17, 17, 20, 23, 26, 32, 35

In (a), (b) and (c) refer to example 11, where the numbers are 3, 4, 4, 5, 6, 7, 9, 10. Call this list L.

(a) These numbers are each 10 more than the numbers in list L. The mean is $10 + 6 = 16$ and the standard deviation does not change because the spread is the same as in list L. So the standard deviation is 2.3.

(b) These numbers are each 9 times the corresponding number in L. The mean is therefore 9 times $6 = 54$.

These numbers are also 9 times as spread out as those in list L. The standard deviation is therefore $9 \times 2.3 = 20.7$.

(c) These numbers have been obtained by muliplying each member of L by 3 and adding 5 to the result each time. Another way of saying this is that each new number y is obtained by multiplying x, the original number, by 3 and adding 5. That is, $y = 3x + 5$.

In this case, the mean is $(3 \times 6) + 5 = 23$. The standard deviation is 3 times as wide, which makes its value $3 \times 2.3 = 6.9$.

■ **In general, if a set of data $x_1, x_2, x_3, \ldots$ with mean μ_x and standard deviation σ_x is transformed into another set $y_1, y_2, y_3, \ldots$ by a relation of the form $y = ax + b$, where a and b are constants, the mean μ_y and the standard deviation σ_y of the transformed set are given by the equations:**

$$\mu_y = a\mu_x + b$$

and

$$\sigma_y = a\sigma_x$$

Shorter methods of calculation

Often calculations can be substantially shortened by using the formulae which follow. You will not be expected to prove these formulae but you will be expected to use them appropriately.

For a set of data $x_1, x_2, x_3, \ldots x_n$ occurring with frequencies $f_1, f_2, f_3, \ldots f_n$ respectively you can tabulate and evaluate:

- **Σf the total frequency,**
- **Σfx the sum of all products $f_1x_1 + f_2x_2 + \ldots f_nx_n$,**
- **Σfx^2 the sum of all the products $f_1x_1^2 + f_2x_2^2 + \ldots f_nx_n^2$**
- **The mean is calculated using the formula:**

$$\text{Mean } \mu = \frac{\Sigma fx}{\Sigma f}$$

- **The variance is calculated using the formula:**

$$\text{Variance } \sigma^2 = \frac{\Sigma fx^2}{\Sigma f} - \left(\frac{\Sigma fx}{\Sigma f}\right)^2$$

- **The standard deviation σ is the square root of the variance.**

Example 13

Find the mean, variance and standard deviation of the data from example 5, page 59.

The table shows the score x and the frequency f for the young professional golfer. First find the values of fx and fx^2 and include them in the table:

Score x	Frequency f	fx	fx^2
4	4	16	64
5	5	25	125
6	3	18	108
7	3	21	147
8	1	8	64
9	0	0	0
10	2	20	200
Totals	$\Sigma f = 18$	$\Sigma fx = 108$	$\Sigma fx^2 = 708$

The mean is found using the formula:

$$\mu = \frac{\Sigma fx}{\Sigma f} = \frac{108}{18} = 6 \text{ (as before in example 6)}$$

The variance is found using the formula:

$$\frac{\Sigma fx^2}{\Sigma f} - \left(\frac{\Sigma fx}{\Sigma f}\right)^2 = \frac{708}{18} - \left(\frac{108}{18}\right)^2 = 39.33 - 36 = 3.33$$

The standard deviation σ is the positive square root of the variance $= \sqrt{3.33} = 1.8$.

So the mean score is 6 and the standard deviation is 1.8.

You can often make the calculations easier by using a 'code' to reduce the numbers involved to more manageable values, as the next example shows.

Example 14

Find estimates for the mean and the standard deviation for the data given in example 8, page 63.

Mid-interval value (M)	Frequency f	$x = \frac{M - 32.5}{2.5}$	fx	fx^2
10	4	−9	−36	324
25	16	−3	−48	144
32.5	12	0	0	0
37.5	10	2	20	40
45	6	5	30	150
55	2	9	18	162
Totals	$\Sigma f = 50$		$\Sigma fx = -16$	$\Sigma fx^2 = 820$

The first two columns are as presented in example 8. Then a new variable x is introduced which is 2.5 times smaller than the original scale, and the origin is taken at 32.5, as shown. The mid-interval value 37.5 is $+2$ steps from the new origin and the mid-interval value 25 is -3 steps from the new origin and so on. You can now estimate the mean and the standard deviation of this new variable x by calculating

$$\text{Mean } \mu = \frac{\Sigma fx}{\Sigma f} = -\frac{16}{50} = -0.32$$

$$\text{Standard deviation } \sigma = \sqrt{\left[\frac{\Sigma fx^2}{\Sigma f} - \left(\frac{\Sigma fx}{\Sigma f}\right)^2\right]}$$

$$= \sqrt{\left[\frac{820}{50} - \left(\frac{256}{2500}\right)\right]}$$

$$= \sqrt{16.3} = 4.04$$

In the original data, the mid-interval values were denoted by M and so we have $M = 2.5x + 32.5$

The mean of the M values is $32.5 - 2.5 \times 0.32 = 31.7$

The standard deviation of the M values $= 2.5 \times 4.04 = 10.1$

The coding method reduces the heavy arithmetic quite dramatically although this is not too much of a problem these days as scientific calculators are in universal use. Finally, we estimate the mean as 31.7 and the standard deviation as 10.1.

Using a calculator to find the mean and standard deviation

You need to know how to use your calculator to find the mean and the standard deviation of a set of data.

Enter each member of the data set into the calculator and press a special key, often marked x_n, for each new entry. After you have completed entering all the members press a key, often marked $\bar{x}$, and the mean will be displayed.

Most calculators have two keys marked σ_{n-1} and σ_n. The standard deviation that you require is given by pressing the key marked σ_n.

Your calculator may have keys marked Σx and Σx^2. These can be of use when calculating the mean and the variance of sample data.

Read the manual supplied with your calculator, then work through the examples on mean and standard deviation in the manual. Then use your calculator to check your own calculations from the formulae.

In an exam you should always show sufficient working for the examiner to see how you arrived at your answer. **Remember:** marks for method can only be awarded when the method can be seen!

Exercise 2C

Find the mean and the standard deviation of the sample data sets in questions 1–6:

1 1, 2, 3, 4, 5, 6, 7

2 10, 20, 30, 40, 50, 60, 70

3 10, 11, 12, 13, 14, 15, 16

State any connection you have found in questions 1, 2, 3.

4

x	1	3	5	7	9	11
f	3	4	5	5	2	1

5

x	13–15	16–18	19–21	22–24	25–27
f	2	5	9	4	1

6

x	0	1	2	3	4	5	6	7
f	1	7	12	19	23	18	16	4

7 A data set consists of the five numbers 3, 4, 7, x and y. The mode and the mean of this set are 4 and 6.5 respectively. Find x, y and the median of the set.

8

x	1	2	3	4	5	6	7	8
f	5	17	25	19	13	12	3	1

For the frequency distribution given in the table, estimate (a) the median (b) the range (c) the IQR.

9

Mark	0	1	2	3	4	5	6	7	8	9	10
Frequency	2	3	3	5	5	6	11	8	3	2	2

The marks obtained by 50 trainee secretaries in a dictation speed test are shown in the table. Calculate (a) the mean (b) the median (c) the range (d) the IQR (e) the variance (f) the standard deviation.

10 The masses in grams of 10 pain-killing tablets were 5.99, 5.95, 5.96, 6.02, 5.97, 5.99, 6.01, 6.01, 5.97, 6.03.
Find the median mass, the range and the IQR.

11 The following scores were made by a group of people playing skittles:

Score	0	1	2	3	4	5	6	7	8	9	10
Frequency	2	1	0	0	2	1	12	6	3	2	1

Calculate the mean score, the variance and the standard deviation.

12 The mass of strawberries, to the nearest lb, picked by each of a group of students was recorded and is shown in the following table:

Mass (lb)	0–19	20–29	30–39	40–49
Frequency	16	26	36	22

Estimate the mean mass and the standard deviation.

13 In a survey about the mileage covered by a brand of radial tyre, a manufacturer recorded the following:

Mileage (000 miles)	10–12	13–16	17–20	21–25
Frequency	24	36	48	60

In this distribution, the mileage covered by each tyre was rounded to the nearest 1000 miles before entry into the table. Estimate (a) the median (b) the range (c) the IQR.

14 The % marks of 200 recruits in an initiative test for the Marines are given in the following cumulative frequency table.

Marks	$\leqslant 20$	$\leqslant 30$	$\leqslant 40$	$\leqslant 60$	$\leqslant 100$
Cumulative frequency	12	32	75	149	200

(a) Rewrite these data as a frequency distribution.

(b) Hence find estimates for the mean and the standard deviation of the marks of the recruits.

15 With the data given in question 14, estimate the median and the IQR of the marks of the recruits.

16 The Neversofar Ladies Athletic Club have an open 1500 m race every August. Last year 40 runners took part with the following results:

Time (s)	240–259	260–269	270–274	275–290
Frequency	8	10	15	7

(a) Estimate the mean time taken for all runners.

(b) Estimate the standard deviation time of all the runners.

17 By considering a cumulative frequency polygon, estimate the median time taken by the runners to complete the race in question 16.

18 All the children in a number of families were together at a party. Each child was asked to state the number of children in his or her family. Two children said 1 child, four children said 2 children and twelve said 3 children.

(a) Explain carefully how you decide that there were 8 families at the party.

(b) Show that the mean number of children per family at the party is 2.25.

19 For two data sets, X and Y, it is known that

$$\text{for set } X,\ \Sigma f = 20,\ \Sigma fx = 100 \text{ and } \Sigma fx^2 = 1000$$

$$\text{for set } Y,\ \Sigma f = 30,\ \Sigma fy = 180 \text{ and } \Sigma fy^2 = 1200$$

(a) Find the means of set X and set Y.

(b) Find the variances of set X and set Y.

The two sets are now combined into one data set. Find the mean and the variance of this combined set.

SUMMARY OF KEY POINTS

1 Data are observations, facts or measurements.

2 Variates (variables) are either qualitative or quantitative; a quantitative variate is either discrete or continuous.

3 A survey of the whole population is called a census.

4 Sample surveys are conducted on samples of the population.

5 The population under consideration is known as the sampling frame and a member of the population is known as a sampling unit.

6 A random sample is a subset of the population in which every member of the whole population had an equal chance of being selected for the sample.

7 Data may be represented graphically by histograms and by cumulative frequency polygons.

8 The mean, median and mode of a data set or of a frequency distribution are known collectively as measures of central position.

9 The range, interpercentile ranges, in particular the interquartile range IQR, the variance and the standard deviation of a data set or of a frequency distribution are all measures of spread or dispersion.

10 Notation and formulae you should memorise are:

The mean: $\mu = \dfrac{\Sigma fx}{\Sigma f}$

The variance $\sigma^2 = \dfrac{\Sigma fx^2}{\Sigma f} - \left(\dfrac{\Sigma fx}{\Sigma f}\right)^2$

The standard deviation σ = the square root of the variance

Exponentials and logarithms

3

In Book P1 you were shown how to add, subtract, multiply and divide numbers that are written in the same bases and raised to various powers. This chapter continues the work on powers.

3.1 Laws of logarithms

You may recall from Book P1 that the words **power**, **index**, **exponent** and **logarithm** are synonymous: they are four different words to describe exactly the same thing.

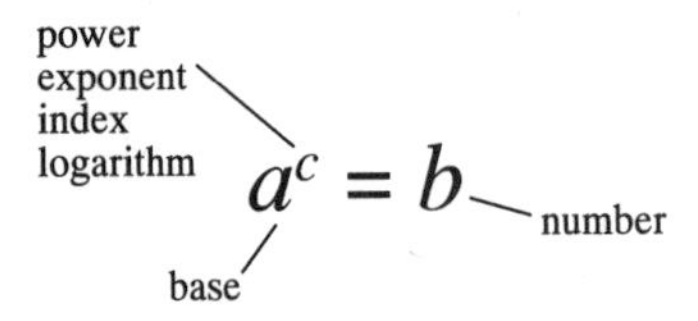

For the time being use the word 'logarithm' and take the base to be positive ($a > 0$). Then b must also be positive. Remember that the graph of $y = a^x (a > 0)$ looks like this:

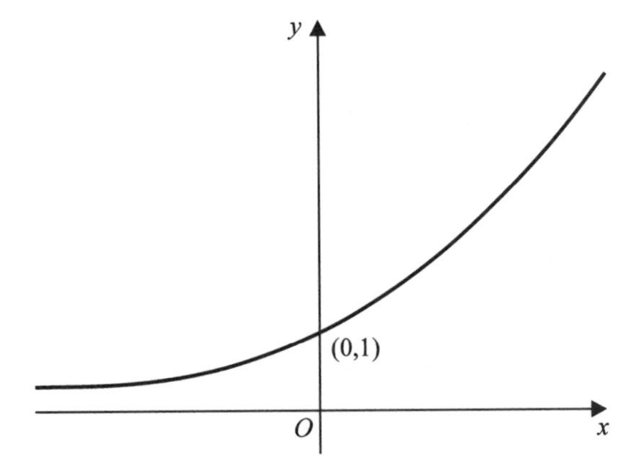

Now the statement $b = a^c$ reads 'the number b is equal to the base a raised to the logarithm c'. Another way of reading the statement is 'c is the logarithm of the number b to the base a'. This can be written

$$c = \log_a b$$

So the statements

$$a^c = b \quad \text{and} \quad c = \log_a b$$

are **identical** and **interchangeable**.

Let $p = \log_a x$ and let $q = \log_a y$, where $a > 0$. Then these two statements can also be written as

$$x = a^p \quad \text{and} \quad y = a^q$$

Now $$xy = a^p \times a^q = a^{p+q}$$

But another way of writing $xy = a^{p+q}$ is

$$p + q = \log_a xy$$

However $$p = \log_a x \text{ and } q = \log_a y$$

■ **So:** $$\mathbf{\log_a xy = \log_a x + \log_a y}$$

Also: $$\frac{x}{y} = \frac{a^p}{a^q} = a^{p-q}$$

But $\frac{x}{y} = a^{p-q}$ can be written

$$p - q = \log_a \frac{x}{y}$$

■ **or** $$\mathbf{\log_a \frac{x}{y} = \log_a x - \log_a y}$$

Now: $$x^n = (a^p)^n = a^{pn}$$

This statement can be written as

$$\log_a x^n = pn$$

But since $p = \log_a x$ then

■ $$\mathbf{\log_a x^n = n \log_a x}$$

If $r = \log_a a$ then

$$a^r = a$$

and this implies that $r = 1$.

■ **So:** $$\mathbf{\log_a a = 1}$$

If $r = \log_a 1$ then

$$a^r = 1$$

But $a^0 = 1$

So: $$r = 0$$

■ **That is** $$\mathbf{\log_a 1 = 0}$$

These five results are very important and you must memorise them.

There are two *common* bases for logarithms, 10 and e.

- **$\log_{10}x$ is usually written $\lg x$.**
- **$\log_e x$ is usually written $\ln x$.**

You should find a 'ln' button on your calculator which will evaluate logarithms to base e. However, many calculators do not have a 'lg' button but instead a button marked 'log'. This is, in fact, the button needed to evaluate logarithms to base 10.

Example 1

Find x if $\log_2 32 = x$.

$$\log_2 32 = x \Rightarrow 2^x = 32$$

But $\qquad 2^5 = 32$

So: $\qquad x = 5$

Example 2

Find the value of $\log_3 243$.

If $y = \log_3 243$

then: $\qquad 3^y = 243$

But $\qquad 3^5 = 243$

So: $\qquad y = 5$

Example 3

Simplify:

$$\log_a 4 + 2\log_a 3 - \log_a 6$$

$$\begin{aligned}
&\log_a 4 + 2\log_a 3 - \log_a 6 \\
&= \log_a 4 + \log_a 3^2 - \log_a 6 \\
&= \log_a 4 + \log_a 9 - \log_a 6 \\
&= \log_a(4 \times 9) - \log_a 6 \\
&= \log_a \frac{36}{6} \\
&= \log_a 6
\end{aligned}$$

Example 4

Express $\log_a \frac{x^3}{y^2 z}$ in terms of $\log_a x$, $\log_a y$ and $\log_a z$.

$$\log_a \frac{x^3}{y^2 z} = \log_a x^3 - \log_a y^2 z$$

$$= 3\log_a x - [\log_a y^2 + \log_a z]$$

$$= 3\log_a x - \log_a y^2 - \log_a z$$

$$= 3\log_a x - 2\log_a y - \log_a z$$

It is possible to express a logarithm written in one base as a logarithm in another base. This is particularly useful when, say, you need to calculate the value of a logarithm in base 7, for which there is no button on your calculator. If you can convert the logarithm from base 7 to base 10 or base e, then you can use your calculator to work out its value.

If $y = \log_a b$

then $a^y = b$

So, taking logarithms to base c gives

$$\log_c(a^y) = \log_c b$$

that is: $y \log_c a = \log_c b$

So: $y = \frac{\log_c b}{\log_c a}$

■ **Thus:** $\boldsymbol{\log_a b = \frac{\log_c b}{\log_c a}}$

Example 5

Calculate, to 3 significant figures, the value of $\log_4 7$.

$$\log_4 7 = \frac{\lg 7}{\lg 4} = \frac{0.845\,09\ldots}{0.602\,05\ldots}$$

$$= 1.40 \text{ (3 s.f.)}$$

Example 6

Calculate, to 3 significant figures, the value of $\log_7 5$.

$$\log_7 5 = \frac{\ln 5}{\ln 7} = \frac{1.6094\ldots}{1.9459\ldots}$$
$$= 0.827 \text{ (3 s.f.)}$$

Exercise 3A

Find the exact value of x, showing your working:

1 $\log_2 8 = x$ **2** $\log_3 27 = x$ **3** $\log_x 125 = 3$
4 $\log_x 36 = 0.5$ **5** $\log_2 x = 4$ **6** $\log_4 64 = x$
7 $\log_5 125 = x$ **8** $\log_x 9 = -2$ **9** $\log_5 x = -1$
10 $\log_9 x = 3\frac{1}{2}$

Find the value of:

11 $\log_3 81$ **12** $\log_4 256$ **13** $\log_3 3$
14 $\log_3 \frac{1}{9}$ **15** $\log_7 343$ **16** $\log_{64} 4$
17 $-\log_2(\frac{1}{8})$ **18** $\log_5(\frac{1}{125})$ **19** $\log_8 2$
20 $\log_{27} 3$

Calculate to 3 significant figures the value of:

21 $\log_4 9$ **22** $\log_5 22$ **23** $\log_6 3$
24 $\log_8 5$ **25** $\log_9 11$

Simplify:

26 $\log_3 7 + \log_3 2$ **27** $\lg 15 - \lg 5$
28 $\ln 5 + \ln 6 - \ln 10$ **29** $2\ln 8 - \ln 5 + 2\ln 10$
30 $2\log_a 3 - 3\log_a 2 + 4\log_a 1$ **31** $3\log_a 4 - \log_a 2 - 3\log_a 6$
32 $2\log_a 7 - 2\log_a a + 2\log_a 3$ **33** $\log_a 5 + \frac{1}{2}\log_a 16 - \log_a 2$
34 $5\log_a a + \frac{1}{3}\log_a 27 + \log_a 2$ **35** $\frac{1}{4}\log_a 81 + 3\log_a(\frac{1}{4}) - 2\log_a(\frac{3}{4})$

Express in terms of $\log_a x$, $\log_a y$ and $\log_a z$:

36 $\log_a\left(\frac{xy}{z}\right)$ **37** $\log_a\left(\frac{x^2y}{z^3}\right)$ **38** $\log_a xy^2z^3$

39 $\log_a \sqrt{(xy^2z)}$ **40** $\log_a \frac{xy}{\sqrt{z^3}}$

3.2 Equations of the form $a^x = b$

Although you can solve equations such as $3^x = 9$, $4^x = 64$, $2^x = 128$, and so on, because the value of x is an integer, in general you cannot solve such equations by inspection. It is time consuming to find a solution to $5^x = 67$ by trial and improvement.

The standard method for solving such equations (other than trial and error) is by taking logarithms.

So if $5^x = 67$

then
$$\lg 5^x = \lg 67$$
$$x \lg 5 = \lg 67$$

and hence
$$x = \frac{\lg 67}{\lg 5}$$

This can now be evaluated with the help of a calculator.

$$x = \frac{1.826\,07\ldots}{0.698\,97\ldots}$$
$$= 2.61 \text{ (3 s.f.)}$$

There is no particular reason to take logarithms to base 10. Natural logarithms would do just as well. These would lead to the solution

$$x = \frac{\ln 67}{\ln 5}$$

So:
$$x = \frac{4.204\,69\ldots}{1.609\,43\ldots}$$
$$= 2.61 \text{ (3 s.f.)}$$

Example 7

Solve the equation $4^{x+2} = 51$.

If $4^{x+2} = 51$

then:
$$\lg 4^{x+2} = \lg 51$$
$$(x+2)\lg 4 = \lg 51$$
$$x + 2 = \frac{\lg 51}{\lg 4}$$
$$x = \frac{\lg 51}{\lg 4} - 2$$
$$= 0.836 \text{ (3 s.f.)}$$

Example 8

Solve the equation $(0.3)^{5x} = 0.51$.

If $(0.3)^{5x} = 0.51$

then:

$$\lg(0.3)^{5x} = \lg 0.51$$
$$5x \lg 0.3 = \lg 0.51$$
$$5x = \frac{\lg 0.51}{\lg 0.3}$$
$$5x = 0.559\,26\ldots$$
$$x = 0.112 \text{ (3 s.f.)}$$

Example 9

Solve the equation $3^{2x} - 6(3^x) + 5 = 0$.

Let $y = 3^x$

Since $3^{2x} = (3^x)^2$ the equation becomes

$$y^2 - 6y + 5 = 0$$
$$(y - 5)(y - 1) = 0$$

So: $y = 1$ or 5

That is $3^x = 1$ or $3^x = 5$

$$3^x = 1 = 3^0 \quad \text{or} \quad \lg 3^x = \lg 5$$
$$x \lg 3 = \lg 5$$
$$x = \frac{\lg 5}{\lg 3}$$
$$x = 0 \text{ or } x = 1.46 \text{ (3 s.f.)}$$

Example 10

Solve the equation $2(5^{2x}) - 5^x = 6$.

Let $y = 5^x$

Then: $2y^2 - y = 6$

or: $2y^2 - y - 6 = 0$

$$(2y + 3)(y - 2) = 0$$
$$y = -1\tfrac{1}{2} \text{ or } y = 2$$

So: $5^x = -1\frac{1}{2}$ or $5^x = 2$

Since 5^x is always positive, $5^x = -1\frac{1}{2}$ gives no real value of x.

$$5^x = 2$$

gives

$$\lg 5^x = \lg 2$$

$$x \lg 5 = \lg 2$$

$$x = \frac{\lg 2}{\lg 5}$$

$$x = 0.431 \text{ (3 s.f.)}$$

Exercise 3B

Solve the equations:

1 $2^x = 7$ **2** $3^x = 19$ **3** $4^x = 11$

4 $5^x = 9$ **5** $7^x = 151$ **6** $4^{-x} = 0.125$

7 $3^{x+1} = 23$ **8** $5^{3x+2} = 43$ **9** $2^{x+1} = 3^x$

10 $5^{x+3} = 3^{x-2}$ **11** $6^{2x-1} = 9^{x+3}$ **12** $4^{3x+2} = 7^{x-3}$

13 $2^{2x} - 5(2^x) + 6 = 0$ **14** $2(3^{2x}) - 9(3^x) + 4 = 0$

15 $3(4^{2x}) + 11(4^x) = 4$ **16** $3^{2x+1} = 3^x + 24$

17 $2^{2x+1} = 5(2^x) + 3$ **18** $4^{2x} + 48 = 4^{x+2}$

19 $2e^x + 2e^{-x} = 5$ **20** $25^x = 5^{x+1} - 6$

3.3 Exponential growth and decay

People such as scientists, sociologists and town planners are often more concerned with the *rate* at which a particular quantity is growing than with its current size. The Director of Education is more concerned with the rate at which the school population is increasing or decreasing than with what the population is now, because he has to plan for the future and ensure that there are enough (and not too many) school places available to meet demand each year. The scientist may need to know the rate at which a colony of bacteria is growing rather than how many of the bacteria exist at this moment, or the rate at which a liquid is cooling rather than the temperature of the liquid now, or the rate at which a radioactive material is decaying rather than how many atoms currently exist.

One thing that each of these populations has in common is that their rate of increase is proportional to the size of the population at any time. Now in Book P1 you were shown that $\frac{dy}{dx}$ represents the rate of change of y with respect to x. So if the size of a population at a given time t is P then the rate of increase of the population is a short way of saying 'the rate of increase of the population as time goes on'. That is, it is the rate of change of P with respect to t, i.e. $\frac{dP}{dt}$. Now if this rate of increase is proportional to the size of the population at any given time, then $\frac{dP}{dt} = kP$, where k is a constant. If you take the equation

$$\frac{dP}{dt} = kP$$

and divide both sides by P you get

$$\frac{1}{P}\frac{dP}{dt} = k$$

If you now integrate both sides with respect to t, you get

$$\int \frac{1}{P}\frac{dP}{dt}\, dt = \int k\, dt$$

or:

$$\int \frac{1}{P}\, dP = \int k\, dt$$

i.e.

$$\ln P = kt + C$$

where C is a constant.

Now

$$\ln P = \log_e P$$

So:

$$\log_e P = kt + C$$

and:

$$P = e^{kt+C} = e^{kt} \cdot e^{C}$$

Since both e and C are constants, it follows that e^C is a constant. Call this constant A.

Then:

$$P = Ae^{kt}$$

This demonstrates that the population grows **exponentially**; that is, P is a function of e^t.

Example 11

(a) Newton's law of cooling states that the rate at which the temperature of a preheated body decreases is proportional to the difference between the temperature of the body and that of the surroundings. Given that θ °C is the excess of the temperature of the body over that of the surroundings at time t minutes after the start, show that the relationship between θ and t is the form $\theta = Ae^{-kt}$ where A and k are constants.

(b) A bowl of water whose temperature is ϕ °C, at time t minutes is placed in a room where the temperature remains constant at 15 °C. Given that $\phi = 100$ at $t = 0$, and $\phi = 40$ at $t = 30$, find the temperature of the water at time (i) $t = 10$ (ii) $t = 45$.
Find also the time when (iii) $\phi = 70$ (iv) $\phi = 53$.

(a) If the temperature of the water at time t minutes is ϕ °C then the rate of change of the temperature of the water is $\frac{d\phi}{dt}$

But $\theta = \phi - M$, where M °C is the constant temperature of the surroundings. Differentiating with respect to t gives

$$\frac{d\theta}{dt} = \frac{d\phi}{dt}$$

since M is a constant.

So the rate of change of the temperature of the water can also be written as $\frac{d\theta}{dt}$.

The rate at which the temperature of the water *decreases* is therefore $-\frac{d\theta}{dt}$. (The minus sign indicates that the temperature is decreasing rather than increasing.)

If this rate is proportional to θ then

$$-\frac{d\theta}{dt} = k\theta$$

where k is a constant.

i.e.
$$\frac{d\theta}{dt} = -k\theta$$

$$\frac{1}{\theta}\frac{d\theta}{dt} = -k$$

$$\int \frac{1}{\theta}\frac{d\theta}{dt}\, dt = \int -k\, dt$$

$$\int \frac{1}{\theta}\, d\theta = \int -k\, dt$$

So:
$$\ln \theta = -kt + C$$

where C is a constant.

$$\theta = e^{-kt+C} = e^{-kt}e^{C}$$

$$\theta = Ae^{-kt}, \text{ where } A = e^{C}$$

The minus sign indicates that the temperature of the water is *decreasing*. This is an example of **exponential decay**.

(b) When $t = 0$, $\phi = 100$, $M = 15$ and

$$\theta = 100 - 15 = 85$$

So: $$85 = Ae^{0} \qquad (1)$$

$$A = 85$$

When $t = 30$, $\phi = 40$ and

$$\theta = 40 - 15 = 25$$

So: $$25 = Ae^{-30k} \qquad (2)$$

Substitute $A = 85$ in (2):

$$25 = 85e^{-30k}$$

$$e^{-30k} = \frac{25}{85}$$

$$\ln e^{-30k} = \ln\frac{25}{85}$$

$$-30k = -1.2237 \text{ (since } \ln e = 1)$$

$$k = 0.04 \text{ (1 s.f.)}$$

So: $$\theta = 85e^{-0.04t}$$

(i) $t = 10 \Rightarrow \theta = 85e^{-0.4} \approx 57$

$$57 \approx \phi - 15$$

$$\phi \approx 72$$

(ii) $t = 45 \Rightarrow \theta = 85e^{-1.8} \approx 14$

$$14 \approx \phi - 15$$

$$\phi \approx 29$$

(iii) $\phi = 70 \Rightarrow \theta = 55$

$$55 = 85e^{-0.04t}$$

$$e^{-0.04t} = \frac{55}{85}$$

$$-0.04t = \ln\frac{55}{85}$$

$$t = -\frac{1}{0.04}\ln\frac{55}{85}$$

$$t \approx 11$$

(iv) $\phi = 53 \Rightarrow \theta = 38$

$$38 = 85e^{-0.04t}$$

$$-0.04t = \ln\frac{38}{85}$$

$$t \approx 20$$

Exercise 3C

1 The law of cooling is $\theta = Ae^{-0.02t}$ where θ °C is the excess of temperature of the water over the room temperature at time t minutes and A is a constant. Given that the constant room temperature is 20 °C and that the temperature of the water is 80 °C when $t = 0$, find the temperature of the water when (a) $t = 10$ (b) $t = 20$ (c) $t = 45$.

2 The law of cooling for a bath of water is $\theta = Ae^{-0.05t}$ where θ is the excess of temperature of the water over the temperature of the bathroom at time t minutes and A is a constant. Given that at time $t = 0$ the temperature of the water is 60 °C and that the bathroom has a constant temperature of 15 °C, calculate the value of t when the temperature of the water is (a) 50 °C (b) 35 °C (c) 27 °C.

3 The population of a country grows according to the law $P = Ae^{0.06t}$ where P million is the population at time t years and where A is a constant. Given that at time $t = 0$ the population is 27.3 millions calculate the population when (a) $t = 10$ (b) $t = 15$ (c) $t = 25$.

4 The population of a country grows according to the law $P = 12e^{kt}$ where P million is the population at time t years and k is a constant. Given that when $t = 7$, $P = 15$, calculate the value of t when the population will be (a) 20 million (b) 30 million (c) 35 million.

5 The rate of increase of a population P million at time t years is proportional to the population at that time. Given that at time $t = 0$, $P = 36.4$ and that at time $t = 10$, $P = 41.2$, find the law for the size of the population in the form $P = f(t)$.

Hence calculate the size of the population when (a) $t = 5$ (b) $t = 20$ (c) $t = 23$.

6 A bowl of water stands in a room of constant temperature 18 °C. The water cools according to Newton's law of cooling and θ °C is the excess of the temperature above 18 °C at time t minutes. Initially the water has a temperature of 75 °C. After 5 minutes it has a temperature of 60 °C. Find the temperature of the water when (a) $t = 15$ (b) $t = 20$ (c) $t = 37$.

7 The mass of a radioactive substance is given by $m = m_0 e^{-kt}$ where m is the mass at time t years, m_0 is the initial mass at time $t = 0$ and k is a constant. Given that $m = 0.8m_0$ when $t = 2$, find the value of k. Hence find the mass in terms of m_0 when (a) $t = 7$ (b) $t = 10$ (c) $t = 17$.

8 The mass of a radioactive substance is given by $m = m_0 e^{-kt}$ where m is the mass at time t years, m_0 is the initial mass and k is a constant. Given that $10m = 9m_0$ when $t = 4$, find the value of k. Hence find the value of t when the mass has reduced to (a) $0.8m_0$ (b) $0.6m_0$ (c) $0.45m_0$.

9 The rate of decay of the mass m of a radioactive substance at time t years is proportional to its mass at that time. Given that initially it has a mass m_0 show that $m = m_0 e^{-kt}$, where k is a constant. Given that $m = 0.85m_0$ when $t = 5$ find
(a) the mass in terms of m_0 when $t = 13$
(b) the time at which $m = 0.5m_0$.

SUMMARY OF KEY POINTS

1 $\log_a xy = \log_a x + \log_a y$

2 $\log_a \frac{x}{y} = \log_a x - \log_a y$

3 $\log_a x^n = n \log_a x$

4 $\log_a a = 1$

5 $\log_a 1 = 0$

6 $\log_a b = \frac{\log_c b}{\log_c a}$

7 An equation of the form $a^x = b$ is solved by taking logarithms of both sides.

8 A population whose rate of increase is proportional to the size of the population at any time obeys a law of the form $P = Ae^{kt}$. This is known as exponential growth.

9 A population whose rate of decrease is proportional to the size of the population at any time obeys a law of the form $P = Ae^{-kt}$. This is known as exponential decay.

Algebra II

4

4.1 Reducing equations to linear form

Scientists, as part of their work, frequently conduct experiments. Once they have collected the scientific data they usually want to see whether they can identify a relationship or law that links two variables in the data. This is not an easy task. It usually involves guessing a relationship and then, by drawing a graph, testing to see whether or not the guess is correct. If the guess is incorrect then the scientist has to modify the original guess and go through the process of drawing another graph.

For example, a scientist collected the following data for the two variables x and y:

x	3.1	5.6	7.0	10.4	12.5
y	2.3	9.7	14.2	24.2	30.4

The scientist guessed that x and y are related by a law of the form $y = ax + b$ and wanted to test this.

Now, you were shown in Book P1 that a straight line has an equation of the form $y = mx + c$. The equation $y = ax + b$ is of the same form, so if the data collected from the experiment *do* satisfy a law of the form $y = ax + b$, then when the corresponding values of x and y are plotted on a graph the five sets of data should lie approximately on a straight line:

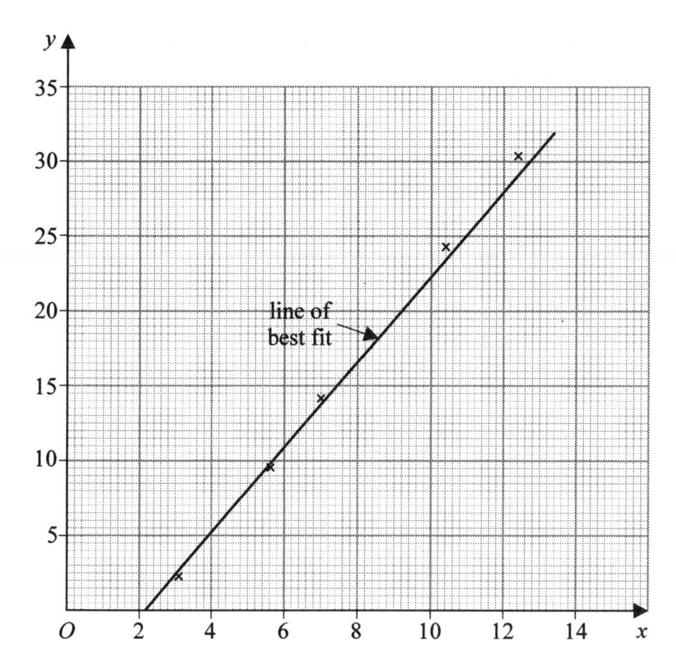

You can see from the graph that the points do, indeed, lie approximately on a straight line. You would not expect them to lie *exactly* on a straight line, since there will always be some experimental error. Because of this, it is not possible to join all the points with a single straight line. Instead, you must draw a **line of best fit**. That is, a straight line that goes as close as possible to each of the plotted points.

You can now find the values of a and b. Since you can see from the graph that the straight line passes through the points (8,16.5) and (11,25) you can obtain the following equations by substituting into the law $y = ax + b$:

$$16.5 = 8a + b \qquad (1)$$

$$25 = 11a + b \qquad (2)$$

(2) – (1): $$8.5 = 3a$$

So: $$a = 2.8 \text{ (1 d.p.)}$$

Substitute in (2):

$$25 = \left(11 \times \frac{8.5}{3}\right) + b$$

$$b = -6.2 \text{ (1 d.p.)}$$

The law that the experimental data obey is therefore approximately

$$y = 2.8x - 6.2$$

There is an alternative, equally valid, approach to finding laws of the form $y = ax + b$, as another set of data illustrates.

Suppose that another scientist collected the following data for two variables x and y:

x	0.7	1.3	3.9	4.4	5.2
y	6.6	5.1	−1.0	−2.3	−4.2

Once again the scientist guessed that x and y are related by a law of the form $y = ax + b$ and plotted corresponding pairs of values of x and y on a graph.

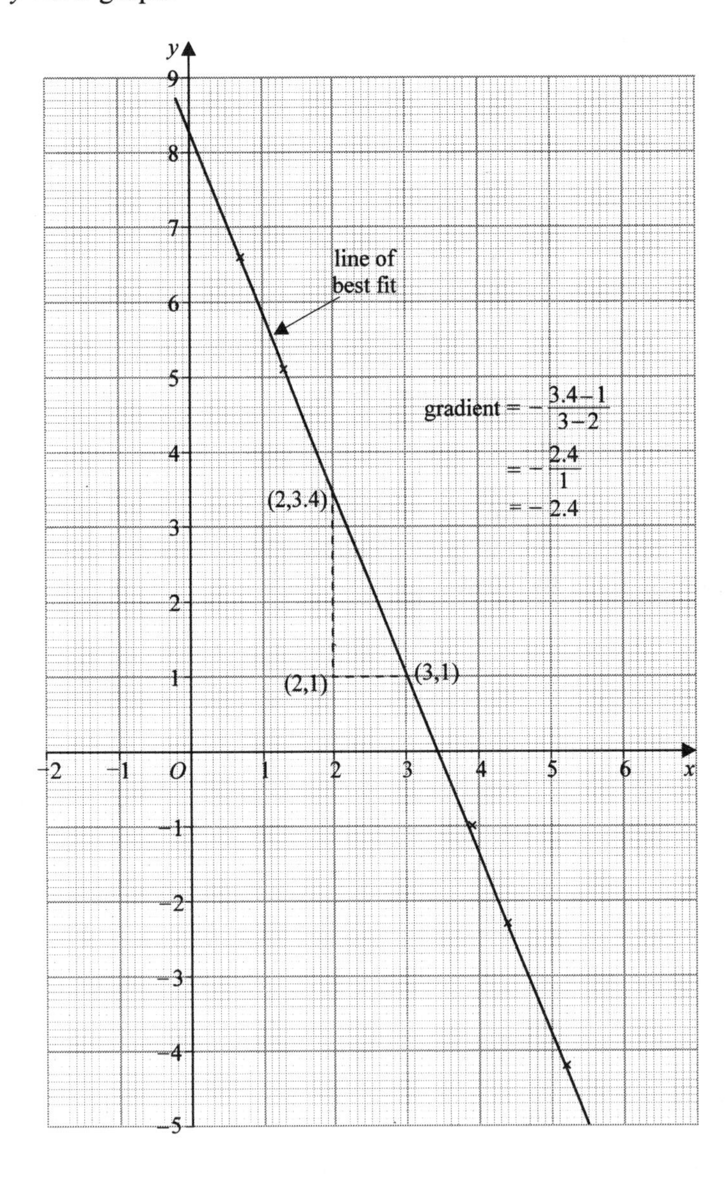

Since the points can be joined approximately by a straight line of best fit, the linear law of the form $y = ax + b$ is confirmed.

Now in Book P1 you were shown that if a straight line has equation $y = mx + c$ then m is the gradient of the line and c is the y-intercept. So if the equation of the line in the above graph is $y = ax + b$ then a is the gradient of the line and b is the y-intercept. Reading from the graph the value where the line cuts the y-axis gives $b = 8.3$.

Draw a right-angled triangle, as shown, to calculate the gradient of the line:

$$a = \frac{-2.4}{1} = -2.4$$

(Remember that the gradient of the line is negative, since it slopes 'downhill'.)

So the linear law that the data satisfy is approximately

$$y = -2.4x + 8.3$$

or:

$$y = 8.3 - 2.4x$$

Either of these two methods is valid. However, you must remember that if you use the first method then you need to choose two points that lie on the line of best fit to produce two simultaneous equations. It is *not* valid to take two values from the given data, unless the line of best fit happens to pass through them once they are plotted on the graph.

This is all very well, but of all the experiments conducted each year only a very small proportion will yield data that fit a linear law. However, many of the laws that we meet, although not linear, can be adjusted so that the data *do* approximately lie on a straight line when they are plotted on a graph. Once the line of best fit has been drawn it is again very easy to proceed to find the equation (or law) that the data fit.

The essential part of all such work on experimental data is trying to guess the equation that the data fit and then, if this equation is not of a linear form, rearranging it into a linear form.

Equations of the form $y = ax^2 + b$

If the data appear to satisfy a law of the form $y = ax^2 + b$, then by letting $Y = y$ and $X = x^2$ the equation becomes $Y = aX + b$, which is the equation of a straight line. So if you plot $Y(= y)$ on the vertical axis against $X(= x^2)$ on the horizontal axis, the points should appear to lie approximately on a straight line. It should be possible to draw a line of best fit and proceed as before.

Example 1

A hosepipe squirts water horizontally and the height, y metres, of the water above a fixed level at a distance, x metres, from the hose is measured as follows:

x	2	4	5	6	7	8
y	6.1	3.6	2.2	−0.1	−2.9	−5.5

It is thought that the law connecting x and y is of the form $y = ax^2 + b$. Show that this is approximately correct and find the values of a and b.

The equation $y = ax^2 + b$ can be transformed into the equation $Y = aX + b$ by letting $Y = y$ and $X = x^2$. The equation $Y = aX + b$ is linear and so should produce approximately a straight line if the law is correct. First, however, it is necessary to produce a *new* table of values for which the two variables are X and Y. Since $Y = y$, it is only necessary to copy down the values of y from the previous table. However, since $X = x^2$, it is necessary to square the values of x from the previous table to find the corresponding values of X for the new table.

X	4	16	25	36	49	64
Y	6.1	3.6	2.2	−0.1	−2.9	−5.5

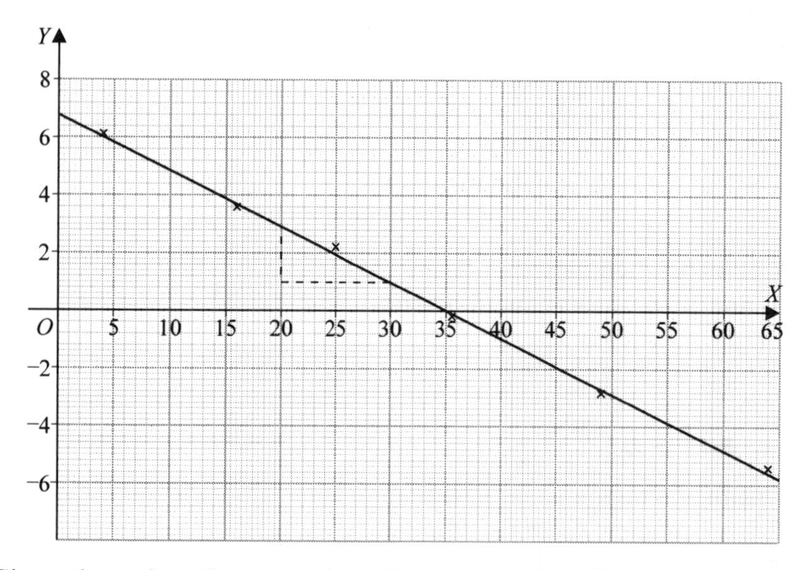

Since the points lie approximately on a straight line and it is thus possible to draw a line of best fit, the data *do* obey a law of the form $y = ax^2 + b$.

In order to find a and b, notice that the line of best fit cuts the y-axis at 6.8. So $b = 6.8$.

The gradient of the line is

$$-\frac{1.9}{10} = -0.19 = -0.2 \text{ (1 d.p.)}$$

So the data fit a law of the form $y = -0.2x^2 + 6.8$.

Equations of the form $y = ax^2 + bx$

If it is thought that the experimental data obey a law of the form $y = ax^2 + bx$ then it is necessary first to rearrange the equation so that it can be written as a linear equation $Y = aX + b$. In this case, dividing by x gives

$$\frac{y}{x} = \frac{ax^2}{x} + \frac{bx}{x}$$

so:
$$\frac{y}{x} = ax + b$$

Now let $Y = \frac{y}{x}$ and $X = x$. The equation becomes $Y = aX + b$ as required.

Example 2

In an experiment a number of readings of the variables x and y were taken. The results are listed in the table below.

x	−2.3	−1.4	−0.3	0.7	1.2	2.9
y	−6	−1.5	0.2	−1.7	−4	−18

It is thought that the data obey a law of the form $y = ax^2 + bx$. Show that this is approximately so and hence find the values of a and b.

If the law is $y = ax^2 + bx$ then

$$\frac{y}{x} = ax + b$$

By letting $Y = \frac{y}{x}$ and $X = x$ you obtain $Y = aX + b$, which is a linear law.

It is now necessary to draw up a *new* table of values. The values of X in this new table will simply be the values of x from the given table, since $X = x$. The values of Y in the new table, however, will be each value of y from the given table divided by its corresponding value of x, since $Y = \frac{y}{x}$.

X	-2.3	-1.4	-0.3	0.7	1.2	2.9
Y	2.6	1.1	-0.7	-2.4	-3.3	-6.2

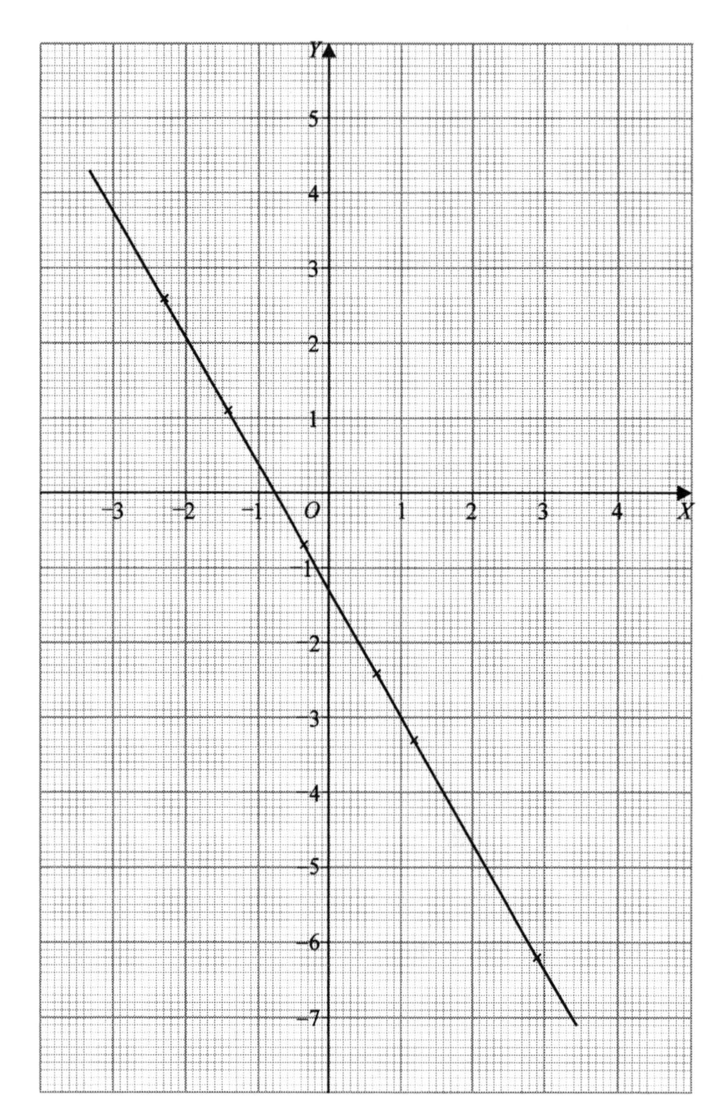

Since the points do lie approximately on a straight line the data do approximately obey a law of the form $y = ax^2 + bx$.

The points $(1, -3)$ and $(-\frac{1}{2}, -\frac{1}{2})$ lie on the line of best fit $Y = aX + b$. So:

$$-3 = a + b \qquad (1)$$

and

$$-\tfrac{1}{2} = -\tfrac{1}{2}a + b \qquad (2)$$

(1) – (2):

$$-2\tfrac{1}{2} = 1\tfrac{1}{2}a$$

$$-\tfrac{5}{2} = \tfrac{3}{2}a$$

or
$$a = -\tfrac{5}{2} \times \tfrac{2}{3}$$
$$a = -\tfrac{5}{3} = -1.7 \text{ (1 d.p.)}$$

Substitute $a = -\frac{5}{3}$ in (1):

$$-3 = -\tfrac{5}{3} + b$$
$$b = -3 + 1\tfrac{2}{3}$$
$$b = -1\tfrac{1}{3} = -1.3 \text{ (1 d.p.)}$$

Equations of the form $\frac{1}{x} + \frac{1}{y} = \frac{1}{a}$

If the experimental law is of the form $\frac{1}{x} + \frac{1}{y} = \frac{1}{a}$, then by letting $X = \frac{1}{x}$ and $Y = \frac{1}{y}$ you obtain $X + Y = \frac{1}{a}$. This can rearranged to

$$Y = -X + \frac{1}{a}$$

which is of the form $y = mx + c$ and so, once again, is linear.

Example 3

Two variables, x and y, are thought to be connected by a law of the form

$$\frac{1}{x} + \frac{1}{y} = \frac{1}{a}$$

In an experiment the following set of data was collected:

x	-25	-0.8	-0.3	0.4	0.8	6.2
y	5.9	0.7	0.3	-0.6	-1	-30

Show that the data do approximately obey such a law and hence find the value of a.

The equation $\frac{1}{x} + \frac{1}{y} = \frac{1}{a}$ can be transformed into a linear equation by putting $X = \frac{1}{x}$ and $Y = \frac{1}{y}$. Then the equation becomes $X + Y = \frac{1}{a}$ or $Y = -X + \frac{1}{a}$.

The new table of values is drawn up by taking the reciprocal of x $\left(\text{i.e. } \frac{1}{x}\right)$ and the reciprocal of y $\left(\text{i.e. } \frac{1}{y}\right)$:

X	−0.04	−1.25	−3.33	2.5	1.25	0.16
Y	0.17	1.43	3.33	−1.67	−1	−0.03

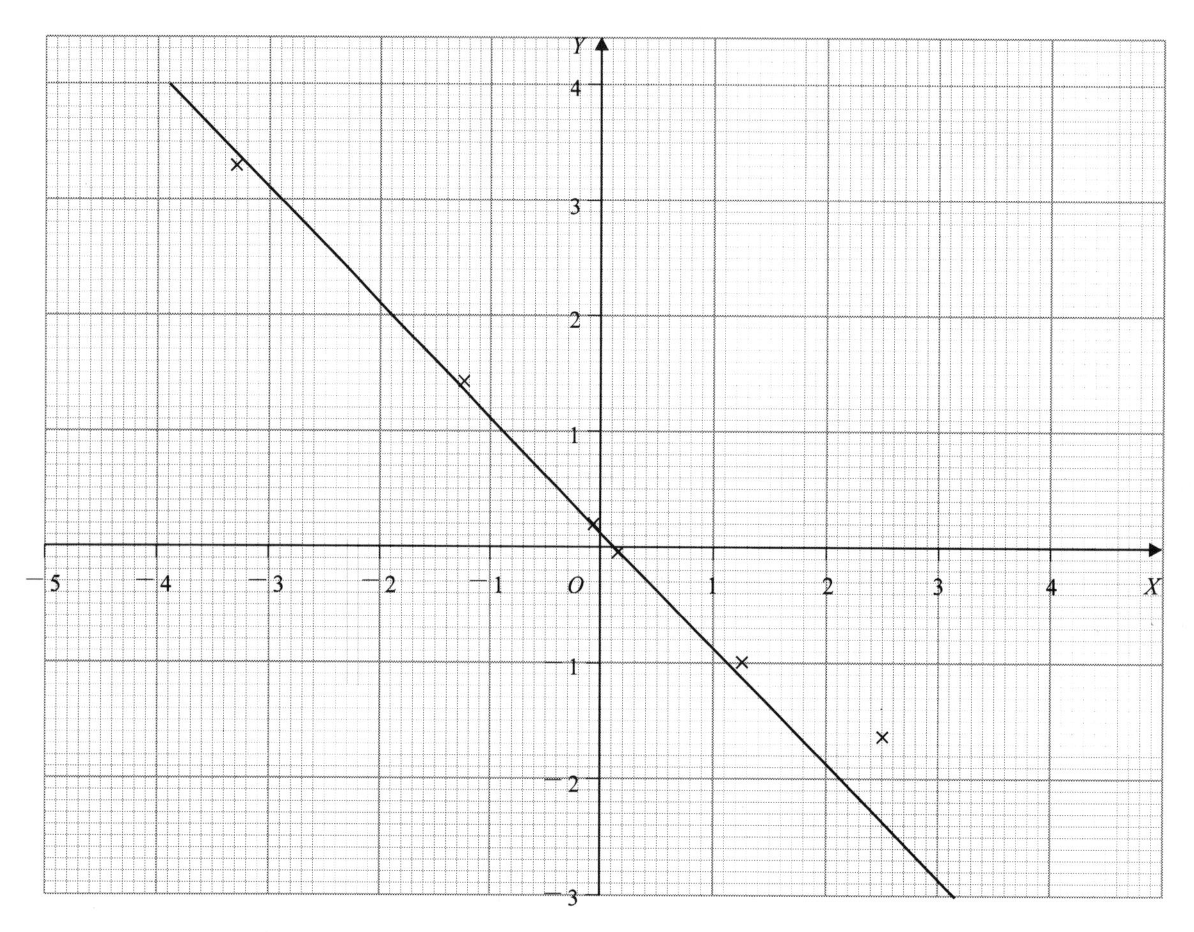

Since the points do lie approximately on a straight line, the data do obey a law of the form $\frac{1}{x} + \frac{1}{y} = \frac{1}{a}$.

The line of best fit cuts the Y-axis where $Y = 0.13$, that is, the Y-intercept is 0.13.

So: $$\frac{1}{a} = 0.13$$

Thus: $$a = \frac{1}{0.13} = 7.7 \text{ (1 d.p.)}$$

Equations of the form $y = ax^n$

Sometimes the law connecting two sets of data takes the form $y = ax^n$. Clearly this is not linear. With this form of law, where a power is involved in the equation, you need to take logarithms of both sides of the equation in order to reduce it to a linear equation.

So:
$$\ln y = \ln(ax^n)$$
$$\ln y = \ln a + \ln x^n$$
$$\ln y = \ln a + n \ln x \qquad \text{(A)}$$

If you let $Y = \ln y$ and $X = \ln x$ then equation (A) becomes
$$Y = \ln a + nX$$
or:
$$Y = nX + \ln a$$

But since a is a constant then $\ln a$ is a constant. So $Y = nX + \ln a$ is a linear equation of the form $y = mx + c$.

Example 4

It is believed that the two variables x and y are connected by a relationship of the form $y = ax^n$ where a and n are constants. Verify the law by drawing a suitable straight line graph and hence calculate estimates of a and n.

x	1.34	3.58	7.60	12.1	14.8
y	208	10.9	1.14	0.283	0.155

Since the law is thought to be of the form $y = ax^n$ you need to take logarithms.

$$y = ax^n$$
$$\ln y = \ln(ax^n)$$
$$\ln y = \ln a + \ln x^n$$
$$\ln y = \ln a + n \ln x$$

Let $Y = \ln y$ and $X = \ln x$. Then the law becomes
$$Y = \ln a + nX$$

In order to draw up a new table of values, you need to find the natural logarithms (ln) of the x-values and the natural logarithms of the y-values:

X	0.29	1.28	2.03	2.49	2.69
Y	5.34	2.39	0.13	−1.26	−1.86

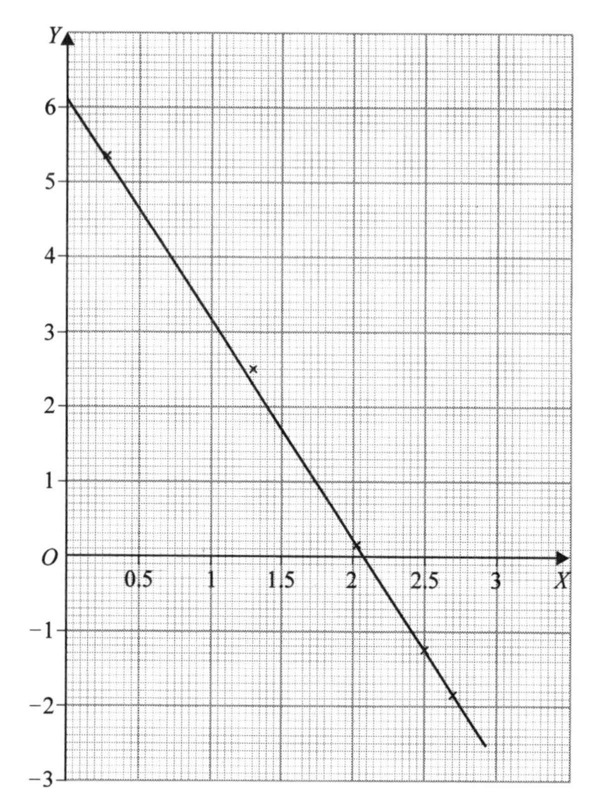

Since the points lie approximately on a straight line the data do approximately satisfy a law of the form $y = ax^n$.

The line of best fit passes through the points (2, 0.2) and (2.5, −1.3). Substitute these into the equation of the line of best fit:

$$Y = \ln a + nX$$

$$0.2 = \ln a + 2n \qquad (1)$$

$$-1.3 = \ln a + 2.5n \qquad (2)$$

(2) − (1): $$-1.5 = 0.5n$$

So: $$n = \frac{-1.5}{0.5} = -3$$

Substitute $n = -3$ in (1):

$$0.2 = \ln a + 2(-3)$$

$$0.2 = \ln a - 6$$

$$\ln a = 6.2$$

So: $$a = \mathrm{e}^{6.2} = 490 \text{ (2 s.f.)}$$

Equations of the form $y = ab^x$

Another possible form of law is $y = ab^x$. Again, when this form of law occurs, that is, when a power is involved, you must take logarithms in order to reduce the law to a linear form.

$$y = ab^x$$

Take logarithms: $$\ln y = \ln (ab^x)$$

$$\ln y = \ln a + \ln b^x$$

So: $$\ln y = \ln a + x \ln b \qquad \text{(B)}$$

Let $Y = \ln y$ and $X = x$. Equation (B) becomes

$$Y = \ln a + (\ln b)X$$

or: $$Y = (\ln b)X + \ln a$$

Since a and b are constants, $\ln a$ and $\ln b$ are also constants. Thus $Y = (\ln b)X + \ln a$ is of the form $y = mx + c$ and is linear.

Example 5

The population P thousands, to one decimal place, of a new town T years after 1975 is summarised in the table.

T	1	2	3	4	5
P	15.4	24.6	47.2	88.3	144.4

It is believed that the above data are connected by a relationship of the form

$$P = kc^T$$

where k and c are constants.

By plotting $\ln P$ against T, obtain estimates for the values of k and c, giving your answers to 2 decimal places. [L]

Since $P = kc^T$ involves a power, you need to take logarithms.

$$\ln P = \ln (kc^T)$$

$$\ln P = \ln k + \ln c^T$$

$$\ln P = \ln k + T \ln c$$

Let $X = T$ and $Y = \ln P$.

Then the equation

$$\ln P = \ln k + T \ln c$$

becomes $$Y = \ln k + X \ln c$$

or: $$Y = (\ln c)X + \ln k$$

which is of the form $y = mx + c$.

To draw up the new table of values the values of X are just copied from the values of T and the values of Y are the logarithms of the values of P.

X	1	2	3	4	5
Y	2.73	3.20	3.85	4.48	4.97

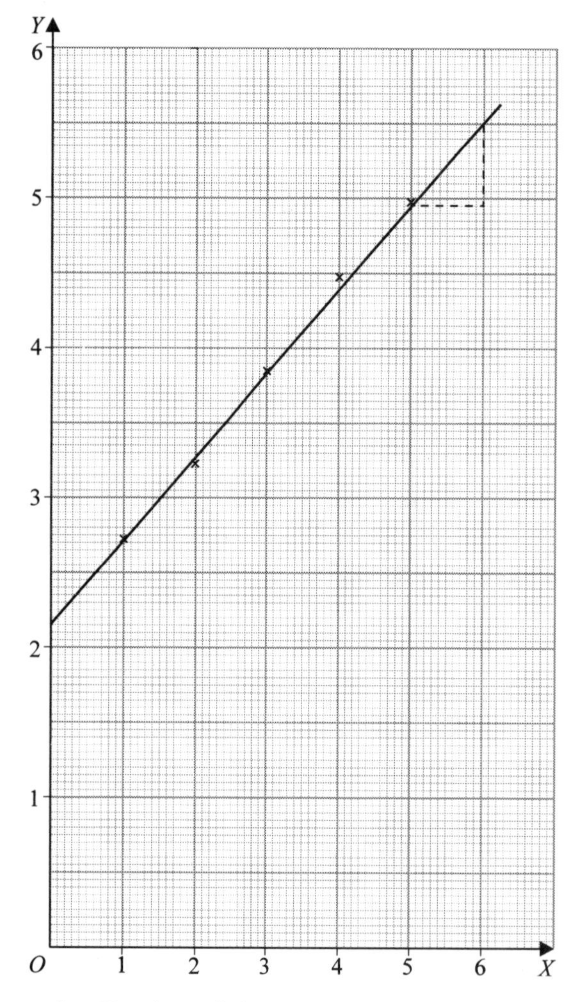

The line cuts the Y-axis at 2.15.

So: $\ln k = 2.15 \Rightarrow k = e^{2.15}$

$$k = 8.58 \text{ (2 d.p.)}$$

The gradient of the line is

$$\frac{0.55}{1} = 0.55$$

So:

$$\ln c = 0.55$$

$$c = e^{0.55}$$

$$= 1.733$$

$$c = 1.73 \text{ (2 d.p.)}$$

Exercise 4A

1 It is believed that two variables x and y satisfy a law of the form $y = ax + b$ where a and b are constants. Verify graphically that this law is approximately correct and hence calculate estimates of a and b.

x	2.2	2.9	3.8	4.8	5.3
y	2.41	3.51	4.81	6.68	7.56

2 In a particular year, a study was made of all the settlements in a Canadian province. The table shows the number N of each type of settlement, and the average spacing S km between settlements of each type.

Classification	Number (N)	Average spacing between settlements (S km)
Hamlet	404	14.5
Village	150	22.1
Town	100	24.6
Small Seat	85	41.4
County Seat	29	64.6
Regional City	9	148
Regional Capital	2	298

It is believed that S and N approximately obey a law of the form

$S = cN^k$ where $c > 0$ and k are constants.

(a) Plot a suitable graph to show that the above data do approximately follow this law.

(b) Use your graph to obtain estimates for the values of c and k, giving your answers to 2 significant figures. [L]

3 Two variables x and y are thought to satisfy a law of the form $y = ab^x$ where a and b are constants. By drawing a straight line graph show that this is approximately correct. Hence calculate estimates for a and b.

x	2.0	2.4	2.8	3.2	3.6
y	69	88	101	131	138

4 In an experiment to find the focal length f of a lens, the following object distances u and the image distances v were recorded.

u	15	20	25	40	50
v	30	20	17	13	12.5

By drawing a straight line graph, find an estimate of f from the formula

$$\frac{1}{u} + \frac{1}{v} = \frac{1}{f}$$

5 The table gives values of x and y which are related by an equation of the form $y = ax^2 + b$. Find estimates of the constants a and b by drawing a suitable straight line graph.

x	1	2	3	4	5
y	12.5	14.0	16.5	20.0	24.5

6 It is thought that the variables s and t satisfy a relation of the form

$$\left(\frac{s}{t}\right)^a = be^{-t}$$

where the constants a and b are positive integers. Show, by drawing a linear graph, that this is supported by the following table of measured values, and find the values of a and b.

t	0.2	0.4	0.6	0.8	1.0
s	1.09	1.96	2.67	3.22	3.64

[L]

7

V	5	10	15	20	25
R	149	175	219	280	359

The table shows corresponding values of the variables V and R obtained in an experiment. By drawing a suitable linear graph, show that these pairs of values may be regarded as approximations to values satisfying a relation of the form $R = a + bV^2$ where a and b are constants.

Use your graph to estimate the values of a and b, giving your answers to 2 significant figures. [L]

8 The expression $y = px^2 + qx$ is an approximation to the relation connecting two variables x and y. Using the values given in the following table draw a suitable graph and from it estimate each of p and q to the nearest whole number.

x	1	2	3	4	5	6
y	74	126	162	172	175	144

[L]

9

x	1.5	2.5	3	4	5
y	−6.0	11.0	6.2	4.1	3.3

The table shows values of a variable y corresponding to certain values of another variable x. By drawing a suitable linear graph, verify that the values of x and y satisfy approximately a relationship of the form

$$\frac{1}{x} + \frac{1}{y} = \frac{1}{a}$$

where a is a constant. Use your graph to estimate the value of a to one decimal place. Estimate also the value of x when $y = 5.1$.

[L]

10 The table shows approximate values of a variable y corresponding to certain values of another variable x. By drawing a suitable linear graph, verify that these values of x and y satisfy approximately a relationship of the form $y = ax^k$. Use your graph to find approximate values of the constants a and k.

x	5	10	15	20	25	30
y	45	63	77	89	100	110

[L]

11 Pairs of values, x and y, are obtained in an experiment as shown in the table.

x	10	20	40	75	100
y	136	334	824	1850	2710

It is believed that x and y are related by a law of the form

$$\lg y = m \lg x + c$$

where m and c are constants.
Write down a table of values for $\lg x$ and $\lg y$.
By drawing a graph, show that the law is approximately valid.
Estimate values of m and c giving your answers to 2 significant figures.
Using your values of m and c, express y in terms of x. [L]

12

x	1.0	1.5	2.0	2.5	3.0
y	12	30	72	173	416

The table shows corresponding values of variables x and y obtained in an experiment. Draw a straight line graph to verify that x and y are connected by a relationship of the form $y = ae^{bx}$, where a and b are constants.
Using your graph obtain estimates of the values of a and b giving your answers to 2 significant figures. [L]

SUMMARY OF KEY POINTS

1 To test whether experimental data satisfy a given law, the law needs to be reduced to the form $Y = aX + b$. The values of X and Y are plotted on a graph, and if they lie approximately on a straight line the proposed law is confirmed.

2 Once it can be seen that the points lie approximately on a straight line, draw by eye the line of best fit.

3 Take two points from the line of best fit and substitute their coordinates into the equation of the line of best fit. The values of a and b can then be found by solving the resultant simultaneous equations.

4 Alternatively, calculate the gradient and the y-intercept of the line of best fit. These are the values of a and b, respectively.

5 The values to plot for various laws are as follows:

Law	Plot
$y = ax + b$	y against x
$y = ax^2 + b$	y against x^2
$y = ax^2 + bx$	$\frac{y}{x}$ against x
$\frac{1}{x} + \frac{1}{y} = \frac{1}{a}$	$\frac{1}{y}$ against $\frac{1}{x}$
$y = ax^n$	$\ln y$ against $\ln x$
$y = ab^x$	$\ln y$ against x

Review exercise 1

1 (a) Given that $(x+1)$ is a factor of the expression $(2x^3 + ax^2 - 5x - 2)$, find the value of the constant a. Show that, with this value of a, $(x-2)$ is another factor of this expression and hence, or otherwise, factorise the expression completely.

(b) When divided by $(x-2)$ the expression $(x^3 + x^2 + 2x + 2)$ leaves a remainder R. Find the value of R. [L]

2 In a psychology experiment, a right-handed student was asked to draw a square 10 times with his left hand to investigate learning effects. The results shown in the table were obtained.

Attempt number (n)	1	2	3	4	5	6	7	8	9	10
Time taken (t seconds)	6.7	6.2	6.0	5.3	4.9	5.0	4.8	4.2	3.9	3.8

The psychologist believes that these data should approximately follow a law of the form

$$t = k\mathrm{e}^{-Vn}$$

where k and V are positive constants.

(a) Plot a suitable graph to show that the above data do approximately follow this law.

(b) Use your graph to obtain estimates for the values of the constants k and V, giving your answers to 1 significant figure. [L]

3 Express

$$\frac{7x-15}{2(x-2)(x-1)} - \frac{7}{2x}$$

as a single fraction in its lowest terms. [L]

4 In an investigation of delays at a roadworks, the times spent by a sample of commuters waiting to pass through the roadworks were recorded to the nearest minute. Shown below is part of a cumulative frequency table resulting from the investigation.

Upper class boundary	2.5	4.5	7.5	8.5	9.5	10.5	12.5	15.5	20.5
Cumulative number of commuters	0	6	21	48	97	149	178	191	200

(a) For how many of the commuters was the time recorded as 11 minutes or 12 minutes?

(b) Estimate (i) the lower quartile

(ii) the 81st percentile, of these waiting times. [L]

5 Given that

$$(x^2 - 2x + 3)(x^2 + kx - 2) \equiv x^4 - 15x^3 + qx^2 + px - 6$$

find the values of k, p and q. [L]

6 Given that, for positive x,

$$\lg x + \lg 25 = \lg(x^3)$$

calculate x, without using a calculator. [L]

7 Given that $(x - 2)$ is a factor of $\mathrm{f}(x)$, where

$$\mathrm{f}(x) \equiv x^3 - x^2 + Ax + B$$

find an equation satisfied by the constants A and B.

Given, further, that when $\mathrm{f}(x)$ is divided by $(x - 3)$ the remainder is 10, find a second equation satisfied by A and B.

Solve your equations to find A and B.

Using your values of A and B, find 3 values of x for which $\mathrm{f}(x) = 0$. [L]

8 Show that for $a \in \mathbb{N}$, $a > 1$ and $p, q > 0$,

$$\log_a p^2, 2\log_a pq \quad \text{and} \quad 2\log_a pq^2$$

are three successive terms of an arithmetic sequence whose common difference is $2\log_a q$.

Given that $pq^2 = a$, show that the sum of the first 5 terms of the arithmetic sequence with first term $\log_a p^2$ and common difference $2\log_a q$ is 10. [L]

9 $$f(x) \equiv \frac{9 - 3x - 12x^2}{(1 - x)(1 + 2x)}$$

Given that $f(x) \equiv A + \frac{B}{1 - x} + \frac{C}{1 + 2x}$, find the values of the constants A, B and C. [L]

10 In an experiment to determine the relationship between the resistance to motion, R newtons, of a plank towed through water and its speed, $V \text{ m s}^{-1}$, the following data were recorded:

V	1.8	3.7	5.5	9.2	11
R	2.01	7.32	15.1	38.9	56.2

Assuming that $R = kV^n$, where n and k are constants,

(a) obtain a relation between $\ln R$ and $\ln V$.

(b) By drawing a graph of $\ln R$ against $\ln V$ estimate, to 2 significant figures, the values of n and k. [L]

11 Solve the equation

$$2^x = 5$$

giving your answer to 3 significant figures.

12 Summarised below are the values of the orders (to the nearest £) taken by a sales representative for a wholesale firm during a particular year.

Value of order (£)	Number of orders
Less than 10	3
10–19	9
20–29	15
30–39	27
40–49	29
50–59	34
60–69	19
70–99	10
100 or more	4

(a) Using interpolation, estimate the median and the semi-interquartile range for these data.

(b) Explain why the median and semi-interquartile range might be more appropriate summary measures for these data than the mean and standard deviation. [L]

13 A notifiable disease spread so that N, the number of people infected t days following January 15th, 1991, $0 \leqslant t \leqslant 25$, obeyed the equation

$$\frac{\mathrm{d}N}{\mathrm{d}t} = kN$$

where k is a positive constant.
On January 15th, 1991, 128 people were infected and 10 days later the number infected had grown to 380. Calculate the value of k to four decimal places. Find how many people were infected 25 days after January 15th, 1991. [L]

14 Given that $\lg 2 = 0.301\,03$, $\lg 3 = 0.477\,12$ and $\lg 5 = 0.698\,97$ calculate, without using a calculator, the value of
(a) $\lg 6$
(b) $\lg 25$
(c) $\lg 1.5$. [L]

15 Given that $x = 2^p$, $y = 4^q$
(a) express $4xy$ as a power of 2
(b) find $\log_2(xy^3)$ in terms of p and q. [L]

16
$$\mathrm{f}(x) \equiv \frac{3x}{(x+2)(x-1)}$$

Express $\mathrm{f}(x)$ in partial fractions.
Hence evaluate $\int_{-1}^{0} \mathrm{f}(x)\,\mathrm{d}x$. [L]

17 Given that $(x+3)$ and $(x-1)$ are both factors of the expression $(x^3 + ax^2 - bx - a)$, calculate the values of the constants a and b.
With these values of a and b, find the values of the remainder when the expression is divided by $(x-2)$. [L]

18 Given that $\lg 8 = 0.903\,09$, find, without using a calculator,
(a) $\lg 2$ (b) $\lg \frac{1}{4}$. [L]

19 Solve

$$2(3^{2x}) + 3^x - 10 = 0$$

giving your answer to 3 significant figures.

20

t seconds	5	10	15	20	25
I amps	212	49.1	10.6	2.31	0.532

The table shows corresponding values of I and t obtained from an experiment. By drawing a suitable graph, show that these values support the hypothesis that I and t are connected by a relationship of the form $I = ae^{kt}$, where a and k are constants.

Use your graph to estimate the values of a and k to 2 significant figures.

Find, to 2 significant figures, the value of $\frac{dI}{dt}$ when $t = 0$.

[L]

21

$$f(x) \equiv \frac{x^2 + 6x + 7}{(x+2)(x+3)}, x \in \mathbb{R}$$

Given that $f(x) \equiv A + \frac{B}{x+2} + \frac{C}{x+3}$,

(a) find the values of the constants A, B and C.

(b) Show that $\int_0^2 f(x)\,dx = 2 + \ln\left(\frac{25}{18}\right)$. [L]

22 The following table shows the time, to the nearest minute, spent reading during a particular day by a group of school children.

Time	Number of children
10–19	8
20–24	15
25–29	25
30–39	18
40–49	12
50–64	7
65–89	5

(a) Represent these data by a histogram.

(b) Comment on the shape of the distribution. [L]

23 The cubic polynomial $f(x)$ is such that $f(2) = f(\frac{1}{2}) = 0$, $f(0) = 6$ and $f(-2) = -20$.
Express $f(x)$ as the product of three linear factors. [L]

24 Express

$$\frac{-5x^2 + 3x + 45}{(x+2)(4x^2+3)}$$

as the sum of partial fractions.

25 Express

$$\frac{3}{2(x+1)} - \frac{3x-1}{2(x-1)(x-2)}$$

as a single fraction, simplifying the numerator as much as possible. [L]

26 A biologist estimated that the number of breeding pairs of a certain rare bird on 1st January 1973 was 138, and that by 1st January 1978 the number had fallen to 109. Suppose the number, n, of breeding pairs of the bird t years after 1st January 1973 is given by,

$$n = Ae^{-kt}$$

where A and k are positive constants and n is treated as a continuous function of t.
(a) Write down the value of A.
(b) Show that, to 3 significant figures, $k = 0.0472$.
(c) Obtain an estimate of the number of breeding pairs of the bird on 1st January 1982. [L]

27 Find the value of each of the constants A, B and C for which

$$\frac{1}{1+x^3} \equiv \frac{A}{(1+x)} + \frac{Bx+C}{(1-x+x^2)}$$ [L]

28 A marketing company always buys new cars on 1st August. Before making any purchases on 1st August 1992, they reviewed their fleet of cars. The following table shows the age, x, in years, of the cars in the fleet.

Age (x)	1	2	3	4	5	6	7	8	9	10	11
Number of cars (f)	14	20	16	14	12	8	6	4	3	2	1

Find

(a) the mode

(b) the median and quartiles

(c) the mean

(d) the standard deviation

of this distribution. [L]

29 Given that $\log_m x = p$, express in terms of p,

(a) $\log_m(x^4)$

(b) $\log_m\left(\frac{1}{x^2}\right)$

(c) $\log_m(mx)$. [L]

30 Solve $6(5^{2x}) - 20(5^x) + 14 = 0$ giving each answer, where appropriate, to 3 significant figures.

31 Given that

$$2\log_2 x = y \text{ and } \log_2(2x) = y + 4$$

find the value of x. [L]

32 Express in its simplest form

$$\frac{5x-3}{2x^2-x-15} + \frac{2x-5}{3x^2-7x-6}$$

33 The population, p, of insects on an island, t hours after midday, is given by

$$p = 1000e^{kt}$$

where k is a constant.

Given that when $t = 0$, the rate of change of the population with respect to time is 100 per hour,

(a) find k.

(b) Find also the population of insects on the island when $t = 6$. Give your answer to 3 significant figures. [L]

34 The periods of revolution, T days, of the planets in the solar system about the sun, and their mean distances from the sun, d (km $\times$ 10^6), are shown in the table.

Planet	Period of revolution T (days)	Mean distance d (km $\times$ 10^6)
Mercury	88	57.9
Venus	225	108.2
Earth	365	149.6
Mars	687	227.9
Jupiter	4 329	778.3
Saturn	10 753	1427
Uranus	30 660	2870
Neptune	60 150	4497
Pluto	90 670	5907

It is believed that these data follow a law of the form

$$T = kd^a$$

where k and a are constants.

By plotting a suitable graph, show that these data are consistent with such a law, and determine an estimate for a to 1 decimal place. [L]

35 Express as the sum of partial fractions

$$\frac{2}{x(x+1)(x+2)}$$

Hence show that

$$\int_2^4 \frac{2}{x(x+1)(x+2)}\,dx = 3\ln 3 - 2\ln 5$$

[L]

36 Given that $f(x) \equiv 2x^3 - 3x^2 - 11x + c$, and that the remainder when $f(x)$ is divided by $x - 2$ is -12, find the value of the constant c.

Show that, with this value of c, $x + 2$ is a factor of $f(x)$, and hence solve the equation $f(x) = 0$. [L]

37 Given that $\log x = p$,

(a) express $\log(x^2)$ in terms of p.

Given also that the sum of the first 20 terms of the series

$$\log x + \log(x^2) + \log(x^3) + \ldots \text{ is } kp$$

(b) calculate the value of k. [L]

38 Telephone calls arriving at a switchboard are answered by the telephonist. The following table shows the time, to the nearest second, recorded as being taken by the telephonist to answer the calls received during one day.

Time to answer (to nearest second)	Number of calls
10–19	20
20–24	20
25–29	15
30	14
31–34	16
35–39	10
40–59	10

Represent these data by a histogram.
Give a reason to justify the use of a histogram to represent these data. [L]

39 Obtain the arithmetic mean and the standard deviation of the set of numbers

1, 2, 3, 4, 5, 6, 7, 8, 9, 10

and deduce the standard deviations of the following sets of numbers:

(a) 14, 15, 16, 17, 18, 19, 20, 21, 22, 23
(b) 3, 6, 9, 12, 15, 18, 21, 24, 27, 30
(c) 7.0, 7.1, 7.2, 7.3, 7.4, 7.5, 7.6, 7.7, 7.8, 7.9

[L]

40 The period of oscillation, T seconds, of a heavy weight attached to a wire of length l metres was investigated for $2 \leqslant l \leqslant 8$. The results obtained are given in the table.

l	2	3	4	5	6	7	8
T	2.81	3.47	4.01	4.49	4.98	5.29	5.63

It is believed that l and T are related by an equation of the form $T = kl^n$, where k and n are positive constants.
Plot values of $\ln T$ against $\ln l$ and hence state, giving reasons, whether or not l and T are related by an equation of the given form.
By taking suitable readings from your graph, estimate the values of k and n to 1 significant figure. [L]

41 The curve with equation $ky = a^x$ passes through the points with coordinates (7,12) and (12,7). Find, to 2 significant figures, the values of the constants k and a. [L]

42 At time $t = 0$ there are 10 000 fish in a lake. At time t days the birthrate of fish is equal to one tenth the number N of fish present, and fish are harvested at a constant rate of 800 per day. Assuming N to be a continuous variable, show that

$$10\frac{\mathrm{d}N}{\mathrm{d}t} = N - 8000$$

Show that $N = a + be^{ct}$, where a, b and c are constants to be determined.

Find the time taken for the population of fish in the lake to increase to 16 000. [L]

43 (a) $$\mathrm{f}(x) \equiv px^3 + qx^2 + rx + s$$

The curve with equation $y = \mathrm{f}(x)$ has gradient 4 at the point with coordinates (0,−5).

(i) Find the values of r and s.

The remainder when $\mathrm{f}(x)$ is divided by $(x - 1)$ is 12, and the remainder when $\mathrm{f}(x)$ is divided by $(x + 2)$ is 15.

(ii) Calculate the values of p and q.

(b) $$\mathrm{g}(x) \equiv 2x^3 - x^2 - 23x - 20$$

(i) Show that $(x + 1)$ is a factor of $\mathrm{g}(x)$.

(ii) Factorise $\mathrm{g}(x)$ completely.

(iii) Solve the equation $\mathrm{g}(x) = 0$. [L]

44 In a certain town an investigation was carried out into accidents in the home to children under 12 years of age. The numbers of reported accidents and the ages of the children concerned are summarised below:

Group	A	B	C	D	E	F
Age of child (years)	0 to <2	2 to <4	4 to <6	6 to <8	8 to <10	10 to <12
Number of accidents	42	52	28	20	18	16

(a) State the modal class.

(b) Calculate, to one decimal place, the mean age and the standard deviation of the distribution of ages.

(c) Draw a cumulative frequency polygon and from it estimate, to the nearest month, the median and the interquartile range for the ages of all children under 12 years of age concerned in reported accidents in the home. State, giving a reason, whether you consider the mode or the median best represents the average age for accidents in the home to children under 12 years of age. [L]

45 (a) Given that $(x + 4)$ is a factor of the expression

$$3x^3 + x^2 + px + 24$$

find the value of the constant p.
Hence, factorise the expression completely.

(b) When divided by $(x + 2)$ the expression $(5x^3 - 3x^2 + ax + 7)$ leaves a remainder of R. When the expression $(4x^3 + ax^2 + 7x - 4)$ is divided by $(x + 2)$ there is a remainder of $2R$. Find the value of the constant a. [L]

46 Express in its simplest form

$$\frac{5}{x^2 + x - 6} - \frac{1}{x^2 + 5x + 6}$$ [L]

47

x	1.0	1.5	2.0	2.5	3.0
y	12	30	72	173	416

The table shows corresponding values of variables x and y obtained in an experiment. Draw a straight line graph to verify that x and y are connected by a relationship of the form $y = ae^{bx}$, where a and b are constants.
Using your graph obtain estimates of the values of a and b giving your answers to 2 significant figures. [L]

48 Given that $\quad \mathrm{f}(x) \equiv \dfrac{2}{(2 - x)(1 + x)^2}$

express $\mathrm{f}(x)$ in the form

$$\frac{A}{(2 - x)} + \frac{B}{(1 + x)} + \frac{C}{(1 + x)^2}$$

where A, B and C are numbers to be found. [L]

49 Newton's law of cooling states that the rate at which a body cools is directly proportional to the excess temperature of the body over the temperature of its immediate surroundings. Given that at time t minutes a body has a temperature T°C and its surroundings a constant temperature θ°C, form a differential equation in terms of T, θ, t, and the constant of proportionality, k, $k > 0$.
Integrate this equation to show that

$$\ln(T - \theta) = -kt + c$$

where c is a constant.
Hence show that

$$T = \theta + Ae^{-kt}$$

where $\ln A = c$.
At 2.23 p.m., the water in a kettle boils at 100 °C in a room of constant temperature 21 °C. After 10 minutes, the temperature of the water in the kettle is 84 °C. Use this information to find the values of k and A, giving your answers to 2 significant figures. Hence find the time, to the nearest minute, when the temperature of the water in the kettle will be 70°C. [L]

50

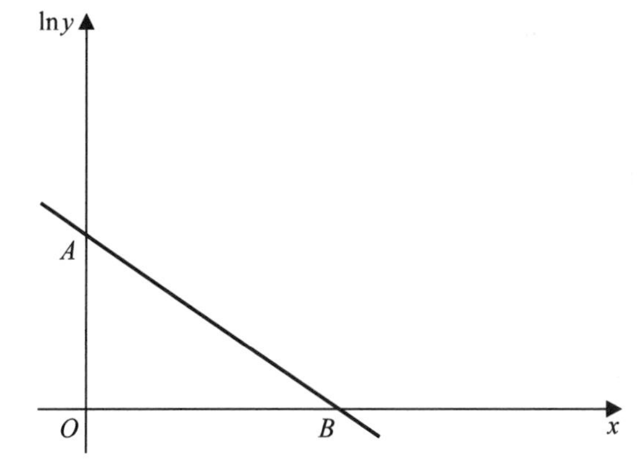

The diagram shows a straight line graph of $\ln y$ against x. The line crosses the axes at points $A(0, 2)$ and $B(3, 0)$. Given that $y = pq^x$, where p and q are constants, find, to one decimal place, the values of p and q. [L]

51 It is given that

$$f(x) \equiv \frac{3x+7}{(x+1)(x+2)(x+3)}$$

(a) Express $f(x)$ as the sum of three partial fractions.

(b) Find $\int f(x)\,dx$.

(c) Hence show that the area of the finite region bounded by the curve with equation $y = f(x)$ and the lines with equations $y = 0$, $x = 0$ and $x = 1$ is $\ln 2$. [L]

52 Write

$$\frac{x^2+8x+9}{(x+1)(x+2)}$$

as the sum of partial fractions.

53 In a borehole the thicknesses, in mm, of the 25 strata are as shown in the table below:

Thickness (mm)	0–	20–	30–	40–	50–	60–
Number of strata	2	5	9	8	1	0

Draw a histogram to illustrate these data. Construct a cumulative frequency table and draw a cumulative frequency polygon. Hence, or otherwise, estimate the median and the interquartile range for these data.
Find the proportion of the strata that are less than 28 mm thick. [L]

54 Given that $f(x) \equiv 3x^3 - 4x^2 - 5x + 2$, show that $(x-2)$ is a factor of $f(x)$.
Express $f(x)$ as a product of three linear factors.
Hence, solve for x, the equation

$$3e^{3x} - 4e^{2x} - 5e^x + 2 = 0, \quad x \in \mathbb{R}.$$ [L]

55 Express

$$\frac{5}{3(x+1)} - \frac{5x-7}{3(x-1)(x-2)}$$

as a single fraction in its lowest terms. [L]

56 (a) Show that $\log_4 3 = \log_2 \sqrt{3}$.

(b) Hence or otherwise solve the simultaneous equations

$$2\log_2 y = \log_4 3 + \log_2 x$$
$$3^y = 9^x$$

given that x and y are positive. [L]

57 Given that $f(x) \equiv \dfrac{11 - 5x^2}{(x+1)(2-x)}$, find constants A and B such that

$$f(x) \equiv 5 + \frac{A}{x+1} + \frac{B}{2-x}$$ [L]

58 The table gives the approximate values of v corresponding to the stated values of u where u and v are variables satisfying an equation of the form

$$v = \frac{a}{u} + \frac{b}{u^3}$$

Estimate to two significant figures values of the constants a and b by drawing a straight line graph relating u^3v and u^2.

u	1.0	1.5	2.0	2.5	3.0
v	10.50	5.33	3.56	2.69	2.17

Estimate the value of u when $u^3v = 23$. [L]

59 Express as a single fraction, in its simplest form,

$$\frac{8}{x^2 - 4} - \frac{2}{x^2 + 3x + 2}$$ [L]

60 The weights of babies born in a hospital maternity unit over a period of 1 month were noted, and are shown in the table.

Weight (kg)	Number of babies
Below 2.50	5
2.50–2.74	14
2.75–2.99	23
3.00–3.24	31
3.25–3.49	28
3.50–3.74	16
3.75–3.99	8
4 and above	2

Plot a cumulative frequency polygon for these data and hence estimate the median and interquartile range for the weights of babies born, giving your answers to 1 decimal place. [L]

61 Carbon 14 is radioactive, and the count rate, N, detected from the carbon 14 content of a piece of wood that is t years old is given by

$$N = N_0 e^{-kt}$$

where N_0 is the count rate of new wood, and k is a positive constant.

Given that the count rate N of a piece of wood is halved after a period of 5600 years, calculate the value of k to 3 significant figures. [L]

62 (a) Sketch the graph of $y = 3^x$.

(b) Solve the equation $3^x = 5$, giving your answer to 3 significant figures.

(c) Use the substitution $y = 3^x$ to solve, to 2 decimal places, the equation

$$2(3^{2x}) - 7(3^x) + 6 = 0$$ [L]

63 Express $\frac{1}{x(x+3)}$ in partial fractions.

Hence show that

$$\int_1^3 \frac{1}{x(x+3)}\, dx = \tfrac{1}{3}\ln 2$$ [L]

64

$$f(x) \equiv px^3 + 11x^2 + 2px - 5$$

The remainder when $f(x)$ is divided by $(x + 2)$ is 15.

(a) Show that $p = 2$.

(b) Factorise $f(x)$ completely.

(c) Deduce the solutions of the equation

$$f(x) = (x + 1)(x + 5)$$ [L]

65 In an experiment to find the focal length f of a lens, the following object distances u and image distances v were recorded:

u	15	20	25	30	35
v	59	31	23	20	18

By plotting the reciprocal of u against the reciprocal of v estimate a value of f from the formula

$$\frac{1}{u} + \frac{1}{v} = \frac{1}{f}$$ [L]

66

x	2.1	2.8	4.7	6.2	7.3
y	13	32	316	2000	7080

The above table shows corresponding values of variables x and y obtained experimentally. By drawing a suitable graph, show that these values support the hypothesis that x and y are connected by a relationship of the form $y = a^x$, where a is a constant. Use your graph to estimate the value of a to 2 significant figures.

67 $f(x) \equiv 2x^2 - px + q + 4$, where p and q are constants.
Given that $f(2) = \frac{1}{2}$ and that the equation $f(x) = 0$ has equal roots, find the two possible values of p and write down the corresponding values of q.
Find, also, the value of the equal roots of the equation $f(x) = 0$ in each case. [L]

68 (a) Write down the common ratio of the geometric series G,

$$e + e^{\frac{1}{2}} + 1 + \dots$$

(b) Calculate, to 3 significant figures, the sum of the first six terms of the series.

(c) Write down, in its simplest form, the common difference of the arithmetic series A,

$$\log_3 2 + \log_3 6 + \log_3 18 + \dots$$

(d) Show that the sum of the first ten terms of A is $10 \log_3 2 + 45$ and evaluate this to 2 decimal places.

(e) One of these two series has a sum to infinity. Calculate, to 2 decimal places, this sum. [L]

69 The rate of decay of a radioactive substance S is proportional to the amount remaining so that

$$\frac{\mathrm{d}x}{\mathrm{d}t} = -kx$$

where x is the amount of S present at time t and k is a positive constant. Given that $x = a$ at time $t = 0$, show that, if the time taken for the amount of S to become $\frac{1}{2}a$ is T, then

$$x = a(\tfrac{1}{2})^{\frac{t}{T}}$$ [L]

70 (a) Solve the equation

$$3y^2 - 16y + 5 = 0$$

(b) By putting $y = 3^x$ into the equation

$$3^{2x+1} - 16(3^x) + 5 = 0$$

show that

$$3y^2 - 16y + 5 = 0$$

(c) Hence, or otherwise, solve, to 3 significant figures where appropriate, the equation

$$3^{2x+1} - 16(3^x) + 5 = 0$$ [L]

71 The following grouped frequency distribution summarises the values of orders taken by the sales representatives employed by a company during one particular year.

Value of order (£)	Number of orders
0–	80
10–	120
20–	226
30–	135
50–	105
100–	40
150–	24
200–	12
500–	8
1000–	0

Let X represent the mid-point of each group.

(a) Using the coding $u = \dfrac{x - 25}{5}$ show that $\Sigma fu = 4355$, where f represents the frequency in each group. Estimate the mean and standard deviation of the value of the orders. (You may use $\Sigma fu^2 = 269\ 975$.)

(b) Explain why these two measures might not be the best ones to use when analysing these data.

(c) Which alternative measures of location and spread would you recommend? [L]

72 The rate of cooling of a metal ball placed in melting ice is proportional to its own temperature T°C. Show that, at time t,

$$T = Ae^{-kt}$$

where A and k are positive constants.
The temperature of the metal ball falls from 50 °C to 40 °C in twenty seconds. Find its temperature at the end of the next twenty seconds. [L]

73 Express f(x), where

$$f(x) \equiv \frac{1+2x}{(1-3x)(1+6x^2)}$$

in partial fractions. [L]

74 Express

$$\frac{7x+6}{2(x+1)(x+2)} - \frac{7}{2(x+3)}$$

as a single fraction in its lowest terms. [L]

75 The table below gives measured values of the variables u and v, which are related by an equation of the form $v = au + \frac{b}{u}$
Find approximately the values of the constants a and b by plotting uv against u^2.

u	1	2	3	4	5
v	12.5	7.0	5.5	5.0	4.9

[L]

76 (a) Use the substitution $y = 3^x$ to find the solution, to 2 decimal places, of the equation

$$3^{2x} - 3(3^x) - 4 = 0$$

Give a reason why there is only one root of this equation.

(b) Sketch the curve with equation

$$y = \log_3 x$$

(c) Given that

$$y = \log_3 x \text{ and } y = \tfrac{1}{2}[1 + \log_3 9x]$$

find the value of x and the corresponding value of y which satisfy these simultaneous equations. [L]

77 (a) Given that $(x + 3)$ is a factor of the expression

$$5x^3 - px^2 - 53x + 84$$

(i) calculate the value of the constant p.

Hence, for this value of p,

(ii) factorise the expression completely.

(b) When the expression $4x^3 - 5x^2 + ax + 4$ is divided by $(x - 2)$ there is a remainder of R. When the expression $5x^3 + ax^2 + 6x + 4$ is divided by $(x - 2)$ there is a remainder of $3R$. Find the value of the constant a. [L]

78 Find the lowest common multiple of

$$(x^2 - 5x + 6),\ (x^2 - 2x) \text{ and } (x^2 - 3x).$$

Express

$$\frac{1}{x^2 - 5x + 6} + \frac{2}{x^2 - 2x} + \frac{3}{x^2 - 3x}$$

as a single fraction in its lowest terms. [L]

79 (a) Solve the equation $\log_3(x + 1) = 2$.

(b) Calculate the value of $\log_7(\frac{1}{5}) + \log_7(\frac{5}{49})$. [L]

80 A controlled experiment was concerned with estimating the number of microbes N present in a culture at time T days after the start of the experiment. Some results from the experiment are shown in the table.

Time T (days)	3	5	10	15	20
Number of microbes (N)	900	2000	5000	9000	16 000

By plotting values of $\lg N$ against the corresponding values of $\lg T$, draw a graph using all the data in the table.

Explain why the graph that you have obtained supports the belief that N and T are related by an equation of the form

$$N = AT^B$$

where A and B are constants.
Use your graph to find an estimate for A, giving your answer to 1 significant figure, and an estimate for B, giving your answer to 2 significant figures. [L]

81 A seed merchant sells a particular seed in packets labelled 'average contents 50 seeds'. A quality control officer samples 100 packets. The results are summarised in the table.

Number of seeds in packet	Number of packets
48	3
49	19
50	27
51	26
52	17
53	6
54	2

Find the median and the mode for the above data.
Calculate also the mean number of seeds in the packets, and show that the standard deviation is 1.32 to 3 significant figures. [L]

82 Express as a single fraction in its lowest terms

$$\frac{5}{x(x-3)} - \frac{25}{x(x-3)(x+2)}$$ [L]

83 Given that the polynomial $\mathrm{P}(x)$ is divisible by $(x-a)^2$, show that $\mathrm{P}'(x)$ is divisible by $(x-a)$.
The polynomial $(x^4 + x^3 - 12x^2 + px + q)$ is divisible by $(x+2)^2$. Find the values of the constants p and q. [L]

84 Given that $\log_8 p = x$ and $\log_8 q = y$, express in terms of x and y,

(a) $\log_8 \sqrt{p} + \log_8 \sqrt{q}$ (b) $\log_8\left(\frac{p^2}{q}\right)$ (c) $\log_8(64pq)$ [L]

85 In a production process, bottles are filled with detergent by means of an automatic pump. The quantities of detergent in a sample of 200 bottles were measured and the following results were obtained.

Quantity in cm^3	900–	902–	904–	906–	908–	910–912
No. of bottles	8	25	45	53	47	22

Taking 905 cm^3 as origin, calculate the arithmetic mean and the standard deviation for this distribution. [L]

86

x	2	3	4	5	6	7
y	39	32	28	25	22.5	20.8

The table shows experimental values of two quantities x and y which are known to satisfy the equation $yx^n = k$, where n and k are constants.

(a) Draw a graph of $\ln y$ against $\ln x$.

(b) Use your graph to estimate values for n and k, giving your answers to 2 significant figures. [L]

87 Express

$$\frac{7}{5(x-3)} - \frac{7x+15}{5(x-1)(x+2)}$$

as a single fraction in its lowest terms. [L]

88

$$f(x) \equiv \frac{5x-1}{(x^2+1)(3x+2)}$$

Express $f(x)$ in partial fractions. [L]

89 (a) Given that $(3x+2)$ is a factor of $3x^3 + Ax^2 - 4x - 4$, show that $A = 5$.

(b) Factorise $3x^3 + 5x^2 - 4x - 4$ completely.

(c) Given that $0° \leqslant t \leqslant 360°$, find the values of t, to the nearest degree, for which

$$3\sin^3 t + 5\sin^2 t - 4\sin t - 4 = 0$$ [L]

90 The first three terms of an arithmetic series are $\lg x$, $\lg 2(x+1)$ and $\lg 4(x+6)$ respectively.

(a) Find the value of x.

(b) Find the value of the common difference of this series. [L]

91 Express as a single fraction

$$\frac{5}{x+2} - \frac{7}{2x+3}$$

simplifying the numerator as much as possible. [L]

92 Calculate the mean and the standard deviation of the odd numbers from 1 to 19 inclusive.

Deduce from your results the mean and the standard deviation of

(a) the numbers 1, 2, 3, . . ., 10

(b) the even numbers from 10 to 28 inclusive. [L]

93 Given that $(x+1)$ and $(x-2)$ are both factors of the expression $ax^3 - x^2 + bx - a$, calculate the values of the constants a and b.

With these values of a and b, calculate the remainder when the expression is divided by $(x-4)$. [L]

94 Two types of bottling machine are each designed to deliver one litre of liquid to every bottle. To test the designs, one machine of each type was used to deliver liquid to 1000 bottles. The data obtained are shown in the table.

Quantity delivered (Q ml)	Number of times quantity was delivered	
	Machine A	Machine B
$998.0 \leqslant Q < 998.5$	10	16
$998.5 \leqslant Q < 999.0$	54	212
$999.0 \leqslant Q < 999.5$	163	548
$999.5 \leqslant Q < 1000.0$	263	193
$1000.0 \leqslant Q < 1000.5$	254	23
$1000.5 \leqslant Q < 1001.0$	175	4
$1001.0 \leqslant Q < 1001.5$	78	3
$1001.5 \leqslant Q < 1002.0$	3	1

(a) Calculate the mean and standard deviation of the volume delivered by each machine, giving your answers to 2 decimal places.

(b) Explain briefly how your results in (a) might influence your choice of type of bottling machine. [L]

95 The times taken by 40 people to complete a puzzle were recorded to the nearest second. A grouped frequency distribution of the recorded results is shown in the table.

Times to complete a puzzle

Time (s)	10–19	20–29	30–39	40–49	50–59	60–69	70–79	80–89
Frequency	4	8	8	9	7	0	3	1

Estimate the mean and the standard deviation of the times taken to complete the puzzle by the 40 people. [L]

96 (a) (i) Given that $\log_3 x = 2$, determine the value of x.

(ii) Calculate the value of y for which

$$2\log_3 y - \log_3(y+4) = 2$$

(iii) Calculate the values of z for which

$$\log_3 z = 4\log_z 3$$

(b) (i) Express $\log_a(p^2q)$ in terms of $\log_a p$ and $\log_a q$.

(ii) Given that $\log_a(pq) = 5$ and $\log_a(p^2q) = 9$, find the values of $\log_a p$ and $\log_a q$. [L]

97 (a) Express

$$\frac{x+3}{2(x+1)(x+5)} - \frac{1}{2(x+2)}$$

as a single fraction in its simplest form.

(b) Find the value of the expression in part (a) if $x = -7$, giving your answer as a single fraction in its lowest terms. [L]

98 A random sample of 120 broad bean seeds was collected. Each seed was weighed to the nearest 0.01 g and the results are summarised in the table.

Weight (g)	Number of beans
1.10–1.29	7
1.30–1.49	24
1.50–1.69	33
1.70–1.89	32
1.90–2.09	14
2.10–2.29	8
2.30–2.49	1
2.50–2.69	1

Calculate an estimate for the mean and standard deviation of the weights of these broad bean seeds, giving your answers to 3 decimal places. [L]

99 (a) Using the result $e^{\ln x} = x$, express in their simplest form

(i) $e^{\ln x + \ln y}$

(ii) $e^{2 \ln x}$

(b) The number N of bacteria in a certain culture at time t hours is given by $N = 600e^{ct}$, where c is a constant. Show that at any instant the number of bacteria is increasing at a rate proportional to the number of bacteria present at that instant.

The number of bacteria increases from 600 when $t = 0$ to 1800 when $t = 2$.

(i) Show that $c = \frac{1}{2}\ln 3$.

(ii) Show that the number of bacteria present at t hours is $600B^{\frac{1}{2}t}$, where B is a constant and state the value of B. [L]

100

Time after formation of protactinium sample, t seconds	80	120	160	200	240
Number of particles detected per second, N	80	67	55	44	38

An isotope of protactinium decays rapidly into a stable isotope of uranium. A particle detector set up to monitor such a decay recorded the levels of activity summarised in the table. It is believed that the number of particles detected per second, N, obeys a law of the form

$$N = N_0 e^{-kt}$$

where t seconds is the time after the formation of the protactinium sample and N_0 and k are positive constants. By plotting a suitable graph, show that the data in the table are consistent with such a law. Use your graph to obtain estimates of N_0 and k to 1 significant figure. [L]

The binomial series for a positive integral index

5

5.1 Binomial expressions

From the algebra covered in Book P1, you will be quite familiar with terms like y^0, $3y^1$, $4p^2$, $-5z^3$, etc., where each term is called a **monomial**. That is, it is an expression consisting of a single term.

Expressions like $1+x$, $2-3x$, $a+bx$ are called **binomial** because they each consist of just two terms joined by a + or a – sign. As you will have read in Book P1, expressions with many terms, such as $3-4x+x^2+x^3$ or x^4-x^2-2x-5, are called **polynomials**.

Often you will find that in the course of your work you need to expand expressions such as $(1+x)^3$ or $(2-3x)^5$, or more generally $(a+b)^n$, where n is a positive integer. We start by considering some simple examples, where binomial expressions are squared and cubed.

Example 1
Expand (a) $(1+x)^2$ (b) $(1+x)^3$.

(a) Removing the brackets gives

$$\begin{aligned}(1+x)^2 &= 1(1+x)+x(1+x)\\ &= 1+x+x+x^2\\ &= 1+2x+x^2\end{aligned}$$

(b) Using the result found in (a):

$$\begin{aligned}(1+x)^3 &= (1+x)(1+x)^2\\ &= (1+x)(1+2x+x^2)\\ &= 1(1+2x+x^2)+x(1+2x+x^2)\\ &= 1+2x+x^2+x+2x^2+x^3\end{aligned}$$

Hence: $$(1+x)^3 = 1+3x+3x^2+x^3$$

Example 2
Expand (a) $(1 - x)^2$ (b) $(1 - x)^3$

These expansions could be done by multiplying out the brackets as in example 1. It is quicker, however, to take the results from example 1 and replace x by $-x$. Then you have:

(a) $(1 - x)^2 = 1 + 2(-x) + (-x)^2 = 1 - 2x + x^2$

(b) $(1 - x)^3 = 1 + 3(-x) + 3(-x)^2 + (-x)^3$

$= 1 - 3x + 3x^2 - x^3$

Example 3
Expand $(1 - 2x)^3$.

This expansion can be done by putting $-2x$ in the place of x in the expansion of $(1 + x)^3$. Then you get:

$$(1 - 2x)^3 = 1 + 3(-2x) + 3(-2x)^2 + (-2x)^3$$
$$= 1 - 6x + 12x^2 - 8x^3$$

Simple bionomial expansions such as $(1 + x)^2$ and $(1 + x)^3$ are obtained by multiplying out brackets. Then other similar binomial expressions may be expanded by using these results, as shown in examples 2 and 3. Although these illustrations are really simple, the methods used are of great importance and you will see them used again both in this chapter and in other contexts.

Exercise 5A

1 By using the expansion of $(1 + x)^2$, expand each of the following:
(a) $(1 + 3x)^2$ (b) $(1 - 5x)^2$ (c) $(1 - x^2)^2$
(d) $(1 + 2x^3)^2$ (e) $[1 - (x - 2)^2]^2$

2 Use the expansion of $(1 + x)^3$ to expand:
(a) $(1 + 2y)^3$ (b) $(1 - x^3)^3$ (c) $(1 + 3x^{-1})^3$

3 By considering the products $(1 + x)(1 + x)^3$ and $(1 + x)(1 + x)^4$ find the expansion of (a) $(1 + x)^4$ (b) $(1 + x)^5$

4 Using the results you obtained in question 3, or otherwise, work out (a) $(1 - 3x)^4$ (b) $(1 + 2x)^5$.

5.2 Pascal's triangular array

From your work in this chapter, you can build up these expansions:

$$(1+x)^1 = 1 + x$$
$$(1+x)^2 = 1 + 2x + x^2$$
$$(1+x)^3 = 1 + 3x + 3x^2 + x^3$$

Also: $$(1+x)^0 = 1$$

You can summarise this by writing down only the coefficients, in this way:

Expansion	Coefficients in ascending powers of x
$(1+x)^0$	1
$(1+x)^1$	1 1
$(1+x)^2$	1 2 1
$(1+x)^3$	1 3 3 1
$(1+x)^4$	1 4 6 4 1
$(1+x)^5$	1 5 10 10 5 1
. . . . . .	. . . . and so on . . . .

This continuing triangular array is known as **Pascal's triangle** and it provides you with a means of determining *any* coefficient in the expansion of $(1+x)^n$ for integral n up to, say, $n = 20$ (or more if you have the time!) but a better method is available for larger n as you will see later in this chapter.

Each entry E in the array, except the 'ones' at each end, is the sum of the two entries on either side of E in the previous line. For reference, call line $(1+x)^0$ line 0, $(1+x)^1$ line 1, $(1+x)^2$ line 2, etc. Each new line of the array is formed from the previous line in this way. For example, the 10 in line 5 comes from adding the 6 and 4 in line 4.

Example 4

Expand $(1+x)^7$ in ascending powers of x. Hence find the first four terms of the expansion of $(1+4x)^7$ in ascending powers of x.

Line 7 of Pascal's triangle is 1 7 21 35 35 21 7 1 and hence the expansion of $(1+x)^7$ in ascending powers of x is

$$1 + 7x + 21x^2 + 35x^3 + 35x^4 + 21x^5 + 7x^6 + x^7$$

If you replace x by $4x$, you find that the first four terms of the expansion of $(1 + 4x)^7$ are:

$$1 + 7(4x) + 21(4x)^2 + 35(4x)^3$$
$$= 1 + 28x + 336x^2 + 2240x^3$$

Example 5

Determine the first four terms of the expansion of $(1 - x)^{13}$ in ascending powers of x. Show that the estimate B of 0.999^{13} obtained by using these four terms of the series is 0.987 077 714.

By considering the next term in your series show that B is *not correct* to 9 decimal places.

The first four entries in line 13 of Pascal's triangle are:

$$1, \quad 13, \quad 78, \quad 286$$

and hence the first four terms in the expansion of $(1 - x)^{13}$ are

$$1 - 13x + 78x^2 - 286x^3$$

Put $x = 0.001$ in this expansion:

$$(1 - 0.001)^{13} = 0.999^{13} \approx 1 - 0.013 + 0.000\,078 - 0.000\,000\,286$$
$$= 0.987\,077\,714, \text{ as required}$$

The fifth Pascal triangle number in line 13 is 715 and the fifth term in the expansion of $(1 + x)^{13}$ is $715x^4$. When $x = 0.001$, this term is 0.000 000 000 715 and so $0.999^{13} \approx 0.987\,077\,714\,715$. So 0.999^{13} is 0.987 077 715 (9 d.p.) and estimate B is *not* correct to 9 decimal places.

5.3 The expansion of $(a + b)^n$, where n is a positive integer

If you multiply directly and remove the brackets, you can easily establish that:

$$(a + b)^2 = a^2 + 2ab + b^2$$
$$(a + b)^3 = a^3 + 3a^2b + 3ab^2 + b^3$$
$$(a + b)^4 = a^4 + 4a^3b + 6a^2b^2 + 4ab^3 + b^4$$
$$(a + b)^5 = a^5 + 5a^4b + 10a^3b^2 + 10a^2b^3 + 5ab^4 + b^5$$

Using Pascal's triangular array and continuing this process, you can produce expansions for $(a + b)^n$ where $n = 6, 7, \ldots$.

Example 6

Expand in descending powers of x: (a) $(2x+3)^4$ (b) $(x-x^{-1})^3$.

(a) Write $a = 2x$ and $b = 3$ in the expansion of $(a+b)^4$. This gives:

$$\begin{aligned}(2x+3)^4 &= (2x)^4 + 4(2x)^3(3) + 6(2x)^2(3)^2 + 4(2x)(3)^3 + (3)^4 \\ &= 16x^4 + 96x^3 + 216x^2 + 216x + 81\end{aligned}$$

(b) Write $a = x$ and $b = -x^{-1}$ in the expansion of $(a+b)^3$. This gives:

$$\begin{aligned}(x-x^{-1})^3 &= x^3 + 3x^2(-x^{-1}) + 3x(-x^{-1})^2 + (-x^{-1})^3 \\ &= x^3 - 3x + 3x^{-1} - x^{-3}\end{aligned}$$

Exercise 5B

1 Expand $(1+5x)^3$.
Use your expansion to simplify $(1+5x)^3 + (1-5x)^3$.

2 Expand $(1-\frac{1}{2}x)^n$, for $n = 2, 3, 4$.

3 Expand $(2-3x)^3$ and $(3+2x)^3$.
Hence express $3(2-3x)^3 - 2(3+2x)^3$ in terms of x.

4 Find in ascending powers of x the first four terms in the expansions of (a) $(1-2x)^6$ (b) $(2-x)^7$.

5 Find in descending powers of y the first three non-zero terms in the expansion of $(1-4y)^9 - (1+4y)^9$.

6 Use the binomial series of $(1-2x)^8$ to evaluate 0.98^8 to 7 decimal places.

7 Use the first four terms in the binomial expansion of $(1+4x)^{12}$ in ascending powers of x to determine an approximation for
(a) 1.004^{12} (b) 0.996^{12}
justifying in each case the accuracy of your approximation.

8 Simplify $(4-3x)^3 - (3+4x)^3$.

9 Find and simplify the first three terms of the expansion of $(5x-2)^4$ in descending powers of x.

10 The first three terms in the expansion of $(A+x)^m$ in ascending powers of x are $64 + 192x + Bx^2$. Find the values of m, A and B.

11 Simplify:
(a) $(1+\sqrt{3})^4 + (1-\sqrt{3})^4$
(b) $(\sqrt{2}+\sqrt{3})^4 + (\sqrt{2}-\sqrt{3})^4$

12 Find the expansion of $(z - \frac{1}{z})^9$ in descending powers of z.

13 Find the coefficients of the terms indicated in the following expansions:

(a) $(1 - 2x)^{14}$; x^2 term

(b) $(2 + 3x)^5$; x^3 term

(c) $(1 - 2x^3)^5$; x^9 term

(d) $(3 - 4x)^7$; x^6 term

14 Find the term that is independent of x in the expansion of

$$(3x - \tfrac{1}{3x})^6$$

15 Evaluate

$$\int_0^1 (2x^2 + 1)^3 \, dx$$

5.4 Expanding $(a + b)^n$ by formula for positive integral n

As you have already observed the first few lines of Pascal's triangular array are:

line

line									
0					1				
1				1		1			
2			1		2		1		
3		1		3		3		1	
4	1		4		6		4		1

. . . . and so on . . . .

The question you may have asked, and, in fact, you *should* have asked is 'What are the entries in the nth line?'. By observation, you can see that the nth line will have $(n + 1)$ numbers, of which the first is 1 and the second is n. The third number is

$$\frac{n(n - 1)}{1 \times 2}$$

and the fourth is

$$\frac{n(n - 1)(n - 2)}{1 \times 2 \times 3}$$

For example in line 4 you have

$$1,\ 4,\ \frac{4 \times 3}{1 \times 2},\ \frac{4 \times 3 \times 2}{1 \times 2 \times 3},\ \frac{4 \times 3 \times 2 \times 1}{1 \times 2 \times 3 \times 4}$$

That is 1, 4, 6, 4, 1.

Test the formula for yourself for different values of n. The general term in the expansion of $(1 + x)^n$ is taken to be the x^r term, where r is any positive integer less than or equal to n.

The coefficient of x^r in the expansion of $(1 + x)^n$ is

$$\frac{n(n-1)(n-2) \times \ldots (n-r+1)}{1 \times 2 \times 3 \times \ldots \times r}$$

The number $1 \times 2 \times 3 \times \ldots \times r$, that is the number obtained when all the integers from 1 to r inclusive are multiplied together, is written in shorthand form as $r!$ and called **'factorial r'**.

If this notation is used, then you find that the coefficient of x^r is:

$$\frac{n(n-1)(n-2)\ldots(n-r+1)}{r!}$$

$$= \frac{n(n-1)(n-2)\ldots(n-r+1)(n-r)(n-r-1)\ldots 2 \times 1}{r!\,(n-r)(n-r-1)\ldots 2 \times 1}$$

$$= \frac{n!}{r!(n-r)!}$$

This is the form in which you should learn and memorise this coefficient. You will not be expected to prove or derive this formula for the coefficient of the x^r term in the expansion of $(1 + x)^n$ but you must be able to apply it soundly.

Notation

You will find that the coefficient

$$\frac{n!}{r!(n-r)!}$$

is often written as

$$\binom{n}{r} \quad \text{or} \quad {}^nC_r$$

There will probably be a button on your calculator marked ${}^nC_r$. Also 'factorial 0', 0!, is defined to be 1.

- **You should remember that $\binom{n}{r} = {}^nC_r = \frac{n!}{r!(n-r)!}$**

Example 7

Evaluate the following coefficients:

(a) $\binom{5}{3}$ (b) $\binom{9}{5}$

Using the formula, you have:

(a) $\binom{5}{3} = \dfrac{5!}{3!(5-3)!} = \dfrac{120}{6 \times 2} = 10$

(b) $\binom{9}{5} = \dfrac{9!}{5!(9-5)!} = \dfrac{362\,880}{120 \times 24} = 126$

The expansion of $(1+x)^n$ in ascending powers of x is

$$(1+x)^n = 1 + \binom{n}{1}x + \binom{n}{2}x^2 + \ldots + \binom{n}{r}x^r + \ldots + x^n$$

Example 8

Determine the first four terms in the expansion of $(1+x)^{21}$ in ascending powers of x. Hence find the coefficient of x^3 in the expansion of $(2-3x)(1+x)^{21}$.

Using the binomial expansion:

$$\begin{aligned}(1+x)^{21} &= 1 + \binom{21}{1}x + \binom{21}{2}x^2 + \binom{21}{3}x^3 + \ldots \\ &= 1 + 21x + 210x^2 + 1330x^3 + \ldots\end{aligned}$$

Now:

$$(2-3x)(1+x)^{21} = (2-3x)(1 + 21x + 210x^2 + 1330x^3 \ldots)$$

The coefficient of $x^3 = 2(1330) - 3(210) = 2030$

Example 9

In the expansion of $(1+x)^n$ in ascending powers of x, the coefficients of x^2 and x^3 are equal. Find the value of n.

The coefficient of x^2 is

$$\binom{n}{2} = \frac{n!}{2!(n-2)!}$$

The coefficient of x^3 is

$$\binom{n}{3} = \frac{n!}{3!(n-3)!}$$

Since these are equal: $\dfrac{n!}{2!(n-2)!} = \dfrac{n!}{3!(n-3)!}$

That is: $2!(n-2)! = 3!(n-3)!$

$$\frac{(n-2)!}{(n-3)!} = \frac{3!}{2!}$$

$$n - 2 = 3$$

So: $n = 5$

Exercise 5C

1 Find the first three terms in the expansion of $(1-x)^{23}$ in ascending powers of x.

2 Given that $(1+x)^{15} = 1 + Ax + Bx^2 + Cx^3 \ldots$, find A, B and C.

3 Find, in descending powers of x, the first three terms in the expansion of $(5x-3)^7$.

4 Simplify $(3+2x)^4 + (3-2x)^4$.

5 Obtain, in ascending powers of y, the first four terms in the expansion of (a) $(1-5y)^8$ (b) $(1-4y)(1-5y)^8$.

6 Obtain, in ascending powers of x, the first four terms in the expansion of $(1+x)^{10}$.
Find an approximation of $(0.998)^{10}$ from your expansion to the best degree of accuracy possible, justifying your decision.

7 Expand and simplify the first three terms, in descending powers of x, in the expansion of $(x+2x^{-1})^{10} - (x-2x^{-1})^{10}$.

8 Given that $(\sqrt{3}-2)^6 = a\sqrt{3} + b$, find the values of a and b.

9 Show that

$$\frac{\sqrt{2}-1}{\sqrt{2}+1} = 3 - 2\sqrt{2}$$

and that

$$\frac{\sqrt{2}+1}{\sqrt{2}-1} = 3 + 2\sqrt{2}$$

Hence find the exact value of

$$\left(\frac{\sqrt{2}-1}{\sqrt{2}+1}\right)^5 + \left(\frac{\sqrt{2}+1}{\sqrt{2}-1}\right)^5$$

10 $\mathrm{P}(x) \equiv \left(x^2 - \frac{3}{x}\right)^n$

Find the term that is independent of x in the expansion of $\mathrm{P}(x)$,
(a) when $n = 3$ (b) when $n = 12$.

11 Use the binomial expansion to find the value of
(a) $(1.001)^{12}$, correct to 8 decimal places
(b) $(9.999)^{15}$, correct to 10 significant figures.

12 The coefficients of the x and x^2 terms in the expansion of $(1 + kx)^n$ are 44 and 924 respectively. Find the values of the constants k and n and the coefficient of the x^3 term in the expansion.

13 Expand $(1 + ax)^7$ in ascending powers of x up to and including the term in x^3.
In the expansion of $(b + x)(1 + ax)^7$ in ascending powers of x, the first and second terms are 5 and $71x$ respectively. Find
(a) the values of the constants a and b
(b) the x^2 and x^3 terms in this expansion.

14 The coefficients of x, x^2 and x^3 in the expansion of $(1 + x)^n$ are the first three terms of an arithmetic series. Show that $n = 7$.

15 The first three terms in the expansion of $(1 + kx)^n$ are 1, $14x$ and $84x^2$ respectively. Find the values of the constants k and n and the coefficients of the x^3 and x^4 terms.

16 Find the coefficient of x^{12} in the expansion of $(x - y)^{18}$. Evaluate this term when $x = 6$ and $y = 3^{-1}$.

17 Find the expansion of $(1 + y)^6$.
By writing $y = x + x^2$, find the first four terms in ascending powers of x of the expansion of $(1 + x + x^2)^6$.
By putting $x = 0.01$ in your four terms, find an approximation for $(1.0101)^6$.

18 Part of the expansion of $(a + bx)^5$ is $32 + 40x + 20x^2$. Find the values of the constants a and b and the remaining terms of the expansion.

19 Find the first four terms in descending powers of x of the expansion of $(x^2 + 1)^{40}$.

20 The first three terms of the expansion of $(1 + kx)^n$ in ascending powers of x are 1, $\frac{17}{4}x$ and $\frac{17}{2}x^2$.
Find the values of the constants k and n and the terms in x^3 and x^4.

SUMMARY OF KEY POINTS

1 Binomial expressions consist of two terms, for example $(a + bx)$.

2 Important simple identities are:

$$(a \pm b)^2 \equiv a^2 \pm 2ab + b^2$$

$$(a \pm b)^3 \equiv a^3 \pm 3a^2b + 3ab^2 \pm b^3$$

3 Pascal's triangular array of binomial coefficients is:

$$\begin{array}{ccccccccc|c} & & & & 1 & & & & & (1+x)^0 \\ & & & 1 & & 1 & & & & (1+x)^1 \\ & & 1 & & 2 & & 1 & & & (1+x)^2 \\ & 1 & & 3 & & 3 & & 1 & & (1+x)^3 \\ 1 & & 4 & & 6 & & 4 & & 1 & (1+x)^4 \end{array}$$

. . . . etc . . . .

4 Factorial n, written $n!$, is $1 \times 2 \times 3 \times 4 \times \ldots \times (n-1) \times n$.
Remember: $0! = 1$

5 $\binom{n}{r} = {}^nC_r = \dfrac{n!}{r!(n-r)!}$

6 $(1+x)^n = 1 + \binom{n}{1}x + \binom{n}{2}x^2 + \ldots + \binom{n}{r}x^r + \ldots + x^n$

7 $(a+b)^n = a^n + \binom{n}{1}a^{n-1}b + \binom{n}{2}a^{n-2}b^2 + \ldots + \binom{n}{r}a^{n-r}b^r + \ldots$
$\ldots + b^r$

Trigonometry I

6

In Book P1 you learned the basics of trigonometry. This chapter will extend your knowledge. Three more trigonometric ratios are introduced, with their graphs. Formulae for compound angles are derived. Finally, identities that relate the various trigonometric ratios to each other are formulated by using Pythagoras' theorem together with the identities for double and half angles. These enable you to solve other sorts of trigonometric equations, in addition to those that you learned to solve in Book P1.

6.1 Secant, cosecant and cotangent

There are six trigonometric ratios in total. You have already met the ratios sine, cosine and tangent. The three others are **secant, cosecant** and **cotangent**. These are defined as follows:

- **secant x (written sec x)** $= \dfrac{1}{\cos x}$
- **cosecant x (written cosec x)** $= \dfrac{1}{\sin x}$
- **cotangent x (written cot x)** $= \dfrac{1}{\tan x}$

So:

$$\begin{aligned}\sec 137^\circ = \frac{1}{\cos 137^\circ} &= -\frac{1}{\cos 43^\circ}\\ &= -\frac{1}{0.731\,35}\\ &= -1.367 \text{ (3 d.p.)}\end{aligned}$$

$$\begin{aligned}\operatorname{cosec} 231^\circ = \frac{1}{\sin 231^\circ} &= -\frac{1}{\sin 51^\circ}\\ &= -\frac{1}{0.777\,14}\\ &= -1.287 \text{ (3 d.p.)}\end{aligned}$$

$$\cot 253^\circ = \frac{1}{\tan 253^\circ} = +\frac{1}{\tan 73^\circ}$$
$$= +\frac{1}{3.27085}$$
$$= +0.306 \text{ (3 d.p.)}$$

6.2 Drawing the graphs of sec *x*, cosec *x* and cot *x*

Section 6.1 shows how to find the secant, cosecant or cotangent of any angle. This section shows you how to draw the graphs of these functions.

Graphing sec *x*

Here is a table of values for the function $y = \sec x$ when $0 \leqslant x \leqslant 360^\circ$. Each value of y is given to two decimal places, where appropriate.

x	0	30°	45°	60°	90°	120°	135°	150°	180°	210°	225°	240°	270°	300°	315°	330°	360°
y	1	1.15	1.41	2	∞	−2	−1.41	−1.15	−1	−1.15	−1.41	−2	∞	2	1.41	1.15	1

If you plot these figures on a graph, it looks like this:

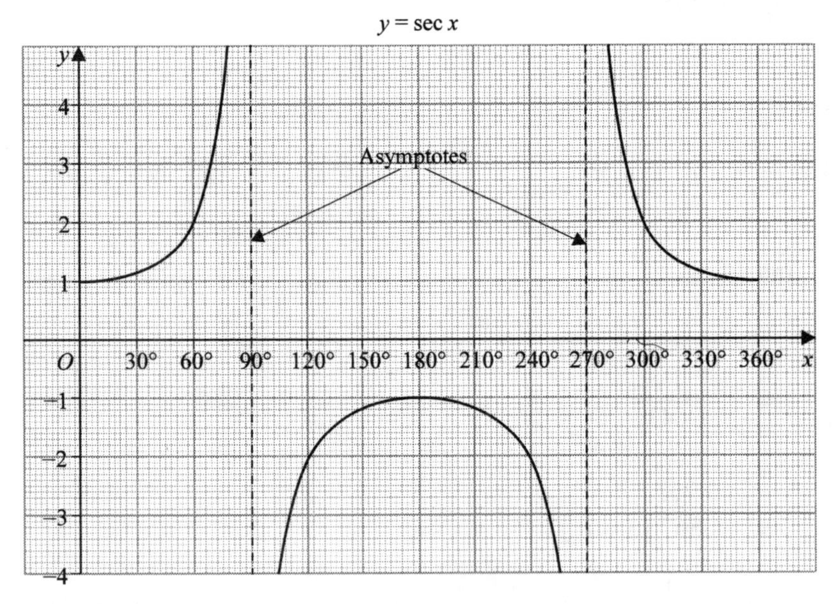

The curve will repeat itself again and again for values of x above 360° and below 0. The function has a local minimum of 1 and a local maximum of −1. The minimum occurs at 0, 360°, 720°, 1080°, etc; in other words, at ±360n°, where n is an integer. The maximum occurs at 180°, 540°, 900°, etc.; in other words, at 180° ± 360n°.

The curve does not meet the x-axis. It has asymptotes at 90°, 270°, 450°, etc.; in other words, at 90° ± 180n°. It has a period of 360°.

Graphing cosec x

Here is a table of values for the function $y = \text{cosec } x$ when $0 \leqslant x \leqslant 360°$. Again each value is given to 2 decimal places, where appropriate.

x	0	30°	45°	60°	90°	120°	135°	150°	180°	210°	225°	240°	270°	300°	315°	330°	360°
y	∞	2	1.41	1.15	1	1.15	1.41	2	∞	−2	−1.41	−1.15	−1	−1.15	−1.41	−2	∞

The curve looks like this:

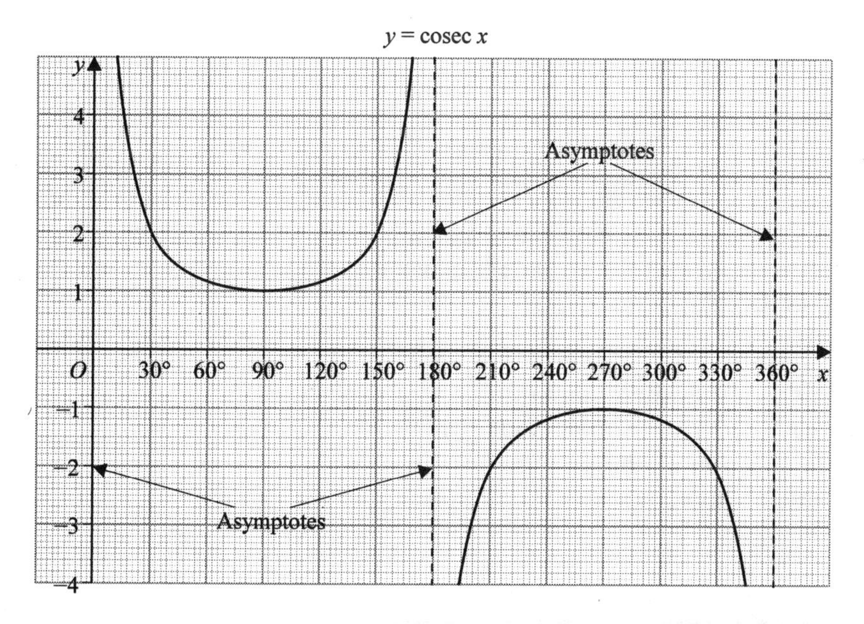

Again the main feature of the curve is that it is periodic. It has a period of 360°. Its local minimum is again 1 and its local maximum −1. But this time the minimum occurs at 90°, 450°, 810°, and so on; in other words, at 90° ± 360n°. The maximum value occurs at 270°, 630°, 990°, and so on; in other words, at 270° ± 360n°.

Again, the graph does not meet the x-axis. It has asymptotes at 0, 180°, 360°, etc.; that is, at ±180n°.

Graphing cot *x*

Here is a table of values for $y = \cot x$ when $0 \leqslant x \leqslant 360°$.

x	0	30°	45°	60°	90°	120°	135°	150°	180°	210°	225°	240°	270°	300°	315°	330°	360°
y	∞	1.73	1	0.58	0	−0.58	−1	−1.73	∞	1.73	1	0.58	0	−0.58	−1	−1.73	∞

The curve looks like this:

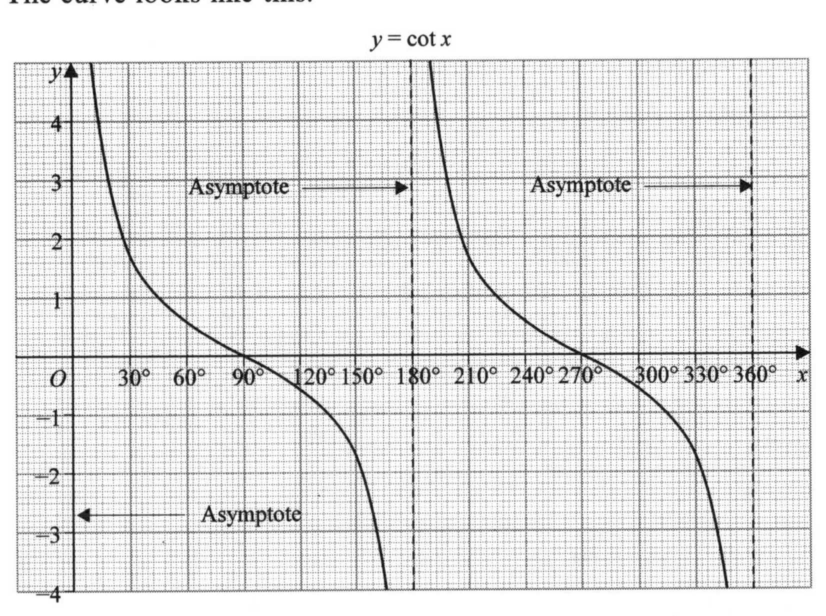

Again, the curve is periodic with period 180°. It cuts the x-axis at 90°, 270°, 450°, and so on; in other words, at $90° \pm 180n°$. It has asymptotes at 0, 180°, 360°, and so on; that is, at $\pm 180n°$.

You must memorise the main features of the curves $y = \sec x$, $y = \operatorname{cosec} x$ and $y = \cot x$ – their shape, their maximum and minimum values and where they occur, the points at which $y = \cot x$ cuts the x-axis, and the position of the asymptotes. For an advanced course in mathematics you must be able to sketch these three curves from memory.

Exercise 6A

1 Find, to 4 significant figures, the value of:

(a) cot 39° (b) sec 47° (c) cosec 63°

(d) sec 23° (e) cot 149° (f) cosec 323°

(g) sec 253° (h) cosec 129° (i) cot 300°

(j) cosec 283°

2 Find, to 4 significant figures, the value of:

(a) $\cot 1.6^{c}$ (b) $\sec 3.4^{c}$
(c) $\operatorname{cosec} 2.3^{c}$ (d) $\cot 4.8^{c}$
(e) $\operatorname{cosec} 5^{c}$ (f) $\sec 6^{c}$
(g) $\cot 3.8^{c}$ (h) $\cot 5.7^{c}$
(i) $\sec 6.83^{c}$ (j) $\operatorname{cosec} 8.23^{c}$

3 Find the value(s) of x, in degrees to $0.1°$, for $0 < x \leqslant 360°$ where:

(a) $\sec x = 1.813$ (b) $\operatorname{cosec} x = -2.164$
(c) $\cot x = -1.23$ (d) $\sec x = -1.114$
(e) $\sec x = -1.132$ (f) $\cot x = 0.147$
(g) $\sec x = 1.614$ (h) $\operatorname{cosec} x = 1.816$
(i) $\cot x = 1.213$ (j) $\operatorname{cosec} x = 1.142$

4 Find the value of x, in radians to 3 decimal places, for $0 < x \leqslant 2\pi$ where:

(a) $\sec x = 1.624$ (b) $\operatorname{cosec} x = -1.624$
(c) $\cot x = 0.718$ (d) $\sec x = -1.934$
(e) $\operatorname{cosec} x = 2.016$ (f) $\cot x = -1.913$
(g) $\sec x = 1.323$ (h) $\operatorname{cosec} x = -1.762$
(i) $\cot x = -0.323$ (j) $\sec x = -2.053$

6.3 The compound angle formulae: sin(*A* ± *B*), cos(*A* ± *B*), tan(*A* ± *B*)

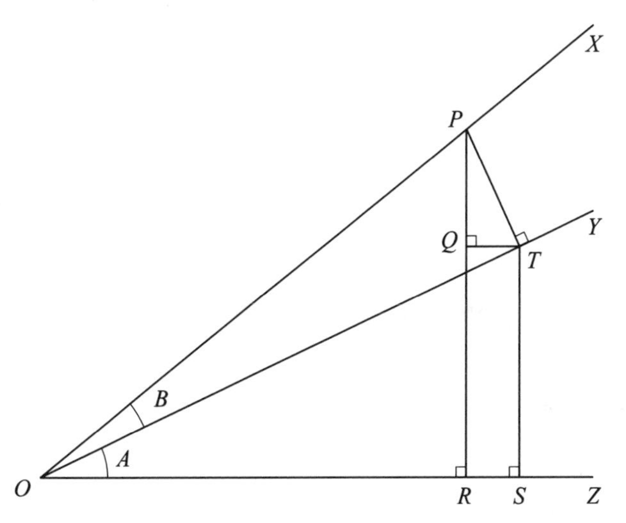

In the diagram the angles $TOR = A$ and $POT = B$ are each acute and the angle $POR = (A + B)$ is also acute. PT is perpendicular to

OY, PR is perpendicular to OZ, TS is perpendicular to OZ and QT is perpendicular to PR.

Since QT is parallel to OS,

$$\angle QTO = \angle TOS = A$$

Since $\angle PTO = 90°$, $\angle PTQ = 90° - A$. Since PQT is a triangle,

$$\angle PQT + \angle PTQ + \angle QPT = 180°$$

That is: $$90° + (90° - A) + \angle QPT = 180°$$

$$180° - A + \angle QPT = 180°$$

$$\angle QPT = A$$

Now:
$$\begin{aligned}\sin(A + B) &= \frac{PR}{OP} \\ &= \frac{PQ + QR}{OP} \\ &= \frac{PQ + TS}{OP} \\ &= \frac{PQ}{OP} + \frac{TS}{OP} \\ &= \left(\frac{PQ}{PT} \times \frac{PT}{OP}\right) + \left(\frac{TS}{OT} \times \frac{OT}{OP}\right) \\ &= \cos A \sin B + \sin A \cos B\end{aligned}$$

■ **So:** $$\mathbf{\sin(A + B) \equiv \sin A \cos B + \cos A \sin B}$$

Similarly:
$$\begin{aligned}\cos(A + B) &= \frac{OR}{OP} \\ &= \frac{OS - RS}{OP} \\ &= \frac{OS - QT}{OP} \\ &= \frac{OS}{OP} - \frac{QT}{OP} \\ &= \left(\frac{OS}{OT} \times \frac{OT}{OP}\right) - \left(\frac{QT}{PT} \times \frac{PT}{OP}\right)\end{aligned}$$

■ $$\mathbf{\cos(A + B) \equiv \cos A \cos B - \sin A \sin B}$$

We have shown that

$$\sin(A + B) \equiv \sin A \cos B + \cos A \sin B$$

and $$\cos(A + B) \equiv \cos A \cos B - \sin A \sin B$$

for A and B acute. However, these identities are, in fact, true for *all* values of A and B.

So replacing B by $-B$ in the identity for $\sin(A+B)$ gives

$$\sin(A-B) \equiv \sin A\cos(-B) + \cos A\sin(-B)$$

or:

■ $$\mathbf{\sin(A-B) \equiv \sin A\cos B - \cos A\sin B}$$

and replacing B by $-B$ in the identity for $\cos(A+B)$ gives

$$\begin{aligned}\cos(A-B) &\equiv \cos A\cos(-B) - \sin A\sin(-B)\\ &= \cos A\cos B - \sin A(-\sin B)\end{aligned}$$

■ $$\mathbf{\cos(A-B) \equiv \cos A\cos B + \sin A\sin B}$$

Now

$$\begin{aligned}\tan(A+B) &\equiv \frac{\sin(A+B)}{\cos(A+B)}\\ &= \frac{\sin A\cos B + \cos A\sin B}{\cos A\cos B - \sin A\sin B}\\ &= \frac{\dfrac{\sin A\cancel{\cos B}}{\cos A\cancel{\cos B}} + \dfrac{\cancel{\cos A}\sin B}{\cancel{\cos A}\cos B}}{\dfrac{\cancel{\cos A}\cancel{\cos B}}{\cancel{\cos A}\cancel{\cos B}} - \dfrac{\sin A\sin B}{\cos A\cos B}}\end{aligned}$$

■ $$\mathbf{\tan(A+B) \equiv \frac{\tan A + \tan B}{1 - \tan A\tan B}}$$

Replace B by $-B$:

$$\begin{aligned}\tan(A-B) &\equiv \frac{\tan A + \tan(-B)}{1 - \tan A\tan(-B)}\\ &= \frac{\tan A - \tan B}{1 - \tan A(-\tan B)}\end{aligned}$$

■ $$\mathbf{\tan(A-B) \equiv \frac{\tan A - \tan B}{1 + \tan A\tan B}}$$

Example 1

Evaluate $\tan 75°$ without the use of a calculator.

$$\begin{aligned}\tan 75° &\equiv \tan(45° + 30°)\\ &= \frac{\tan 45° + \tan 30°}{1 - \tan 45° \tan 30°}\\ &= \frac{1 + \frac{1}{\sqrt{3}}}{1 - \left(1 \times \frac{1}{\sqrt{3}}\right)}\\ &= \frac{\sqrt{3} + 1}{\sqrt{3} - 1}\end{aligned}$$

Example 2

Given that angles A and B are acute and $\sin A = \frac{4}{5}$ and $\cos B = \frac{7}{25}$, find, without the use of a calculator, the value of

(a) $\cos(A + B)$ (b) $\tan(A - B)$.

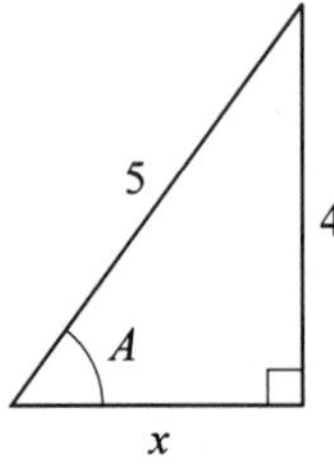

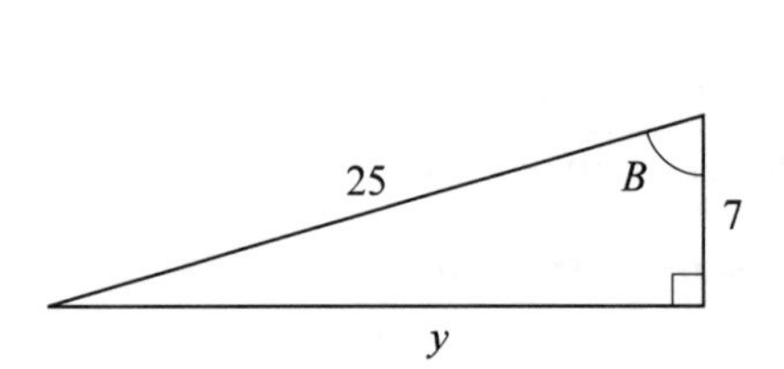

Using Pythagoras' theorem:

$$\begin{aligned}x^2 + 16 &= 25 \\ x^2 &= 9 \\ x &= 3\end{aligned}$$

$$\begin{aligned}y^2 + 49 &= 625 \\ y^2 &= 576 \\ y &= 24\end{aligned}$$

(a)
$$\begin{aligned}\cos(A + B) &\equiv \cos A \cos B - \sin A \sin B\\ &= \left(\tfrac{3}{5} \times \tfrac{7}{25}\right) - \left(\tfrac{4}{5} \times \tfrac{24}{25}\right)\\ &= \tfrac{21}{125} - \tfrac{96}{125}\\ &= -\tfrac{75}{125} = -\tfrac{3}{5}\end{aligned}$$

(b)
$$\begin{aligned}\tan(A - B) &\equiv \frac{\tan A - \tan B}{1 + \tan A \tan B}\\ &= \frac{\frac{4}{3} - \frac{24}{7}}{1 + \left(\frac{4}{3} \times \frac{24}{7}\right)}\\ &= \frac{\frac{28}{21} - \frac{72}{21}}{1 + \frac{32}{7}}\end{aligned}$$

$$= \frac{-\frac{44}{21}}{\frac{39}{7}}$$

$$= -\frac{44}{_3\cancel{21}} \times \frac{\cancel{7}}{39}$$

$$= -\frac{44}{117}$$

Example 3

Find the exact value of

$$\sin 157^\circ \cos 97^\circ - \cos 157^\circ \sin 97^\circ$$

$$\sin 157^\circ \cos 97^\circ - \cos 157^\circ \sin 97^\circ = \sin(157^\circ - 97^\circ)$$

$$= \sin 60^\circ = \frac{\sqrt{3}}{2}$$

Example 4

Given that $\sin A = \frac{12}{13}$ and $90^\circ < A < 180^\circ$, and $\tan B = \frac{4}{3}$ and $180^\circ < B < 270^\circ$, find the value of (a) $\sin(A + B)$ (b) $\cos(A - B)$.

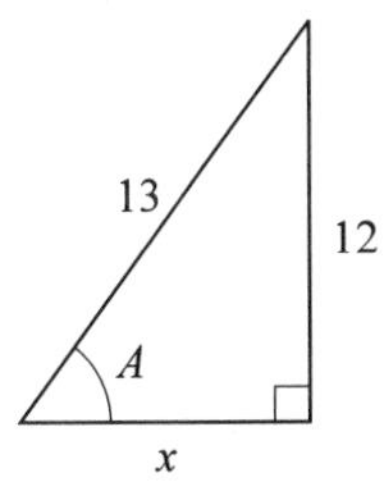

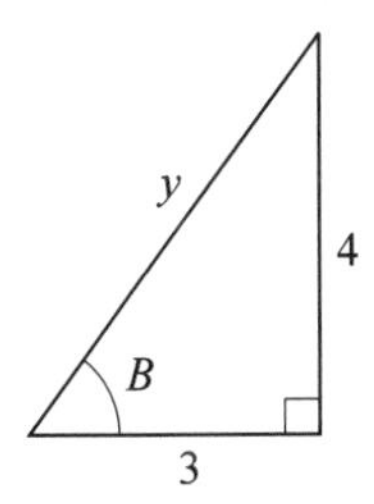

If A and B were acute, then by Pythagoras' theorem:

$$x^2 + 12^2 = 13^2 \qquad y^2 = 4^2 + 3^2$$

$$x^2 + 144 = 169 \qquad y^2 = 16 + 9$$

$$x^2 = 25 \qquad y^2 = 25$$

$$x = 5 \qquad y = 5$$

If A were acute then $\sin A$ would be $\frac{12}{13}$ and $\cos A$ would be $\frac{5}{13}$. But since $90^\circ < A < 180^\circ$

$$\sin A = \frac{12}{13} \quad \text{and} \quad \cos A = -\frac{5}{13}$$

Similarly, if B were acute then $\sin B$ would be $\frac{4}{5}$ and $\cos B$ would be $\frac{3}{5}$. However, $180^\circ < B < 270^\circ$. So

$$\sin B = -\frac{4}{5} \quad \text{and} \quad \cos B = -\frac{3}{5}$$

(a)

$$\sin(A + B) \equiv \sin A \cos B + \cos A \sin B$$

$$= \frac{12}{13}(-\tfrac{3}{5}) + (-\tfrac{5}{13})(-\tfrac{4}{5})$$

$$= -\frac{36}{65} + \frac{20}{65}$$

$$= -\frac{16}{65}$$

(b)
$$\cos(A-B) \equiv \cos A \cos B + \sin A \sin B$$
$$= (-\tfrac{5}{13})(-\tfrac{3}{5}) + (\tfrac{12}{13})(-\tfrac{4}{5})$$
$$= \tfrac{15}{65} - \tfrac{48}{65}$$
$$= -\tfrac{33}{65}$$

Exercise 6B

1 Find, without using a calculator, the exact value of:

(a) $\sin 75°$ (b) $\cos 75°$

(c) $\tan 105°$ (d) $\sin 15°$

(e) $\cos 105°$ (f) $\tan 15°$

(g) $\sin 165°$ (h) $\cos 165°$

2 Find, without using a calculator, the exact value of:

(a) $\sin 40° \cos 50° + \cos 40° \sin 50°$

(b) $\cos 70° \cos 10° + \sin 70° \sin 10°$

(c) $\dfrac{\tan 80° + \tan 40°}{1 - \tan 80° \tan 40°}$

(d) $\cos 20° \cos 40° - \sin 20° \sin 40°$

(e) $\sin 50° \cos 20° - \cos 50° \sin 20°$

(f) $\dfrac{\tan 47° - \tan 17°}{1 + \tan 47° \tan 17°}$

3 Simplify:

(a) $\sin\theta \cos 3\theta + \cos\theta \sin 3\theta$

(b) $\cos 5\theta \cos 2\theta + \sin 5\theta \sin 2\theta$

(c) $\dfrac{\tan 4\theta - \tan 2\theta}{1 + \tan 4\theta \tan 2\theta}$

(d) $3\sin 7\theta \cos 2\theta - 3\cos 7\theta \sin 2\theta$

(e) $4\sin 6\theta \sin 4\theta + 4\cos 6\theta \cos 4\theta$

(f) $2\sin\theta \sin 4\theta + 2\cos\theta \cos 4\theta$

4 If A and B are acute angles such that $\sin A = \frac{3}{5}$ and $\sin B = \frac{12}{13}$, find the exact value of:

(a) $\sin(A+B)$ (b) $\cos(A-B)$

(c) $\tan(A+B)$ (d) $\cot(A-B)$

5 If $\sin A = \frac{3}{5}$ and $\cos B = \frac{5}{13}$ where A is obtuse and B is acute, find the exact value of:

(a) $\sin(A - B)$ (b) $\sec(A + B)$
(c) $\tan(A - B)$ (d) $\cot(A + B)$

6 If $\cos A = \frac{5}{13}$ and $\sin B = \frac{24}{25}$ where A is acute and B is obtuse, find the exact value of:

(a) $\sin(A + B)$ (b) $\operatorname{cosec}(A - B)$
(c) $\cos(A - B)$ (d) $\sec(A + B)$
(e) $\tan(A - B)$ (f) $\cot(A + B)$

7 Find, without using calculus, (i) the greatest, (ii) the least value that the following can take, and state the value of θ, $0 < \theta \leqslant 360°$, for which these values occur.

(a) $\sin\theta\cos 70° + \cos\theta\sin 70°$
(b) $\sin\theta\cos 20° - \cos\theta\sin 20°$
(c) $\sin 25°\cos\theta + \cos 25°\sin\theta$

8 Find, without using calculus, the least value of the following and the value of θ, $0 < \theta \leqslant 360°$, for which this value occurs.

(a) $\cos 50°\cos\theta + \sin 50°\sin\theta$
(b) $\sin\theta\sin 43° + \cos\theta\cos 43°$
(c) $3\cos\theta\cos 105° - 3\sin\theta\sin 105°$

6.4 Pythagoras' theorem

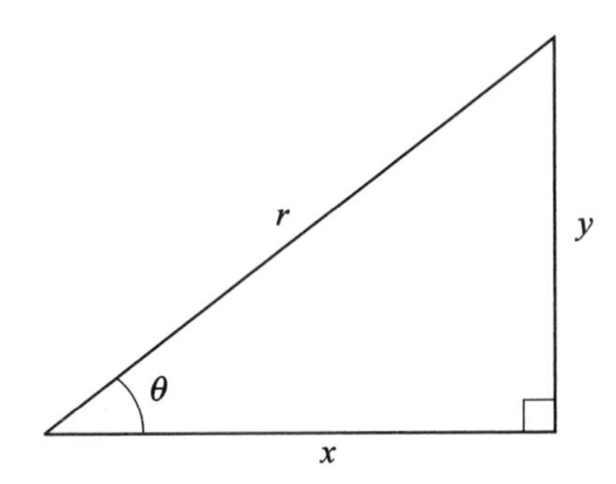

In this triangle

$$\frac{y}{r} = \sin\theta \quad \text{and} \quad \frac{x}{r} = \cos\theta$$

Now Pythagoras' theorem states:

$$x^2 + y^2 = r^2$$

Dividing by r^2 gives

$$\frac{x^2}{r^2}+\frac{y^2}{r^2}=\frac{r^2}{r^2}$$

or:

$$\frac{x^2}{r^2}+\frac{y^2}{r^2}=1$$

That is:

$$\left(\frac{x}{r}\right)^2+\left(\frac{y}{r}\right)^2=1$$

Using the facts that $\frac{y}{r}=\sin\theta$ and $\frac{x}{r}=\cos\theta$ gives:

$$\cos^2\theta+\sin^2\theta\equiv 1$$

or, more usually:

■ $$\mathbf{\sin^2\theta+\cos^2\theta\equiv 1}$$

Divide both sides by $\cos^2\theta$:

$$\frac{\sin^2\theta}{\cos^2\theta}+\frac{\cancel{\cos^2\theta}}{\cancel{\cos^2\theta}}\equiv\frac{1}{\cos^2\theta}$$

$$\left(\frac{\sin\theta}{\cos\theta}\right)^2+1\equiv\left(\frac{1}{\cos\theta}\right)^2$$

That is: $$\tan^2\theta+1\equiv\sec^2\theta$$

or, more usually:

■ $$\mathbf{\sec^2\theta\equiv 1+\tan^2\theta}$$

Again, starting from $\sin^2\theta+\cos^2\theta\equiv 1$, divide both sides by $\sin^2\theta$:

$$\frac{\cancel{\sin^2\theta}}{\cancel{\sin^2\theta}}+\frac{\cos^2\theta}{\sin^2\theta}\equiv\frac{1}{\sin^2\theta}$$

$$1+\left(\frac{\cos\theta}{\sin\theta}\right)^2\equiv\left(\frac{1}{\sin\theta}\right)^2$$

That is: $$1+\cot^2\theta\equiv\operatorname{cosec}^2\theta$$

or, more usually:

■ $$\mathbf{\operatorname{cosec}^2\theta\equiv 1+\cot^2\theta}$$

Although these three identities have been proved for θ acute, they are, in fact, true for all values of θ. You will not be asked to prove them, but you must learn them and use them confidently. They are very useful in the study of Advanced level mathematics. It is therefore *very* important that you remember them. You will be expected to use them in proving other simple identities, in the solution of equations and in calculus.

The general strategy when you are asked to prove an identity is to take the left-hand side (LHS) of the identity and try to rearrange it to obtain the right-hand side (RHS) of the identity, or vice versa if this looks as though it may prove more fruitful.

Example 5
Prove that

$$\tan^2\theta - \cot^2\theta \equiv (\sec\theta - \operatorname{cosec}\theta)(\sec\theta + \operatorname{cosec}\theta)$$

$$\begin{aligned}\text{LHS} &= \tan^2\theta - \cot^2\theta \\ &= (\sec^2\theta - 1) - (\operatorname{cosec}^2\theta - 1) \\ &= \sec^2\theta - 1 - \operatorname{cosec}^2\theta + 1 \\ &= \sec^2\theta - \operatorname{cosec}^2\theta \\ &= (\sec\theta - \operatorname{cosec}\theta)(\sec\theta + \operatorname{cosec}\theta) \\ &= \text{RHS}\end{aligned}$$

Example 6
Prove that

$$2 - \tan^2 A \equiv 2\sec^2 A - 3\tan^2 A$$

$$\begin{aligned}\text{RHS} &= 2\sec^2 A - 3\tan^2 A \\ &= 2(1 + \tan^2 A) - 3\tan^2 A \\ &= 2 + 2\tan^2 A - 3\tan^2 A \\ &= 2 - \tan^2 A \\ &= \text{LHS}\end{aligned}$$

Exercise 6C

Prove the identities:

1 $\cos^2\theta + 3\sin^2\theta \equiv 3 - 2\cos^2\theta$

2 $\operatorname{cosec}\theta - \sin\theta \equiv \cos\theta\cot\theta$

3 $\cot^2\theta + \cos^2\theta \equiv (\operatorname{cosec}\theta - \sin\theta)(\operatorname{cosec}\theta + \sin\theta)$

4 $\sec^2\theta - \sin^2\theta \equiv \tan^2\theta + \cos^2\theta$

5 $\sec^4\theta - \tan^4\theta \equiv 2\sec^2\theta - 1$

6 $(\cos\theta + \sec\theta)^2 \equiv \tan^2\theta + \cos^2\theta + 3$

7 $\cos^4\theta - \sin^4\theta \equiv \cos^2\theta - \sin^2\theta$

8 $\operatorname{cosec}^2\theta(\tan^2\theta - \sin^2\theta) \equiv \tan^2\theta$

9 $\left(\frac{1+\sin\theta}{\cos\theta}\right)^2 + \left(\frac{1-\sin\theta}{\cos\theta}\right)^2 \equiv 2 + 4\tan^2\theta$

10 $(\tan\theta + \sec\theta)(\cot\theta + \operatorname{cosec}\theta) \equiv (1 + \sec\theta)(1 + \operatorname{cosec}\theta)$

6.5 Double angle formulae

In section 6.3 you were introduced to the identity

$$\sin(A+B) \equiv \sin A\cos B + \cos A\sin B$$

Now let $B = A$; then:

$$\sin(A+A) \equiv \sin A\cos A + \cos A\sin A$$

That is: $\sin 2A \equiv \sin A\cos A + \sin A\cos A$

or:

■ $$\mathbf{\sin 2A \equiv 2\sin A\cos A}$$

Similarly:

$$\cos(A+B) \equiv \cos A\cos B - \sin A\sin B$$

Again, let $B = A$:

$$\cos(A+A) \equiv \cos A\cos A - \sin A\sin A$$

That is:

■ $$\mathbf{\cos 2A \equiv \cos^2 A - \sin^2 A}$$

Now $\sin^2 A + \cos^2 A \equiv 1$

or $\cos^2 A \equiv 1 - \sin^2 A$

So: $\cos 2A \equiv (1 - \sin^2 A) - \sin^2 A$

■ $$\mathbf{\cos 2A \equiv 1 - 2\sin^2 A}$$

Also, since $\sin^2 A + \cos^2 A \equiv 1$,

$$\sin^2 A \equiv 1 - \cos^2 A$$

So if $\cos 2A \equiv \cos^2 A - \sin^2 A$, then

$$\begin{aligned}\cos 2A &\equiv \cos^2 A - (1 - \cos^2 A)\\ &\equiv \cos^2 A - 1 + \cos^2 A\end{aligned}$$

or:

■ $$\mathbf{\cos 2A \equiv 2\cos^2 A - 1}$$

Finally, you learned in section 6.3 that

$$\tan(A+B) \equiv \frac{\tan A + \tan B}{1 - \tan A \tan B}$$

Let $B = A$; then:

$$\tan(A+A) \equiv \frac{\tan A + \tan A}{1 - \tan A \tan A}$$

So:

■ $$\mathbf{\tan 2A \equiv \frac{2\tan A}{1 - \tan^2 A}}$$

These five results are known as the **double angle formulae**, because they allow you to convert from double angles ($2A$) to single angles (A). Once again, you must learn these.

6.6 Half angle formulae

From the double angle formulae you can find the half angle formulae.

$$\cos 2A \equiv 1 - 2\sin^2 A$$

So: $$2\sin^2 A \equiv 1 - \cos 2A$$

or: $$\sin^2 A \equiv \tfrac{1}{2}(1 - \cos 2A)$$

Let $A = \frac{1}{2}\theta$. Then:

■ $$\mathbf{\sin^2 \tfrac{1}{2}\theta \equiv \tfrac{1}{2}(1 - \cos\theta)}$$

Also: $$\cos 2A \equiv 2\cos^2 A - 1$$

So: $$2\cos^2 A \equiv \cos 2A + 1$$

or: $$\cos^2 A \equiv \tfrac{1}{2}(\cos 2A + 1)$$

Again, let $A = \frac{1}{2}\theta$. Then:

■ $$\mathbf{\cos^2 \tfrac{1}{2}\theta \equiv \tfrac{1}{2}(1 + \cos\theta)}$$

So, given the value of $\cos\theta$, you can now find the value of $\sin\frac{1}{2}\theta$ and $\cos\frac{1}{2}\theta$. This will be particularly useful in integration and in the solution of equations.

Example 7

Evaluate exactly $2\sin 15^\circ \cos 15^\circ$.

Since $2\sin A\cos A \equiv \sin 2A$,

then:
$$2\sin 15^\circ \cos 15^\circ \equiv \sin(2 \times 15^\circ)$$
$$= \sin 30^\circ$$
$$= \tfrac{1}{2}$$

Example 8

Given that θ is acute and $\sin\theta = \frac{5}{13}$, find the exact value of $\sin 2\theta$.

By Pythagoras' theorem:

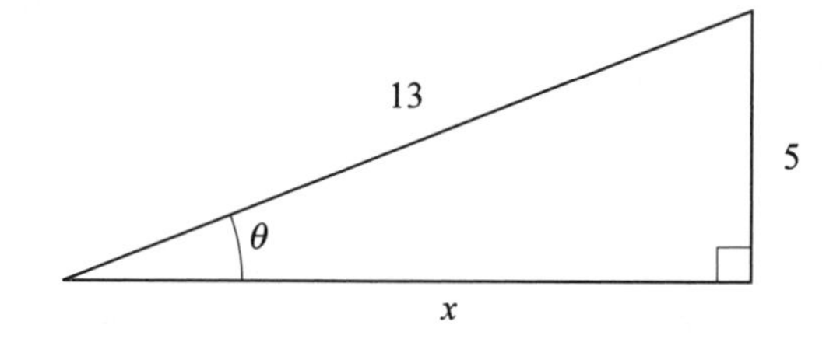

$$x^2 + 5^2 = 13^2$$
$$x^2 + 25 = 169$$
$$x^2 = 169 - 25 = 144$$
$$x = 12$$

So:
$$\cos\theta = \tfrac{12}{13}$$
$$\sin 2\theta \equiv 2\sin\theta\cos\theta$$
$$= 2 \times \tfrac{5}{13} \times \tfrac{12}{13}$$
$$= \tfrac{120}{169}$$

Example 9

Prove the identity: $\cot A - \tan A \equiv 2\cot 2A$

$$\text{RHS} \equiv 2\cot 2A$$
$$= \frac{2}{\tan 2A}$$
$$= 2\left(\frac{1 - \tan^2 A}{2\tan A}\right)$$
$$= 2\left(\frac{1}{2\tan A} - \frac{\tan^2 A}{2\tan A}\right)$$
$$= \frac{\cancel{2}}{\cancel{2}\tan A} - \frac{\cancel{2}\tan^2 A}{\cancel{2}\tan A}$$
$$= \frac{1}{\tan A} - \tan A$$
$$= \cot A - \tan A$$
$$= \text{LHS}$$

Example 10

Eliminate θ from $x = \cos 2\theta - 1$, $y = 2\sin\theta$.

$$x = \cos 2\theta - 1$$
$$= (1 - 2\sin^2\theta) - 1$$
$$x = -2\sin^2\theta$$

Now $y = 2\sin\theta$

$$\frac{y}{2} = \sin\theta$$
$$\left(\frac{y}{2}\right)^2 = \sin^2\theta$$

So:

$$x = -2\left(\frac{y}{2}\right)^2$$
$$= \frac{-2y^2}{4}$$
$$4x + 2y^2 = 0$$

or:

$$2x + y^2 = 0$$

Exercise 6D

1 Find, without using a calculator, the exact value of:

(a) $2\sin 75^\circ \cos 75^\circ$ (b) $\cos^2 22\tfrac{1}{2}^\circ - \sin^2 22\tfrac{1}{2}^\circ$

(c) $\dfrac{2\tan 67\frac{1}{2}^\circ}{1 - \tan^2 67\frac{1}{2}^\circ}$ (d) $1 - 2\sin^2 15^\circ$

(e) $2\cos^2 105^\circ$ (f) $\sec 112\tfrac{1}{2}^\circ \operatorname{cosec} 112\tfrac{1}{2}^\circ$

(g) $\dfrac{1 - \tan^2 15^\circ}{\tan 15^\circ}$ (h) $\sin^2 75^\circ$

2 If $\cos\theta = \frac{3}{5}$ and θ is acute, find the exact value of:

(a) $\sin 2\theta$ (b) $\cos 2\theta$ (c) $\tan 2\theta$

3 If $\tan\theta = \frac{12}{5}$ and θ is acute, find the exact value of:

(a) $\sin 2\theta$ (b) $\cos 2\theta$ (c) $\tan 2\theta$

4 Given that $\sin\theta = \frac{7}{25}$ and that θ is obtuse, find the exact value of:

(a) $\sin 2\theta$ (b) $\sec 2\theta$ (c) $\cot 2\theta$

5 Given that $\tan\theta = -\frac{3}{4}$ and that θ is obtuse, find the exact value of:

(a) $\sec 2\theta$ (b) $\operatorname{cosec} 2\theta$ (c) $\tan 2\theta$

6 Given that $\sin 2\theta = 1$ and that θ is acute, find the exact value of:

(a) $\sin\theta$ (b) $\cos\theta$ (c) $\tan\theta$

7 Find the possible values of $\tan\frac{1}{2}\theta$ when $\tan\theta = \frac{3}{4}$.

8 Find the possible values of $\tan\frac{1}{2}\theta$ when $\tan\theta = \frac{7}{24}$.

9 Obtain an expression for $\sin 3\theta$ in terms of $\sin\theta$ and hence find the value of $\sin 3\theta$ if $\sin\theta = \frac{1}{8}$.

10 Obtain an expression for $\cos 3\theta$ in terms of $\cos\theta$ and hence find the value of $\cos 3\theta$ if $\cos\theta = \frac{1}{4}$.

11 Eliminate θ from the following pairs of equations:

(a) $x = \sin 2\theta + 2, \quad y = \cos\theta$

(b) $x = 1 + \cos\theta, \quad y = 3\cos 2\theta$

(c) $x = \tan 2\theta, \quad y = \tan\theta$

(d) $x = 3 + \cos 2\theta, \quad y = \sec\theta$

(e) $x = \operatorname{cosec}\theta, \quad y = 1 - \cos 2\theta$

Prove the following identities:

12 $\dfrac{\cos 2A}{\cos A - \sin A} \equiv \cos A + \sin A$

13 $2\operatorname{cosec} 2\theta \equiv \sec\theta \operatorname{cosec}\theta$

14 $\cot\theta - \tan\theta \equiv 2\cot 2\theta$

15 $\tan 2\theta \sec\theta \equiv 2\sin\theta \sec 2\theta$

16 $\sin 2\theta \equiv \dfrac{2\tan\theta}{1 + \tan^2\theta}$

17 $\dfrac{1}{\cos\theta - \sin\theta} - \dfrac{1}{\cos\theta + \sin\theta} \equiv 2\sin\theta \sec 2\theta$

18 $\dfrac{\sin 2\theta}{1 - \cos 2\theta} \equiv \cot\theta$

19 $\tan 3\theta \equiv \dfrac{3\tan\theta - \tan^3\theta}{1 - 3\tan^2\theta}$

20 $\dfrac{\sin 2\theta + \sin\theta}{\cos 2\theta + \cos\theta + 1} \equiv \tan\theta$

6.7 How to solve more complicated trigonometric equations

In chapter 7 of Book P1 you were shown how to solve trigonometric equations such as:

$$\sin 2x = \tfrac{1}{2} \quad \text{for} \quad 0 < x \leqslant 2\pi$$

and :

$$\tan(3x - 40^\circ) = 0.61 \quad \text{for} \quad 0 < x \leqslant 360^\circ$$

Now that you know a number of trigonometric identities, you can solve some further trigonometric equations. In all such equations, the strategy is to reduce the equation to one trigonometric ratio only, using the identities proved in this chapter.

Example 11

Solve the equation

$$6\cos^2\theta + \sin\theta - 5 = 0 \quad \text{for} \quad 0 < \theta \leqslant 360^\circ$$

$$6\cos^2\theta + \sin\theta - 5 = 0$$

$$6(1 - \sin^2\theta) + \sin\theta - 5 = 0$$

$$6 - 6\sin^2\theta + \sin\theta - 5 = 0$$

$$-6\sin^2\theta + \sin\theta + 1 = 0$$

or:

$$6\sin^2\theta - \sin\theta - 1 = 0$$

$$(3\sin\theta + 1)(2\sin\theta - 1) = 0$$

So either $3\sin\theta + 1 = 0$ or $2\sin\theta - 1 = 0$

that is: $\sin\theta = -\tfrac{1}{3}$ or $\sin\theta = \tfrac{1}{2}$

Thus: $\theta = 199.5^\circ, 340.5^\circ$ or $\theta = 30^\circ, 150^\circ$

So $\theta = 30^\circ, 150^\circ, 199.5^\circ, 340.5^\circ$.

Example 12

Solve the equation

$$\cos\theta\cos 30^\circ - \sin\theta\sin 30^\circ = \tfrac{1}{2}$$

for $-180^\circ < \theta < 180^\circ$.

$$\cos\theta\cos 30^\circ - \sin\theta\sin 30^\circ = \tfrac{1}{2}$$

$$\Rightarrow \quad \cos(\theta + 30^\circ) = \tfrac{1}{2}$$

That is: $\theta + 30° = -60°,\ 60°$

So: $\theta = -90°$ or $30°$

Example 13

Solve the equation

$$\sec^2\theta = 3 - \tan\theta$$

for $0 \leqslant \theta < 360°$.

Since $\sec^2\theta \equiv 1 + \tan^2\theta$,

$$\sec^2\theta = 3 - \tan\theta$$

$\Rightarrow$ $$1 + \tan^2\theta = 3 - \tan\theta$$

$$\tan^2\theta + \tan\theta - 2 = 0$$

$$(\tan\theta - 1)(\tan\theta + 2) = 0$$

$$\tan\theta = 1 \quad \text{or} \quad \tan\theta = -2$$

$$\theta = 45°,\ 225° \quad \text{or} \quad \theta = 116.6°,\ 296.6°$$

So: $\theta = 45°,\ 116.6°,\ 225°,\ 296.6°$

Example 14

Solve the equation

$$\cos(\theta + 60°) = \sin\theta$$

for $0 < \theta \leqslant 360°$.

$$\cos(\theta + 60°) = \sin\theta$$

$\Rightarrow$ $$\cos\theta\cos 60° - \sin\theta\sin 60° = \sin\theta$$

So: $$(\cos\theta \times \tfrac{1}{2}) - (\sin\theta \times \tfrac{\sqrt{3}}{2}) = \sin\theta$$

That is: $$\cos\theta - \sqrt{3}\sin\theta = 2\sin\theta$$

$$\cos\theta = 2\sin\theta + \sqrt{3}\sin\theta$$

$$\cos\theta = (2 + \sqrt{3})\sin\theta$$

$$\frac{1}{2+\sqrt{3}} = \frac{\sin\theta}{\cos\theta}$$

$$\tan\theta = \frac{1}{2+\sqrt{3}}$$

So: $\theta = 15°$ or $195°$

Example 15

Solve $\cos 2\theta = \tan 2\theta$ for $0 \leqslant \theta \leqslant \pi$.

Since $\tan 2\theta = \dfrac{\sin 2\theta}{\cos 2\theta}$,

$$\cos 2\theta = \tan 2\theta$$

$$\Rightarrow \qquad \cos 2\theta = \frac{\sin 2\theta}{\cos 2\theta}$$

$$\cos^2 2\theta = \sin 2\theta$$

$$1 - \sin^2 2\theta = \sin 2\theta$$

$$\sin^2 2\theta + \sin 2\theta - 1 = 0$$

$$\sin 2\theta = \frac{-1 \pm \sqrt{(1+4)}}{2}$$

$$= \frac{-1 \pm \sqrt{5}}{2}$$

$$= 0.618\,03 \text{ or } -1.618\,03$$

Reject $\sin 2\theta = -1.618\,03$(sin x has a minimum value of -1)

$$2\theta = 0.666\,23 \text{ or } 2.475\,35$$

$$\theta = 0.333\,11 \text{ or } 1.237\,67$$

So: $\theta = 0.333^c$ or 1.238^c (3 d.p.)

Exercise 6E

Solve these equations for $0 \leqslant \theta \leqslant 360°$, giving θ to 1 decimal place where appropriate:

1 $\sin\theta\cos 15° + \cos\theta\sin 15° = 0.4$

2 $\cos\theta\cos 33° - \sin\theta\sin 33° = -0.2$

3 $\dfrac{\tan 47° - \tan\theta}{1 + \tan 47°\tan\theta} = 1.5$

4 $\cos(\theta - 45°) = \frac{1}{\sqrt{3}}\cos\theta$

5 $\sin(\theta + 30°) = 2\cos\theta$

6 $2\cos(\theta - 60°) = \sin\theta$

7 $\sin(\theta + 15°) = 3\cos(\theta - 15°)$

8 $2\sin\theta\cos\theta = 1 - 2\sin^2\theta$

9 $\sin 2\theta + \sin\theta - \tan\theta = 0$

10 $\sin 2\theta + \sin\theta = 0$

11 $\cos 2\theta = 2\sin\theta$

12 $\tan 2\theta + \tan\theta = 0$

13 $3\cos^2\theta - 2\sin\theta - 2 = 0$

14 $3\sec^2\theta + 1 = 8\tan\theta$

15 $2 + \cos\theta\sin\theta = 8\sin^2\theta$

16 $\sin 2\theta = \cos\theta$

17 $3\cos 2\theta - 7\cos\theta - 2 = 0$

18 $\sec^2\theta = 4\tan\theta$

19 $2\sin 2\theta = \tan\theta$

20 $\sin 2\theta - \tan\theta = 0$

21 $\operatorname{cosec}^2\theta = 3\cot\theta - 1$

22 $3\cos 2\theta + 2\sin^2\theta + 1 = 2\sin\theta$

23 $(\sin\theta + \cos\theta)^2 = \frac{1}{2}$

24 $2\tan\theta + \sin 2\theta\sec\theta = 1 + \sec\theta$

25 $\sin(2\theta - 30°) = \cos(2\theta + 30°)$

26 $\cot\theta + 3\cot 2\theta - 1 = 0$

27 $\tan(3\theta - 20°) = \sqrt{2} - 1$

28 $3\tan\theta + 4\sin\theta = 0$

SUMMARY OF KEY POINTS

1. $\sec x = \dfrac{1}{\cos x}$ $\qquad \operatorname{cosec} x = \dfrac{1}{\sin x}$ $\qquad \cot x = \dfrac{1}{\tan x}$
2. $\sin(A \pm B) \equiv \sin A\cos B \pm \cos A\sin B$
3. $\cos(A \pm B) \equiv \cos A\cos B \mp \sin A\sin B$
4. $\tan(A \pm B) \equiv \dfrac{\tan A \pm \tan B}{1 \mp \tan A\tan B}$
5. $\sin^2 A + \cos^2 A \equiv 1$
6. $\sec^2 A \equiv 1 + \tan^2 A$
7. $\operatorname{cosec}^2 A \equiv 1 + \cot^2 A$
8. $\sin 2A \equiv 2\sin A\cos A$
9. $\cos 2A \equiv \cos^2 A - \sin^2 A$
10. $\cos 2A \equiv 2\cos^2 A - 1$
11. $\cos 2A \equiv 1 - 2\sin^2 A$
12. $\tan 2A = \dfrac{2\tan A}{1 - \tan^2 A}$
13. $\sin^2\frac{1}{2}A = \frac{1}{2}(1 - \cos A)$
14. $\cos^2\frac{1}{2}A = \frac{1}{2}(1 + \cos A)$

Trigonometry II

7

In chapter 6 you were introduced to a number of new trigonometrical results. This enabled you, among other things, to prove trigonometrical identities and solve further trigonometrical equations. This chapter goes back to the more practical side of trigonometry, which you encountered in Book P1 when you tried to find the lengths of sides and the sizes of angles in triangles. The knowledge that you gained in the first book only allowed you to do this for right-angled triangles. This chapter introduces you to the sine rule and cosine rule which, in turn, will allow you to find the lengths of sides and the sizes of angles in triangles that do *not* have a right angle. It also shows you an alternative way of calculating the area of a triangle.

After that you will learn to apply trigonometry to practical situations in three dimensions, and the chapter ends with an introduction to using cartesian coordinates in three dimensions.

7.1 The sine rule

In this chapter we shall frequently use a triangle (△) ABC:

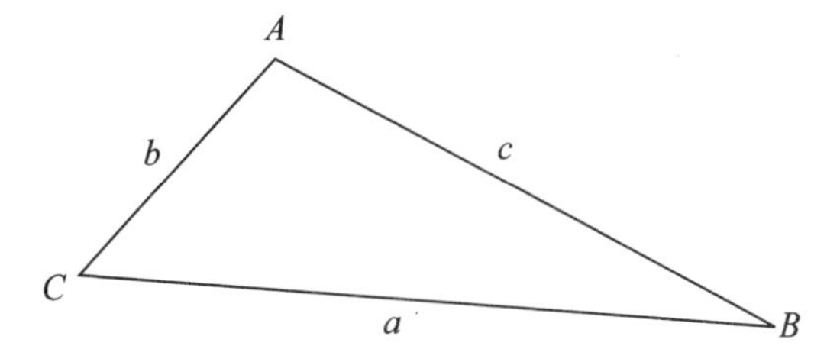

For brevity, $\angle BAC$ will simply be called angle A, $\angle ABC$ will be called $\angle B$ and $\angle ACB$ will be called $\angle C$. The side BC will be called a, because it is the side opposite $\angle A$, the side AC will be called b and the side AB will be called c. This is standard notation.

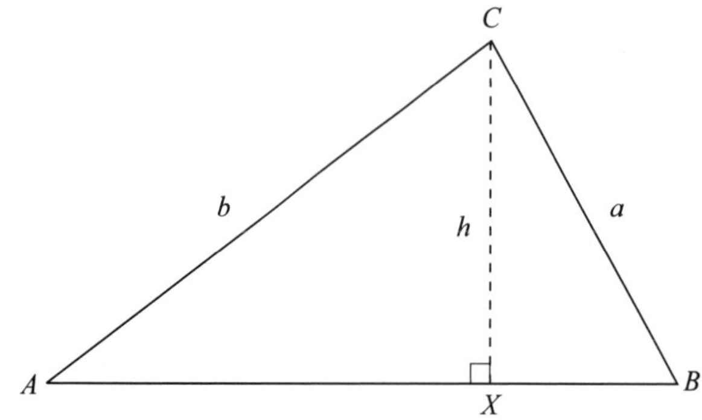

In this $\triangle ABC$, both angle A and angle B are acute. CX is the perpendicular from C to the side AB.

Let $CX = h$

In $\triangle AXC$,

$$\frac{h}{b} = \sin A$$

So: $\quad h = b \sin A$

In $\triangle BXC$,

$$\frac{h}{a} = \sin B$$

So: $\quad h = a \sin B$

Thus $\quad a \sin B = b \sin A$

$\div \sin A$: $\quad \frac{a \sin B}{\sin A} = \frac{b \cancel{\sin A}}{\cancel{\sin A}}$

$\div \sin B$: $\quad \frac{a \cancel{\sin B}}{\sin A \cancel{\sin B}} = \frac{b}{\sin B}$

So: $\quad \frac{a}{\sin A} = \frac{b}{\sin B}$

In a similar way, by drawing the perpendicular fom B to the line AC you can show that

$$\frac{a}{\sin A} = \frac{c}{\sin C}$$

Putting these two results together, you get:

$$\frac{a}{\sin A} = \frac{b}{\sin B} = \frac{c}{\sin C}$$

Now suppose that $\angle B$ is obtuse:

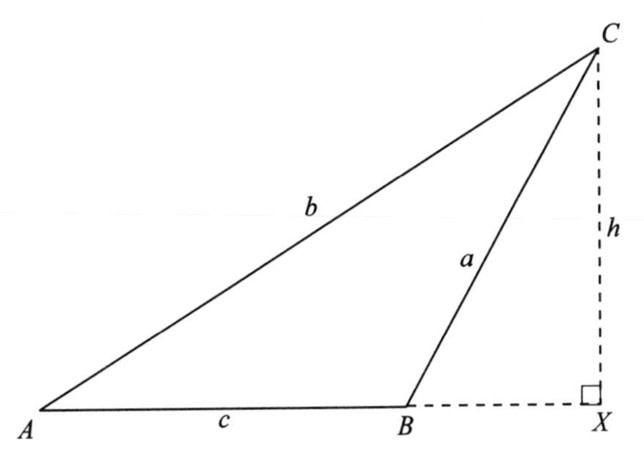

CX is the perpendicular from C to the side AB. (This time the side AB has to be extended.)

Let $CX = h$, as before.

In $\triangle AXC$,

$$\frac{h}{b} = \sin A$$

So: $$h = b \sin A$$

In $\triangle BXC$,

$$\frac{h}{a} = \sin(180° - B)$$

But $$\sin(180° - B) = \sin B$$

So: $$\frac{h}{a} = \sin B$$

or $$h = a \sin B$$

Thus, as before: $$a \sin B = b \sin A$$

and so: $$\frac{a}{\sin A} = \frac{b}{\sin B}$$

Again, if you draw the perpendicular from B to the side AC then you can show that

$$\frac{a}{\sin A} = \frac{c}{\sin C}$$

Consequently, once again you have

■ $$\frac{\mathbf{a}}{\mathbf{\sin A}} = \frac{\mathbf{b}}{\mathbf{\sin B}} = \frac{\mathbf{c}}{\mathbf{\sin C}}$$

This is known as the **sine rule**.

Remember, although these results have been proved by using a construction line that is perpendicular to one of the sides of the triangle, in neither case was the triangle ABC right-angled. The sine rule applies to *any* triangle.

You can use the sine rule, in general, if the given triangle contains two known angles and one known side and you need to find another side. It can also be used where the given triangle contains two known sides and one known angle, *which is not the angle between the two sides* (often called the 'non-included' angle), and where you need to find another angle.

Example 1

Calculate, in cm to 3 significant figures, the length of the side AB of the triangle ABC in which $\angle ACB = 62°$, $\angle ABC = 47°$ and $AC = 7\,\text{cm}$.

By the sine rule:

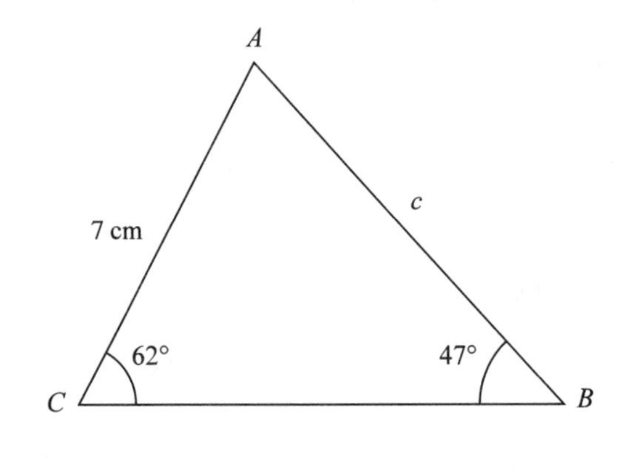

$$\frac{c}{\sin 62°} = \frac{7}{\sin 47°}$$

So:

$$\frac{c}{\cancel{\sin 62°}} \times \cancel{\sin 62°} = \frac{7}{\sin 47°} \times \sin 62°$$

$$c = \frac{7 \sin 62°}{\sin 47°}$$
$$= 8.4509$$
$$= 8.45 \text{ cm (3 s.f.)}$$

Example 2

Calculate, in cm to 3 significant figures, the length of the side AC of the triangle ABC in which $BC = 6.4\,\text{cm}$, $\angle ACB = 43°$ and $\angle BAC = 71°$.

Since the sum of the three angles in any triangle is 180°,

$$71° + 43° + \angle ABC = 180°$$

So:

$$\angle ABC = 180° - 114° = 66°$$

By the sine rule:

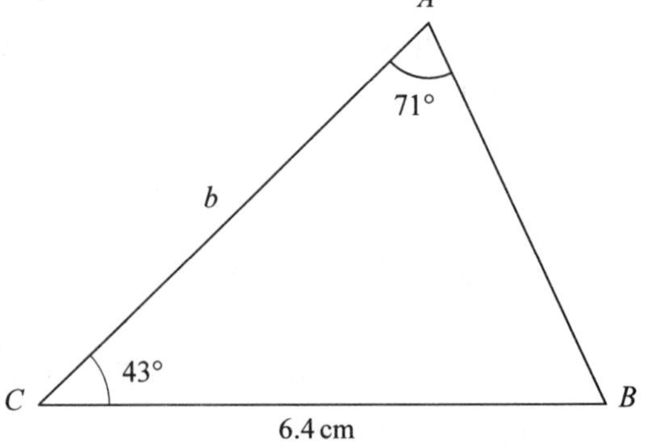

$$\frac{b}{\sin 66°} = \frac{6.4}{\sin 71°}$$

$$\frac{b}{\cancel{\sin 66°}} \times \cancel{\sin 66°} = \frac{6.4}{\sin 71°} \times \sin 66°$$

$$b = \frac{6.4 \sin 66°}{\sin 71°}$$
$$= 6.183 \text{ cm}$$
$$= 6.18 \text{ cm (3 s.f.)}$$

When the information given about the triangle is two sides and the non-included angle, you must be careful. It is sometimes possible from such information to obtain *two* solutions. That is, it is sometimes possible to find two *different* triangles with the same given data, as the next example demonstrates.

Example 3

Calculate, in degrees to 1 decimal place, the size of the angles CAB and ACB in the triangle ABC, where $AC = 4$ cm, $BC = 5$ cm and $\angle ABC = 42°$.

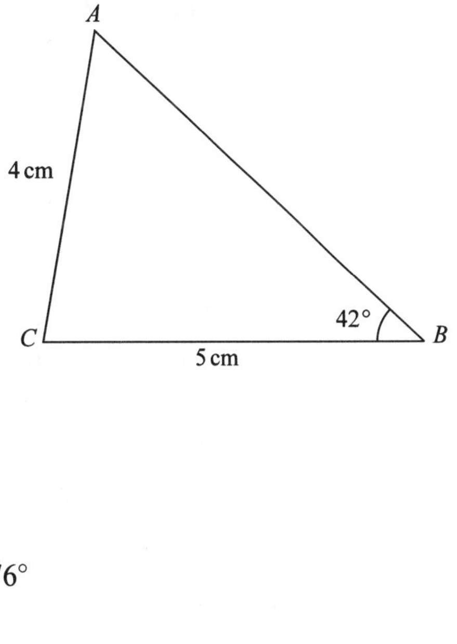

By the sine rule, $$\frac{5}{\sin A} = \frac{4}{\sin 42°}$$

That is: $$\frac{\sin A}{5} = \frac{\sin 42°}{4}$$

So: $$\frac{\sin A}{\not{5}} \times \not{5} = \frac{\sin 42°}{4} \times 5$$

$$\sin A = \frac{5 \sin 42°}{4}$$

$$\sin A = 0.8364$$

So: $A = 56.76°$ and $C = 180° - 42° - 56.76°$

$= 56.8°$ (1 d.p.) $= 81.2°$ (1 d.p.)

OR (since sine is also positive in the second quadrant):

$$A = 180° - 56.76° = 123.24°$$
$$= 123.2° \text{ (1 d.p.)}$$

and: $$C = 180° - 42° - 123.24°$$

$$= 14.8° \text{ (1 d.p.)}$$

Thus there are two possible triangles that can be drawn in this case with the given information, as shown below.

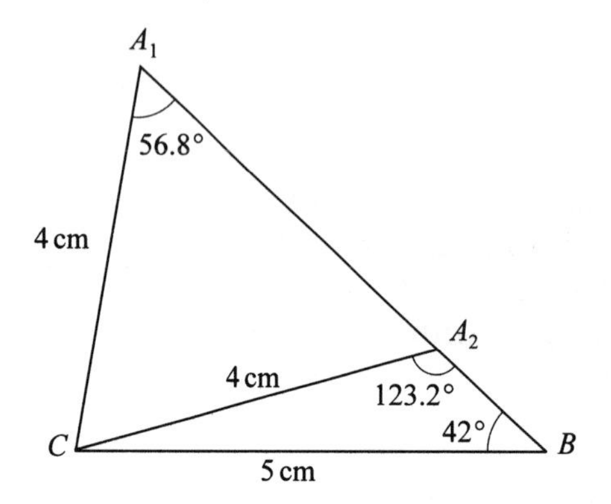

Example 4

Find the length of AB and the sizes of $\angle ACB$ and $\angle BAC$ for the triangle ABC in which $AC = 6.3$ cm, $BC = 4.8$ cm and $\angle ABC = 53°$.

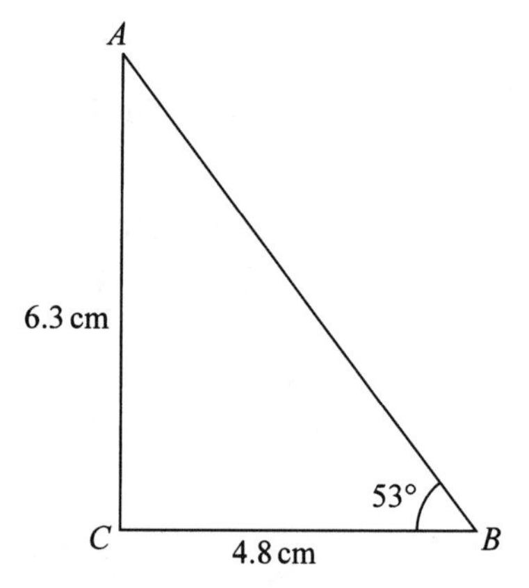

$$\frac{4.8}{\sin A} = \frac{6.3}{\sin 53°}$$

So:

$$\frac{\sin A}{4.8} = \frac{\sin 53°}{6.3}$$

$$\frac{\sin A}{\cancel{4.8}} \times \cancel{4.8} = \frac{\sin 53°}{6.3} \times 4.8$$

$$\sin A = \frac{4.8 \sin 53°}{6.3}$$

$$= \frac{4.8 \times 0.7986}{6.3}$$

$$\sin A = 0.6084$$

So:

$$A = 37.47°$$

$$= 37.5° \text{ (1 d.p.)}$$

and:

$$C = 180° - 37.47° - 53°$$

$$= 89.5° \text{ (1 d.p.)}$$

OR

$$A = 180° - 37.47°$$

$$= 142.5°$$

which is impossible because

$$142.5° + 53° = 195.5°$$

which is more than 180°, and so C cannot exist. So, to find AB we write

$$\frac{AB}{\sin 89.5^\circ} = \frac{6.3}{\sin 53^\circ}$$

$$\frac{AB}{\cancel{\sin 89.5^\circ}} \times \cancel{\sin 89.5^\circ} = \frac{6.3}{\sin 53^\circ} \times \sin 89.5^\circ$$

$$AB = \frac{6.3 \sin 89.5^\circ}{\sin 53^\circ}$$
$$= 7.89\,\text{cm (3 s.f.)}$$

So in this case it is only possible to draw *one* triangle from the given information.

Exercise 7A

Calculate the lengths of the unknown sides and the sizes of the unknown angles in these triangles:

1 Triangle ABC: $AB = 7.3$ cm, angle $A = 51°$, angle $B = 38°$

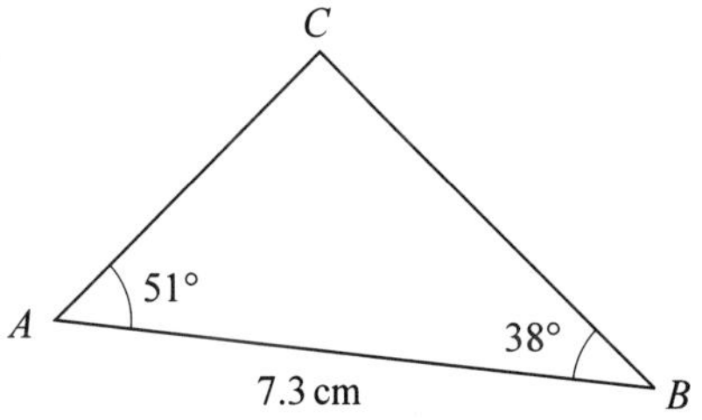

2 Triangle ABC: $AC = 12.4$ cm, angle $A = 43.4°$, angle $B = 69.2°$

3 Triangle ABC: $AC = 19.3$ cm, angle $C = 61.5°$, angle $B = 49°$

4 Triangle ABC: $BC = 8.9$ cm, angle $A = 29.3°$, angle $C = 71.4°$

5 Triangle ABC: $AC = 5.3$ cm, angle $A = 55°$, angle $C = 75°$

6 Triangle ABC: $AC = 6.5$ cm, angle $A = 48.5°$, angle $C = 75.3°$

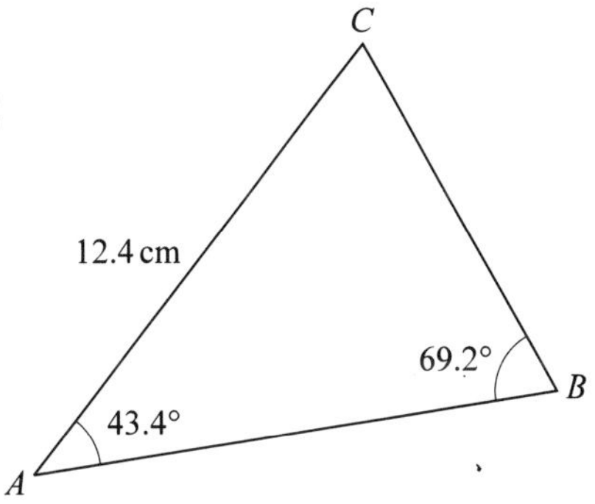

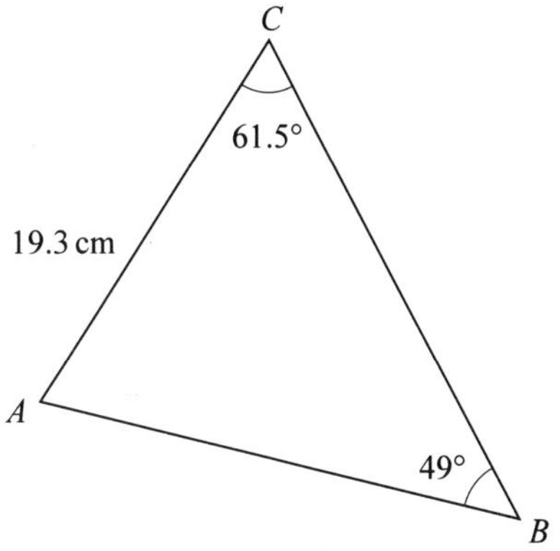

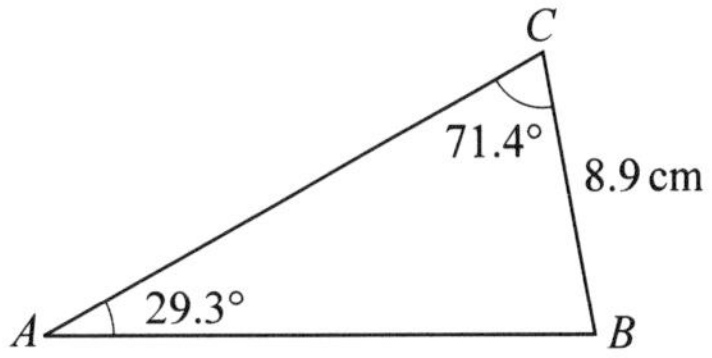

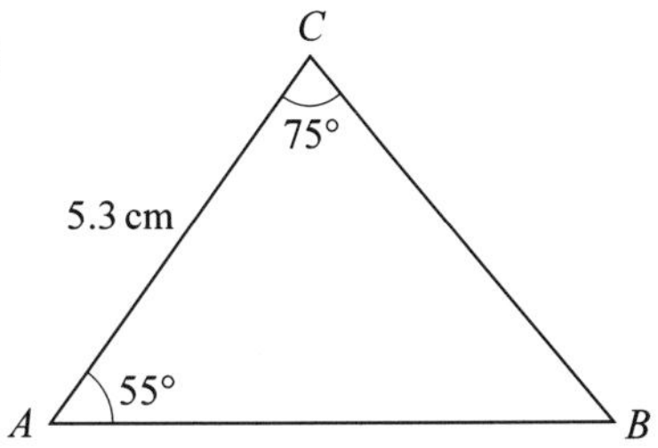

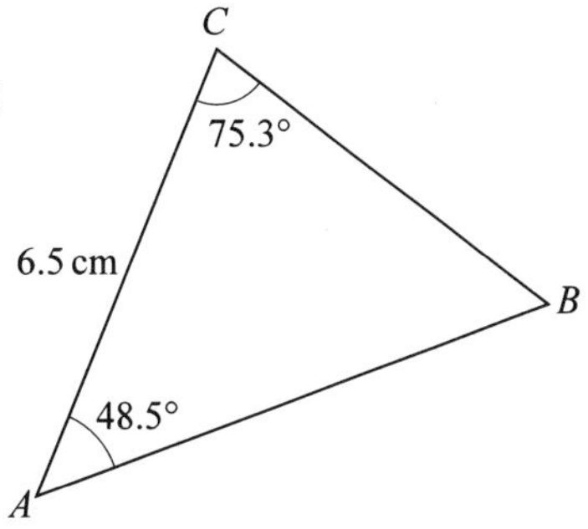

7

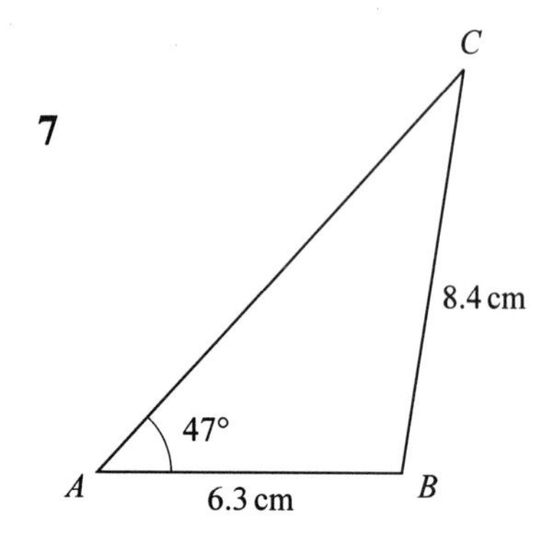

8

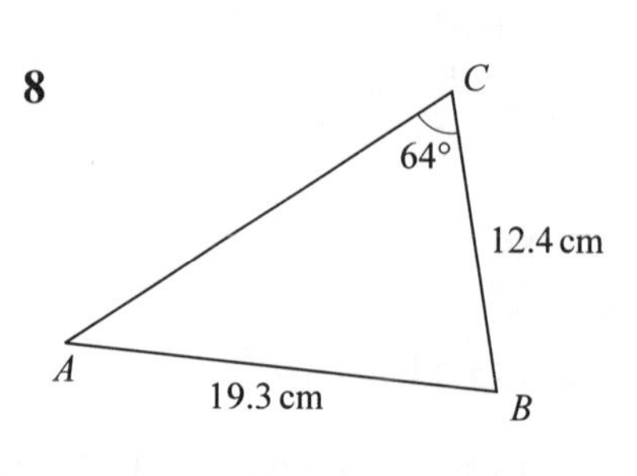

9

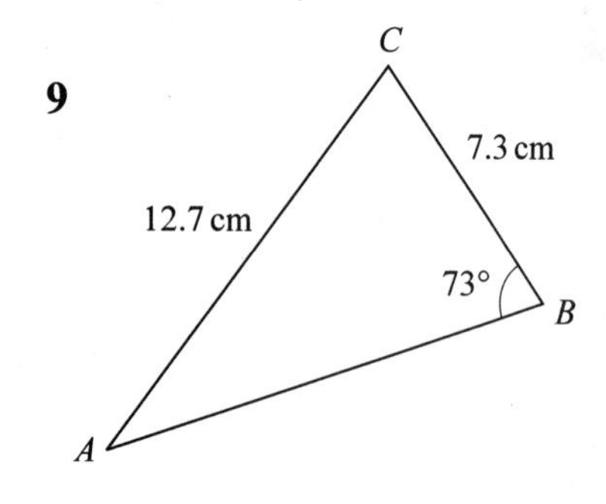

10

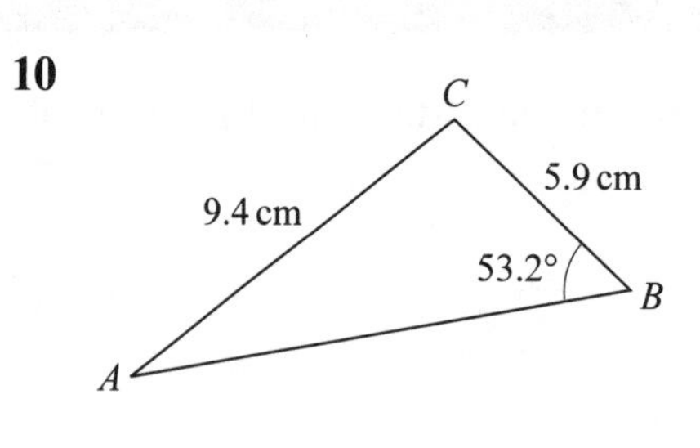

11

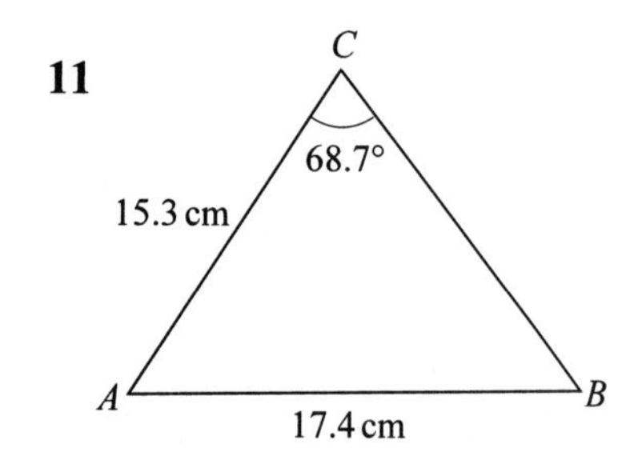

12

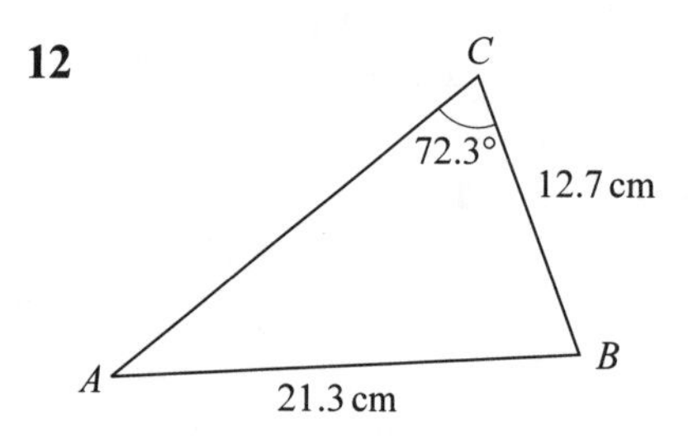

13

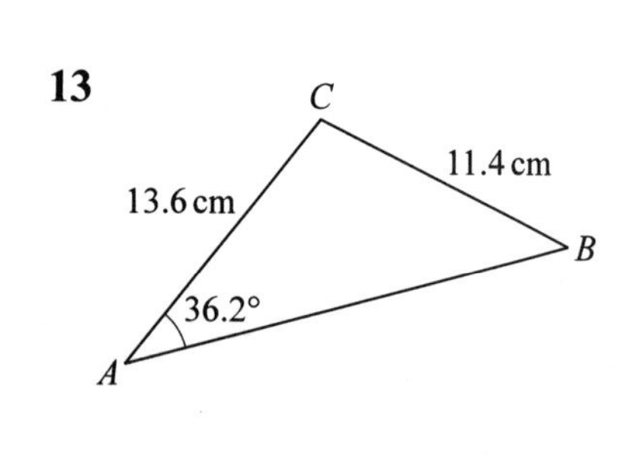

14

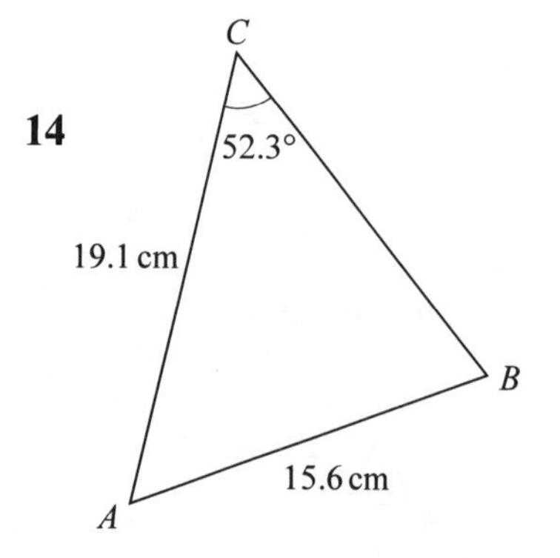

15

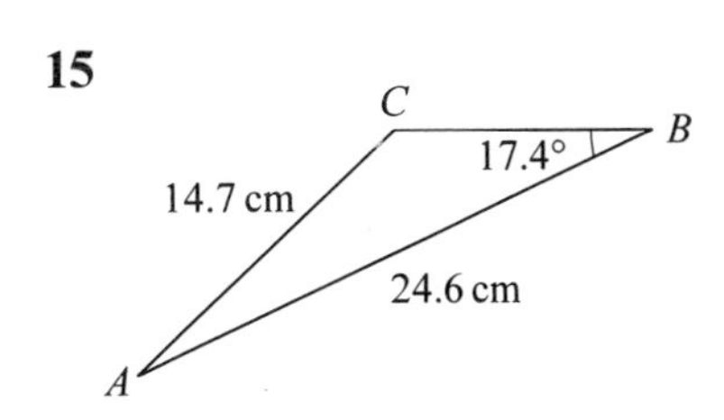

16

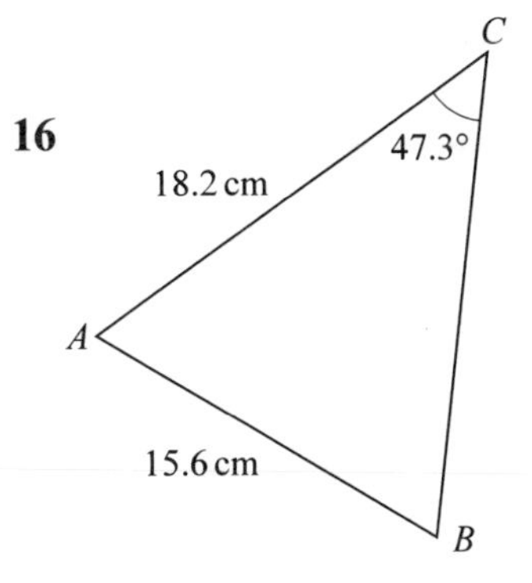

17

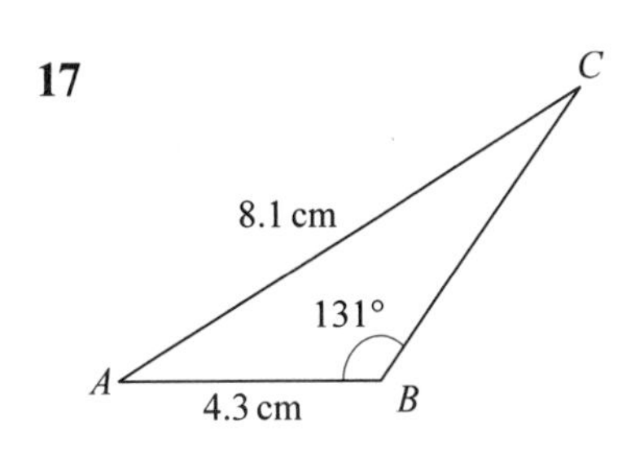

18

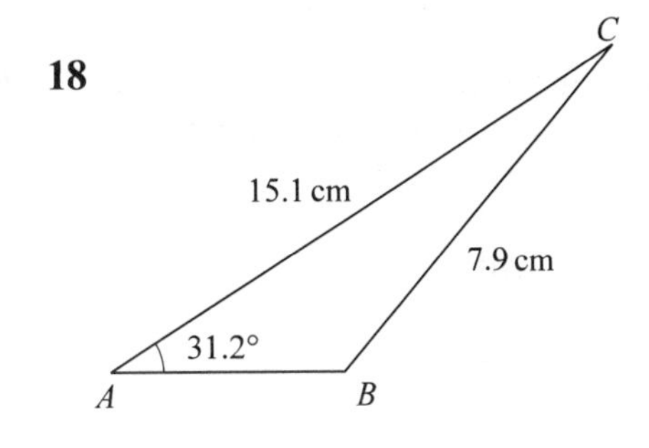

19

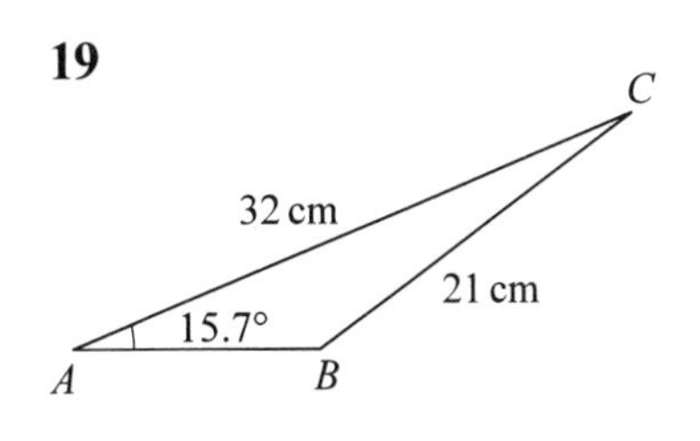

20

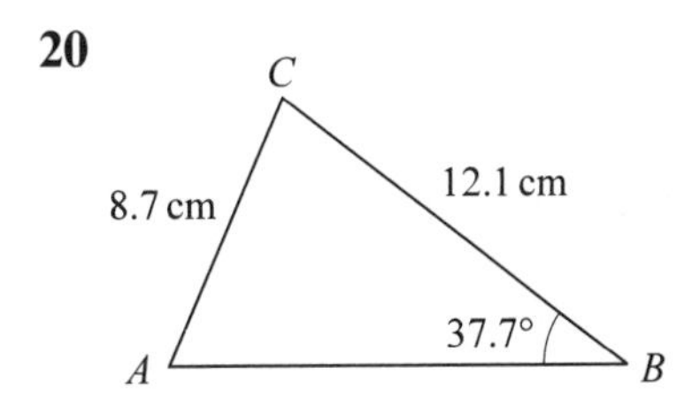

7.2 The cosine rule

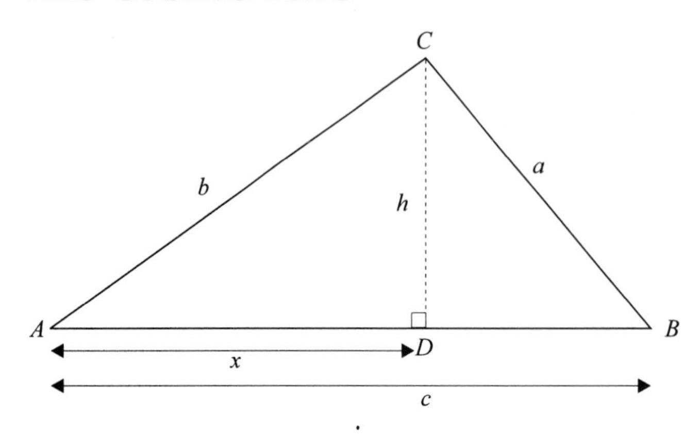

In $\triangle ABC$, CD is the perpendicular from C to the side AB, $CD = h$ and $AD = x$. Angle B is acute. In $\triangle ADC$ you can use Pythagoras' theorem to write

$$h^2 + x^2 = b^2 \quad (1)$$

In $\triangle BCD$, you can use Pythagoras' theorem to write

$$h^2 + DB^2 = a^2$$

But: $\quad DB = c - x$

So: $\quad h^2 + (c - x)^2 = a^2 \quad (2)$

From (1): $\quad h^2 = b^2 - x^2$

Substitute for h^2 in (2):

$$b^2 - x^2 + (c - x)^2 = a^2$$

That is: $\quad b^2 - x^2 + c^2 - 2cx + x^2 = a^2$

or $\quad a^2 = b^2 + c^2 - 2cx \quad (3)$

But in $\triangle ADC$,

$$\frac{x}{b} = \cos A$$

or:

$$x = b \cos A$$

Substitute this into equation (3):

$$a^2 = b^2 + c^2 - 2c(b \cos A)$$

or $\quad a^2 = b^2 + c^2 - 2bc \cos A$

Now consider the case when A is obtuse:

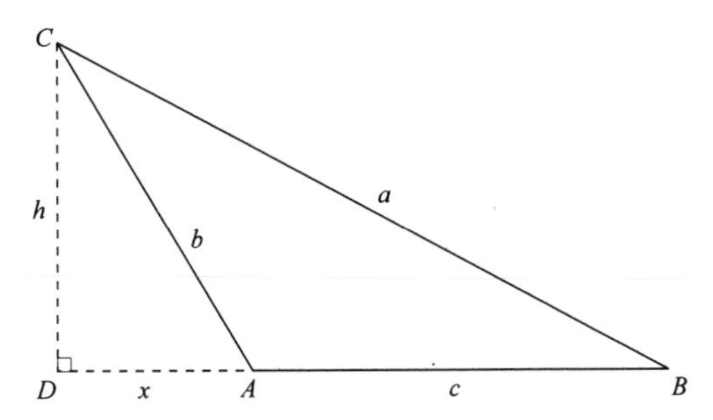

In $\triangle ABC$, $AB = c$, $BC = a$, $AC = b$, CD is the perpendicular from C to the line BA (which is extended or **produced**, as it is called in mathematics), $DA = x$ and $CD = h$.

In $\triangle ADC$, you can use Pythagoras' theorem, to write:

$$h^2 + x^2 = b^2 \qquad (1)$$

In $\triangle BCD$, you can use Pythagoras' theorem, to write

$$h^2 + (x + c)^2 = a^2 \qquad (2)$$

From (1): $\quad h^2 = b^2 - x^2$

Substitute this into equation (2):

$$b^2 - x^2 + (x + c)^2 = a^2$$

So: $\quad b^2 - x^2 + x^2 + 2cx + c^2 = a^2$

or $\quad a^2 = b^2 + c^2 + 2cx \qquad (3)$

Now in $\triangle ADC$, $\quad \dfrac{x}{b} = \cos CAD$

So:

$$\begin{aligned} x &= b\cos CAD \\ &= b\cos(180^\circ - A) \\ &= -b\cos A \end{aligned}$$

By substituting this into equation (3) you get:

$$a^2 = b^2 + c^2 + 2c(-b\cos A)$$

or $\quad a^2 = b^2 + c^2 - 2bc\cos A$

as before.

This is the **cosine rule**. It can written as:

■ $\quad \boldsymbol{a^2 = b^2 + c^2 - 2bc\cos A}$

OR $\quad \boldsymbol{b^2 = a^2 + c^2 - 2ac\cos B}$

OR $\quad \boldsymbol{c^2 = a^2 + b^2 - 2ab\cos C}$

Now if

$$a^2 = b^2 + c^2 - 2bc\cos A$$

then: $$a^2 - b^2 - c^2 = -2bc\cos A$$

So: $$-a^2 + b^2 + c^2 = 2bc\cos A$$

or $$\frac{-a^2 + b^2 + c^2}{2bc} = \cos A$$

So: $$\cos A = \frac{b^2 + c^2 - a^2}{2bc}$$

This is another form of the cosine rule. You can get two other similar formulae if you rearrange the other two forms of the cosine rule. So you can get:

■ $$\boldsymbol{\cos A = \frac{b^2 + c^2 - a^2}{2bc}}$$

OR $$\boldsymbol{\cos B = \frac{a^2 + c^2 - b^2}{2ac}}$$

OR $$\boldsymbol{\cos C = \frac{a^2 + b^2 - c^2}{2ab}}$$

You can use the cosine rule when you are given the lengths of two sides in a triangle and the size of the angle between them. You can then find the length of the third side of the triangle, using either

$$a^2 = b^2 + c^2 - 2bc\cos A$$

or $$b^2 = a^2 + c^2 - 2ac\cos B$$

or $$c^2 = a^2 + b^2 - 2ab\cos C$$

You can also use the cosine rule when you are given the lengths of all three sides of a triangle and want to find the sizes of its angles. In this case you can find the angles using

$$\cos A = \frac{b^2 + c^2 - a^2}{2bc}$$

or $$\cos B = \frac{a^2 + c^2 - b^2}{2ac}$$

or $$\cos C = \frac{a^2 + b^2 - c^2}{2ab}$$

Example 5

Find the length of the side BC of the triangle ABC in which $AB = 7\,\text{cm}$, $AC = 9\,\text{cm}$ and $\angle BAC = 71°$.

By the cosine rule:

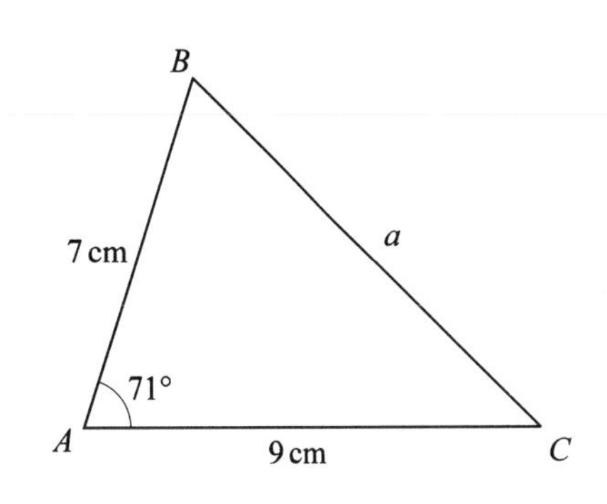

$$a^2 = b^2 + c^2 - 2bc\cos A$$

$$= 9^2 + 7^2 - 2 \times 9 \times 7\cos 71°$$

$$= 81 + 49 - 126\cos 71°$$

$$= 130 - 126 \times 0.3255$$

$$= 130 - 41.02$$

$$a^2 = 88.98$$

So: $\quad a = \sqrt{88.98} = 9.43\,\text{cm (3 s.f.)}$

Example 6

Find the length of the side AB of the triangle ABC in which $BC = 15.3\,\text{cm}$, $AC = 9.4\,\text{cm}$ and $\angle ACB = 121°$.

By the cosine rule:

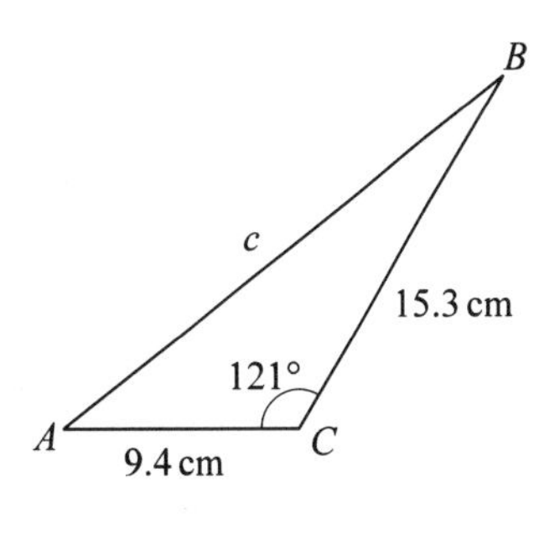

$$c^2 = a^2 + b^2 - 2ab\cos C$$

$$= 15.3^2 + 9.4^2 - 2 \times 15.3 \times 9.4\cos 121°$$

$$= 234.09 + 88.36 - 287.64\cos 121°$$

$$= 322.45 - 287.64(-\cos 59°)$$

$$= 322.45 + 287.64\cos 59°$$

$$= 322.45 + 287.64 \times 0.5150$$

$$= 322.45 + 148.14$$

$$c^2 = 470.59$$

So: $\quad c = 21.7\,\text{cm (3 s.f.)}$

Example 7

Calculate the size of $\angle ABC$ of the triangle ABC in which $AB = 3.6\,\text{cm}$, $BC = 5.2\,\text{cm}$ and $CA = 4.3\,\text{cm}$.

By the cosine rule:

$$\cos B = \frac{a^2 + c^2 - b^2}{2ac}$$

$$= \frac{5.2^2 + 3.6^2 - 4.3^2}{2 \times 5.2 \times 3.6}$$

$$= \frac{27.04 + 12.96 - 18.49}{37.44}$$

$$\cos B = \frac{21.51}{37.44} = 0.5745$$

So: $B = 54.9^\circ$ (1 d.p.)

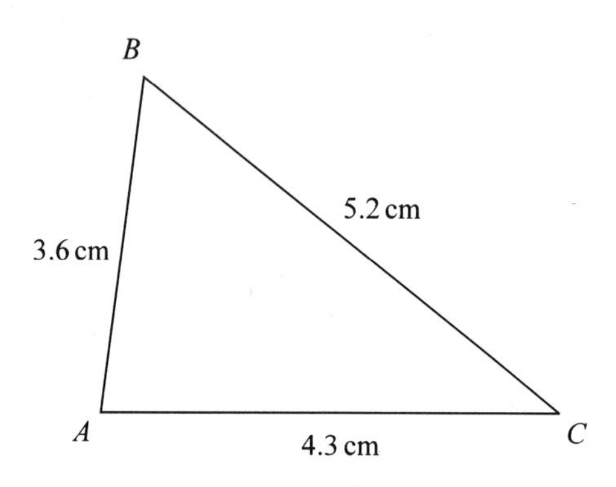

Example 8

Find the size of $\angle ABC$ of the triangle ABC in which $AB = 13.7\,\text{cm}$, $BC = 12.1\,\text{cm}$ and $AC = 19.3\,\text{cm}$.

By the cosine rule:

$$\cos B = \frac{a^2 + c^2 - b^2}{2ac}$$

$$= \frac{12.1^2 + 13.7^2 - 19.3^2}{2 \times 12.1 \times 13.7}$$

$$= \frac{146.41 + 187.69 - 372.49}{331.54}$$

$$= \frac{-38.39}{331.54}$$

$$\cos B = -0.1157$$

So: $B = 96.6^\circ$ (1 d.p.)

B
13.7 cm
12.1 cm
A
19.3 cm
C

Frequently you will be faced with questions that ask you to 'solve' the triangle. This means that you need to find *all* the sides and angles that are not given in the question.

If you are given a triangle and told either

(i) the lengths of all three sides or

(ii) the lengths of two sides and the size of the angle between them

then you can generally solve the triangle by using the cosine rule again and again. However, you will often find that when you have used the cosine rule once, you will then be able to find the other sides and/or angles using the sine rule. As the sine rule is much easier to use than the cosine rule, this is the way that you are recommended to proceed. Nevertheless, if you are one of those masochists who roam the earth, you can stay with the cosine rule!

Example 9

Solve the triangle ABC in which $AC = 12\,\text{cm}$, $BC = 19\,\text{cm}$ and $\angle ACB = 48°$.

We need to calculate c, and the sizes of $\angle CAB$ and $\angle ABC$.

By the cosine rule:

$$\begin{aligned} c^2 &= a^2 + b^2 - 2ab\cos C \\ &= 19^2 + 12^2 - 2 \times 19 \times 12\cos 48° \\ &= 361 + 144 - 456\cos 48° \\ &= 505 - 456 \times 0.669\,13 \\ &= 505 - 305.12 \\ c^2 &= 199.87 \end{aligned}$$

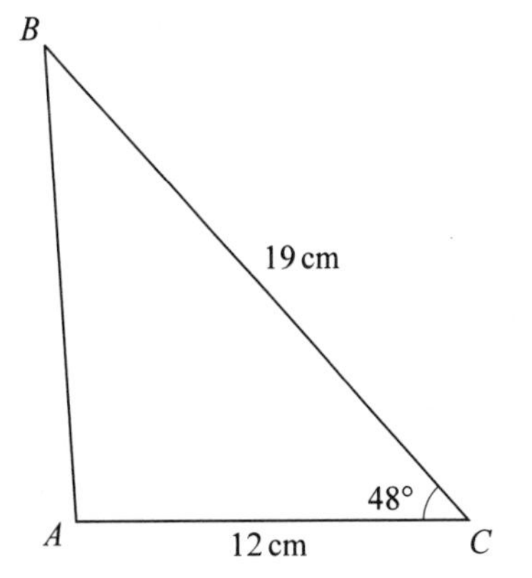

So: $c = 14.137 = 14.1\,\text{cm}$ (3 s.f.)

By the sine rule:

$$\frac{a}{\sin A} = \frac{c}{\sin C}$$

$$\frac{19}{\sin A} = \frac{14.137}{\sin 48°}$$

So: $$\frac{\sin A}{19} = \frac{\sin 48°}{14.137}$$

and $$\sin A = \frac{19 \times 0.743\,14}{14.137}$$

$$A = 87.16° = 87.2° \text{ (1 d.p.)}$$

Consequently: $$B = 180° - 48° - 87.2°$$

$$B = 44.8°$$

Example 10

Solve the triangle ABC in which $AC = 9.5$ cm, $BC = 5.5$ cm and $\angle ACB = 145°$.

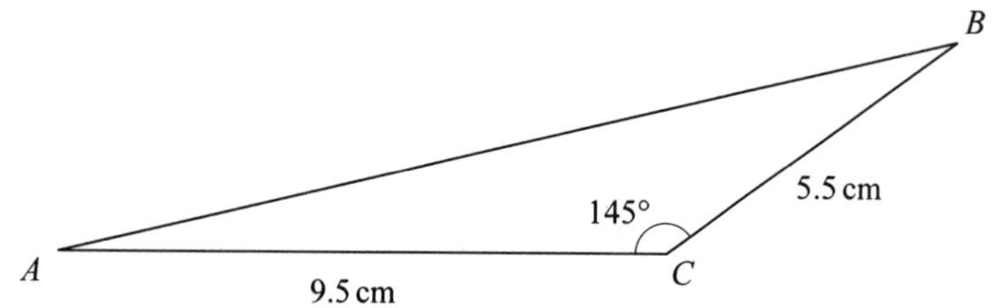

We need to calculate c and the sizes of $\angle BAC$ and $\angle ABC$.

By the cosine rule:

$$\begin{aligned} c^2 &= a^2 + b^2 - 2ab\cos C \\ &= 5.5^2 + 9.5^2 - 2 \times 5.5 \times 9.5 \cos 145° \\ &= 30.25 + 90.25 - 104.5(-\cos 35°) \\ &= 120.5 + 104.5 \times 0.819\,15 \\ c^2 &= 206.10 \end{aligned}$$

So: $$c = 14.356 = 14.4 \text{ cm (3 s.f.)}$$

By the sine rule:

$$\frac{a}{\sin A} = \frac{c}{\sin C}$$

$$\frac{5.5}{\sin A} = \frac{14.35}{\sin 145°}$$

So: $$\frac{\sin A}{5.5} = \frac{\sin 145°}{14.35}$$

That is: $$\sin A = \frac{5.5 \sin 145°}{14.35}$$

$$A = 12.69° = 12.7° \text{ (1 d.p.)}$$

Consequently: $$B = 180° - 145° - 12.7°$$

$$B = 22.3°$$

7.3 The area of a triangle

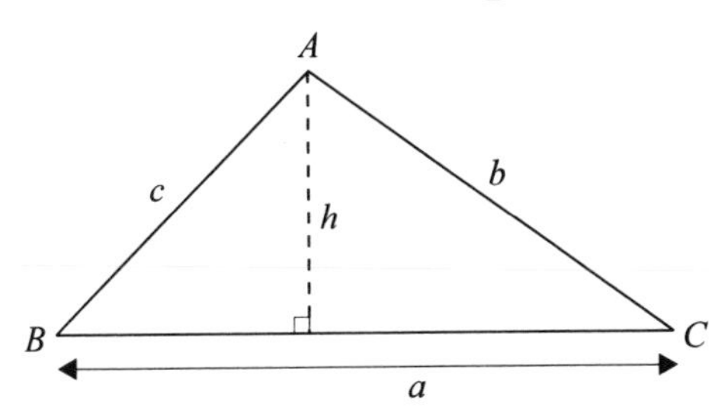

You should know that one formula for the area of $\triangle ABC$ is $\frac{1}{2}ah$, where h is the height of the triangle. This formula is not always the most useful one.

In the diagram,

$$\frac{h}{b} = \sin C$$

So: $$h = b \sin C$$

Thus another useful formula for the area of a triangle is:

$$\tfrac{1}{2}a(b \sin C) = \tfrac{1}{2}ab \sin C$$

This has two other forms:

$$\tfrac{1}{2}bc \sin A$$

and $$\tfrac{1}{2}ac \sin B$$

Exercise 7B

1 Calculate x, in cm to 3 significant figures, in the following triangles:

(a)

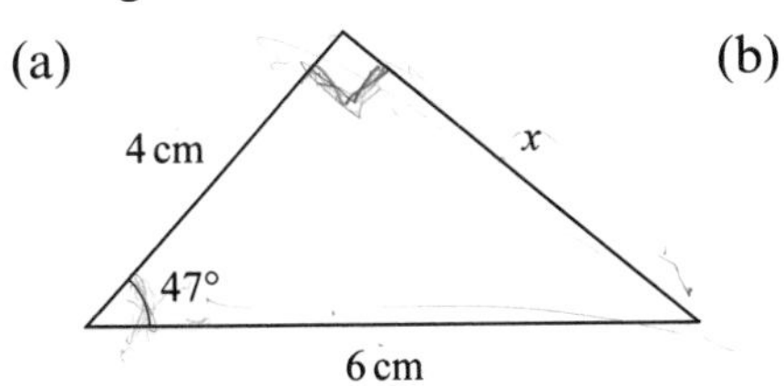

(b)

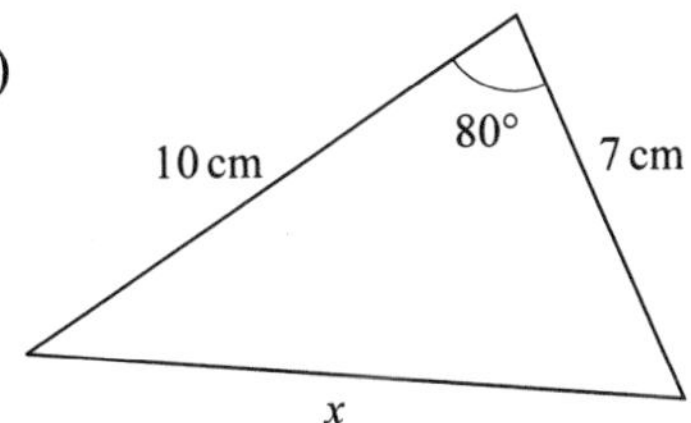

(c)

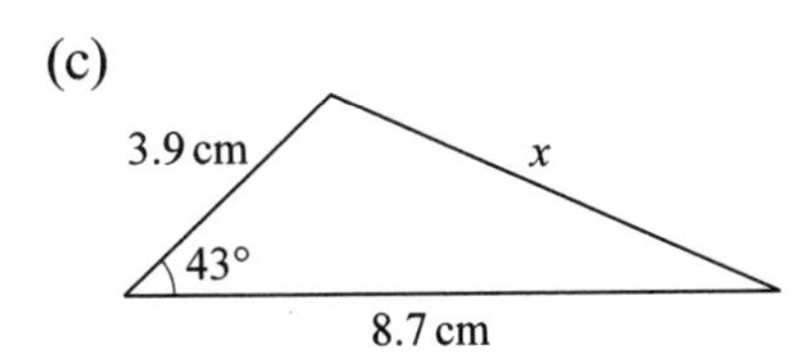

(d)

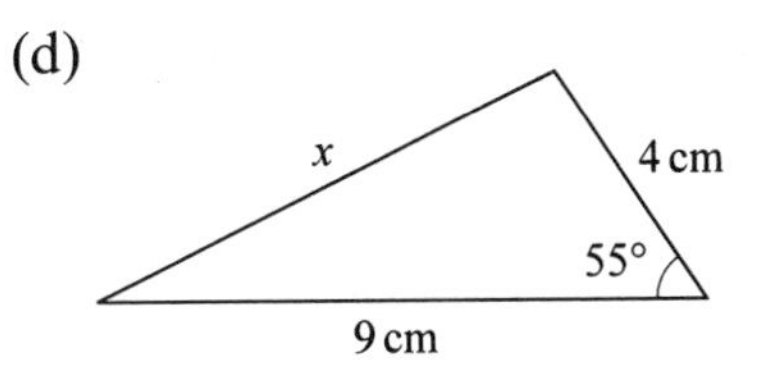

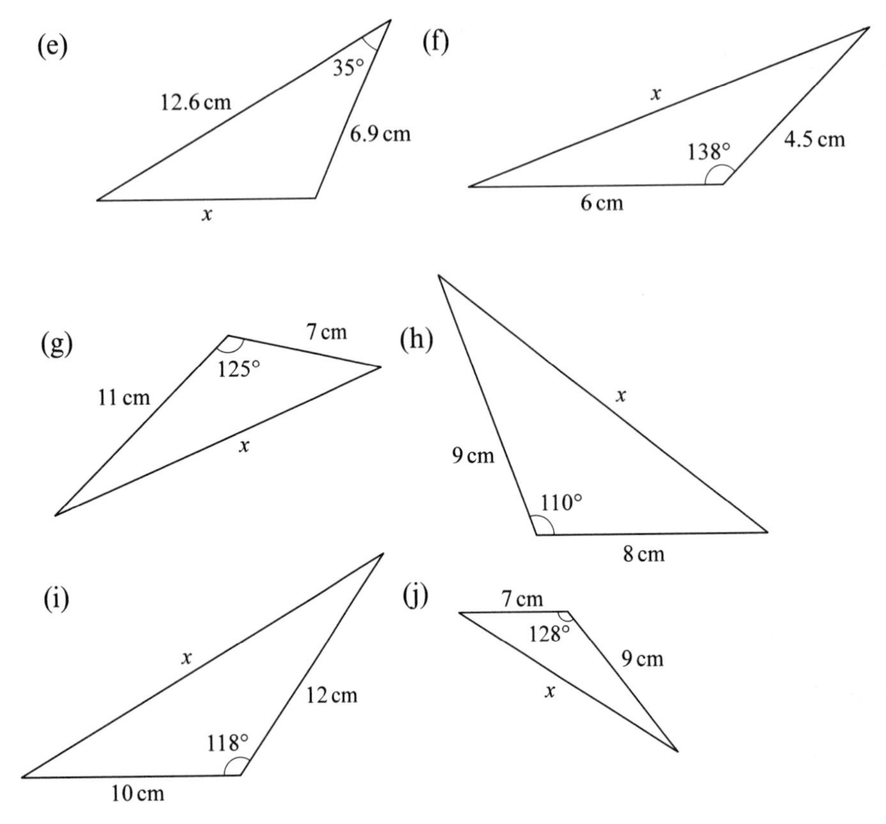

2 Calculate the value of θ, to 0.1°, in the following triangles:

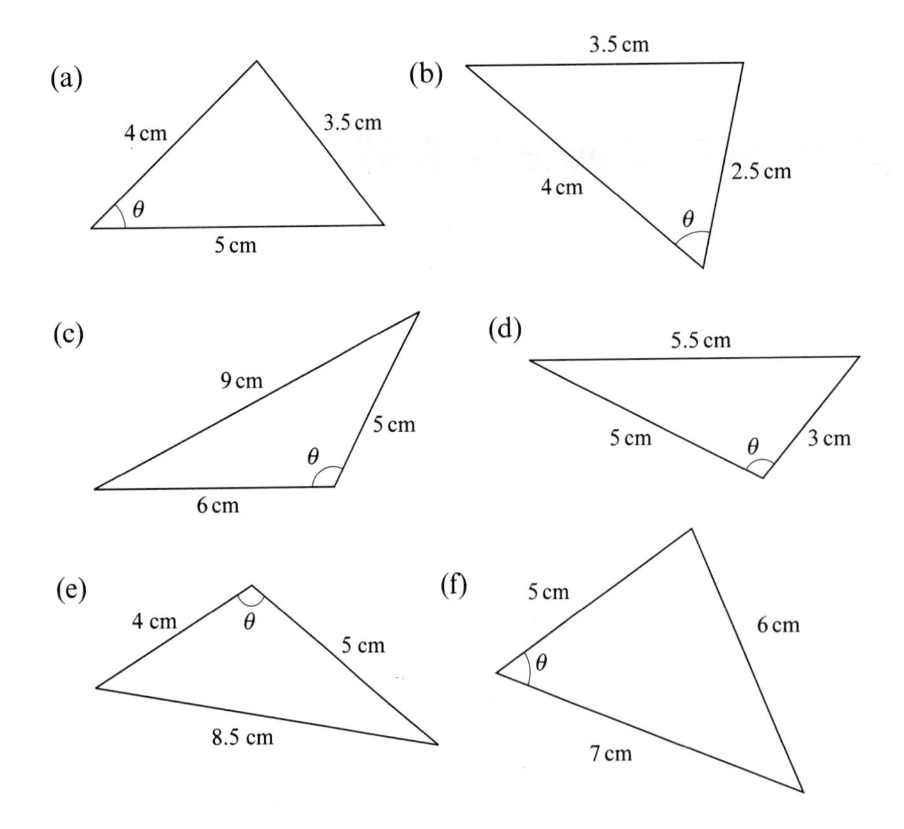

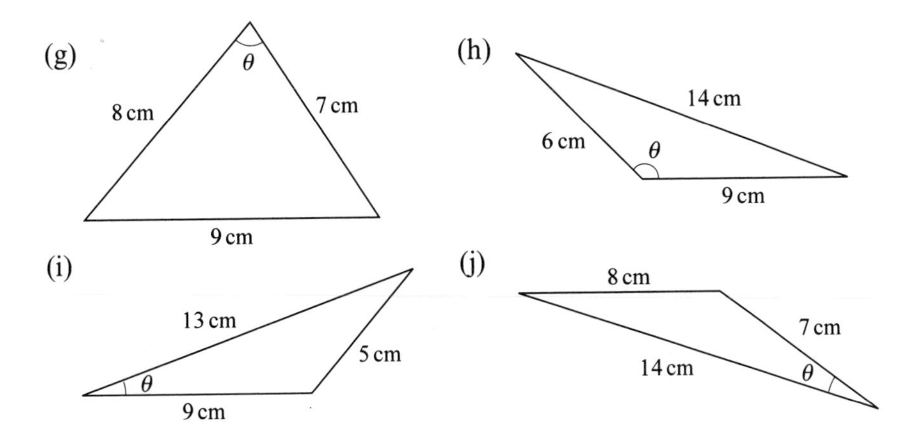

3 Solve the following triangles, giving lengths in cm to 3 significant figures and angles to 0.1°. Find also the area, in cm^2 to 3 significant figures, of each triangle.

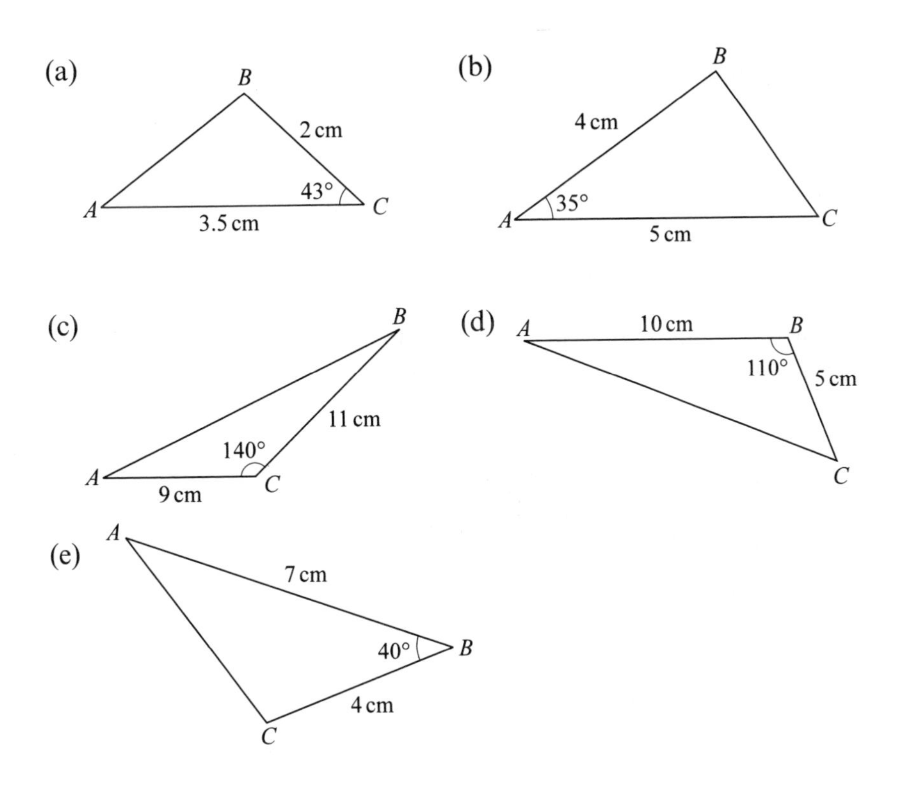

7.4 Problems involving bearings

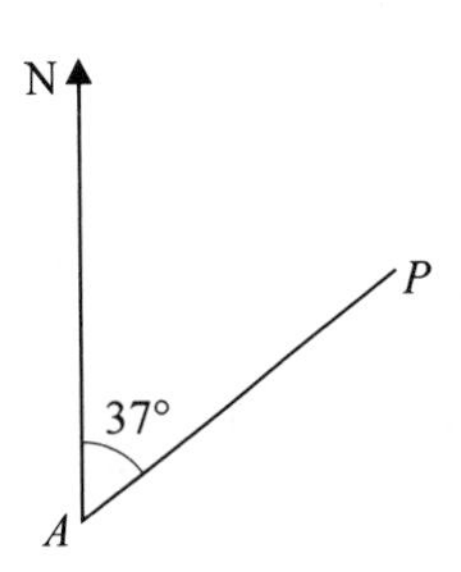

You should have already met bearings in your study of mathematics for GCSE. So you should know that there are two ways of giving a bearing. The first is by giving the angle, measured from the north or south towards the east or west.

So, for example, the bearing of P from A is N37°E.

Likewise, the bearing of Q from B is $S41°W$:

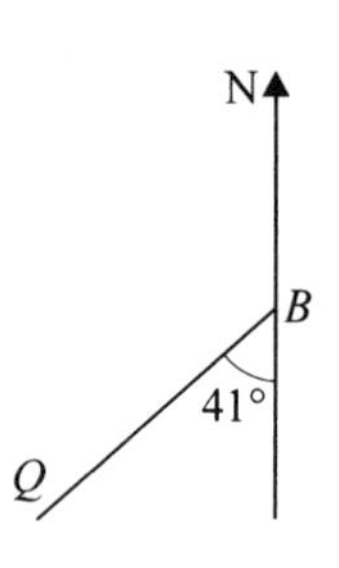

The alternative way, which will be used in this book is to give the bearing as a three-figure angle measured from North in a clockwise direction. So the bearing of P from A in this system is 037°.

Likewise, the bearing of Q from B in this system is $180° + 41°$, that is, 221°:

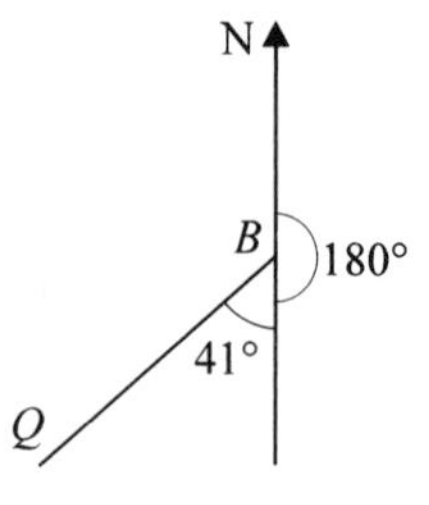

In the same way, the bearing of R from C is $360° - 23°$, that is, 337°:

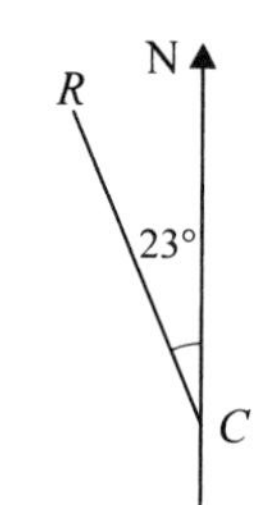

When you are faced with a problem involving bearings, the first thing to do *always* is to draw a diagram. On the diagram you must draw a North line at each point from which a bearing is given. Once you have drawn the diagram the problem will usually reduce to finding lengths and/or angles in a triangle using ratios in a right-angled triangle or the sine rule and/or the cosine rule.

Example 11

Three points P, Q and S are on the same level. The bearings of S from P and Q are 062° and 307°. The distance SP is 190 m and the distance QS is 85 m. Calculate the distance PQ and the bearing of Q from P.

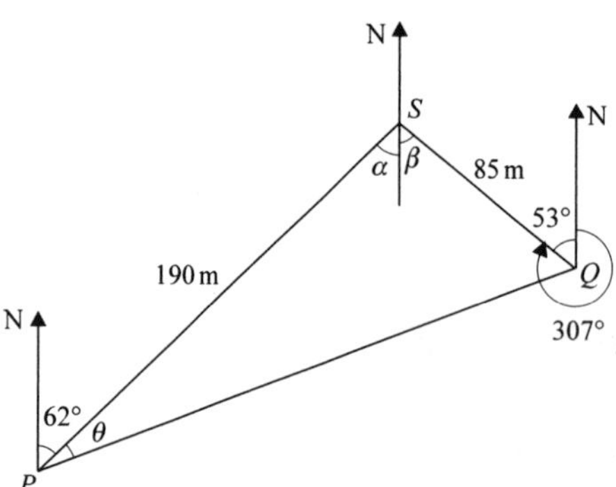

The bearing of S from Q is 307°, so the angle between SQ and the North line is $360° - 307° = 53°$.

Also, because the two North lines at S and Q are parallel, $\beta = 53°$ (alternate angles). Similarly, because the two North lines at P and S are parallel, $\alpha = 62°$.

So: $\angle PSQ = 53° + 62° = 115°$

By the cosine rule:

$$\begin{aligned} s^2 &= p^2 + q^2 - 2pq\cos S \\ &= 85^2 + 190^2 - 2 \times 85 \times 190\cos 115° \\ &= 7225 + 36\,100 - 32\,300(-\cos 65°) \\ &= 43\,325 + 13\,650.5 \\ s^2 &= 56\,975.5 \end{aligned}$$

So: $s = 238.69 = 239\,\text{m}$ (3 s.f.)

By the sine rule:

$$\frac{\sin\theta}{85} = \frac{\sin 115°}{238.7}$$

So:

$$\begin{aligned} \sin\theta &= \frac{85\sin 115°}{238.7} \\ &= \frac{85\sin 65°}{238.7} \\ \sin\theta &= 0.3227 \end{aligned}$$

So: $\theta = 18.8°$

So the angle between the North line at P and PQ is $18.8° + 62° = 80.8°$. That is, the bearing of Q from P is 081° (nearest degree).

Example 12

Two coastguard stations A and B are 5 km apart and B is due east of A. From A the bearings of two ships P and Q are 025° and 061° respectively and from B the bearings are 290° and 338° respectively. Find the distance between the ships.

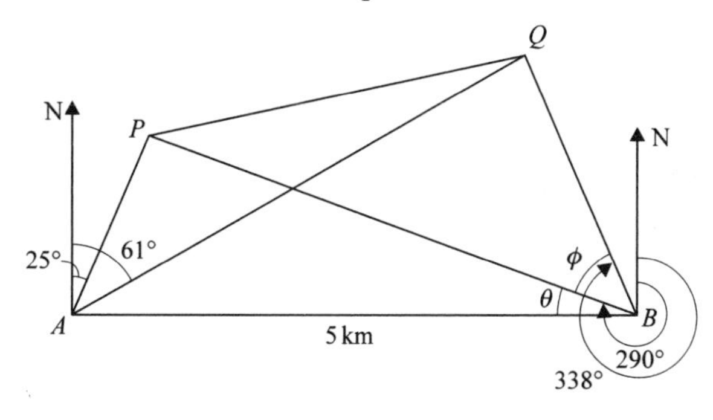

From the diagram you can see that

$$\angle PAQ = 61^\circ - 25^\circ = 36^\circ$$

and $$\angle PAB = 90^\circ - 25^\circ = 65^\circ$$

also: $$\theta = 290^\circ - 270^\circ = 20^\circ$$

and $$\theta + \phi = 338^\circ - 270^\circ = 68^\circ$$

So: $$\phi = 68^\circ - 20^\circ = 48^\circ$$

In $\triangle PAB$, $\angle PAB = 65^\circ$, $\angle PBA = 20^\circ$

So: $$\angle APB = 180^\circ - 65^\circ - 20^\circ = 95^\circ$$

Use the sine rule in $\triangle APB$:

$$\frac{AP}{\sin 20^\circ} = \frac{5}{\sin 95^\circ}$$

$$\begin{aligned} AP &= \frac{5 \sin 20^\circ}{\sin 95^\circ} \\ &= \frac{5 \sin 20^\circ}{\sin 85^\circ} \\ &= 1.716 \text{ km} \end{aligned}$$

In $\triangle AQB$, $\angle ABQ = 68^\circ$ and $\angle QAB = 90^\circ - 61^\circ = 29^\circ$.

So: $$\angle AQB = 180^\circ - 68^\circ - 29^\circ = 83^\circ$$

Use the sine rule in $\triangle AQB$:

$$\frac{AQ}{\sin 68^\circ} = \frac{5}{\sin 83^\circ}$$

$$\begin{aligned} AQ &= \frac{5 \sin 68^\circ}{\sin 83^\circ} \\ &= 4.670 \text{ km} \end{aligned}$$

Use the cosine rule in $\triangle APQ$:

$$\begin{aligned} PQ^2 &= 1.716^2 + 4.670^2 - 2 \times 1.716 \times 4.670 \cos 36^\circ \\ &= 2.944 + 21.80 - 12.966 \end{aligned}$$

$$PQ^2 = 11.78$$

So: $$PQ = 3.43 \text{ km (3 s.f.)}$$

Exercise 7C

1 Three landmarks P, Q and R are on the same horizontal level. Landmark Q is 3 km and on a bearing of 328° from P, landmark R is 6 km and on a bearing of 191° from Q. Calculate the distance and bearing of R from P, giving your answer in km to one decimal place and in degrees to the nearest degree. [L]

2 From a ship, the following distances and bearings are obtained simultaneously:

Lighthouse A 15 nautical miles, 032°

Lighthouse B 10 nautical miles, 272°

Calculate, to the nearest tenth of a nautical mile, the distance between the lighthouses. [L]

3 Two points A and B on a straight coast are 2 km apart, B being due east of A. A ship is observed on bearings 169° and 211° from A and B respectively. Calculate the distance of the ship from the coast.

4 A man walking due north on a straight road sees a mast on a bearing of 035°. After walking 600 m the bearing of the mast is 071° from the man's new position. Find the distance of the mast from the road.

5 A ship sails 3 nautical miles from P to Q on a bearing 071° and then 4 nautical miles from Q to R on a bearing 292°. Calculate the distance PR and the bearing of R from P.

6 From a lighthouse A at a given time, two ships P and Q lie on bearings 300° and 020° respectively. Given that P is 6 km from the lighthouse and Q is 3.5 km from the lighthouse, calculate the distance between the ships at the given time.

7 Three points A, B and C lie on level ground and B is due south of A. The point C lies 350 m from A on a bearing 065° and the distance BC is 450 m. Calculate the bearing of B from C.

8 A ship P is 3 km from a lightship and on a bearing of 345° from the lightship. At the same time a ship Q is 5 km from the lightship and on a bearing 067° from the lightship. Calculate the distance and bearing of P from Q.

9 A ship A is 9 km from a lighthouse P on a bearing 075° and a ship B is 5.5 km from the lighthouse on a bearing 200°. Find the distance and bearing of A from B.

10 Two lighthouses A and B are 7 km apart and B is due east of A. From A the bearings of two ships are 031° and 073° respectively and from B the bearings are 275° and 343° respectively. Find the distance between the ships.

7.5 Applying trigonometry to problems in three dimensions

One of the major problems that you will have with work in three dimensions is identifying which angles or which lengths you need to calculate. It is therefore of the utmost importance that you sketch good diagrams. A good diagram is one that is positioned in such a way that you *can* see the angle or length which you are trying to calculate.

Once you have identified the angle or length that you are trying to calculate, you will find that it always lies in a triangle. Once again, it is important that you sketch this triangle separately; from this point on you can work from a two-dimensional figure, just as you have been doing so far in this chapter.

If you follow these simple steps then problems in three dimensions will be simplified and become much more straightforward.

The angle between a line and a plane

One of the two most frequent calculations that you will have to do is to calculate the angle between a line and a plane.

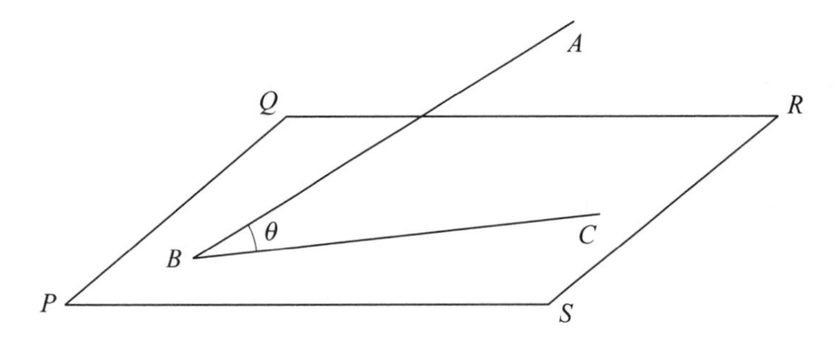

The angle between the line AB and the plane $PQRS$ is defined to be the angle between the line and the projection of the line on the plane. The **projection** of the line AB is the shadow that it casts on the plane when a light is held above it pointing at the plane and at right-angles to the plane. In the example shown, the line AB casts the shadow BC on the plane. So the angle between the line AB and the plane $PQRS$ is defined to be the angle between the line AB and the line BC, that is, the angle θ.

Example 13

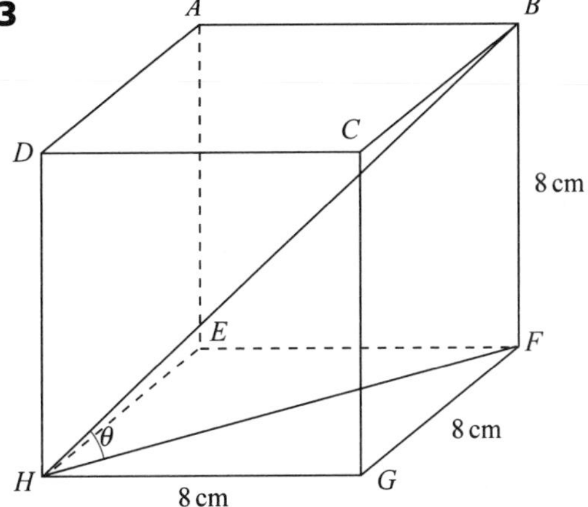

The cube $ABCDEFGH$ has sides 8 cm long. Calculate the angle between the diagonal BH and the plane $EFGH$.

The projection of the line BH on the plane $EFGH$ is the line FH. So the required angle is $\angle BHF = \theta$. This lies in the triangle BFH.

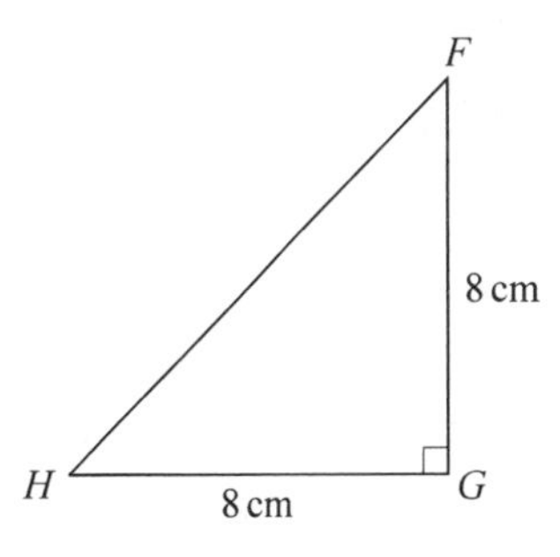

Before you can calculate θ in $\triangle BFH$ you need to know the lengths of two sides. At the moment only one is known: $BF = 8$ cm. So you need to calculate either BH or FH.

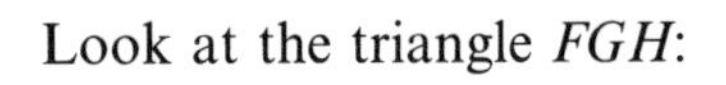

Look at the triangle FGH:

The triangle has a right angle at G and $FG = HG = 8$ cm.

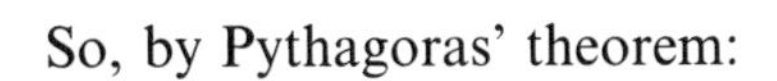

So, by Pythagoras' theorem:

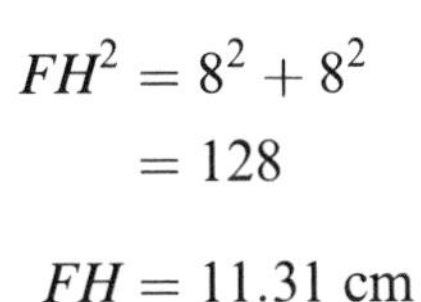

$$FH^2 = 8^2 + 8^2$$
$$= 128$$
$$FH = 11.31 \text{ cm}$$

Now look at $\triangle BFH$:

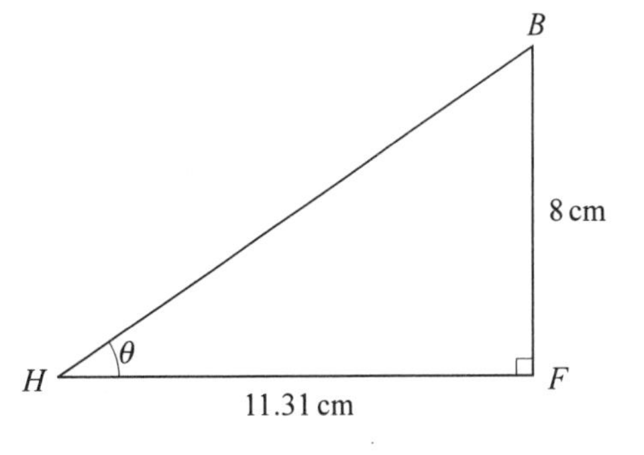

$$\tan\theta = \frac{8}{11.31}$$
$$= 0.7071$$
$$\theta = 35.27^\circ = 35.3^\circ \text{ (1 d.p.)}$$

Example 14

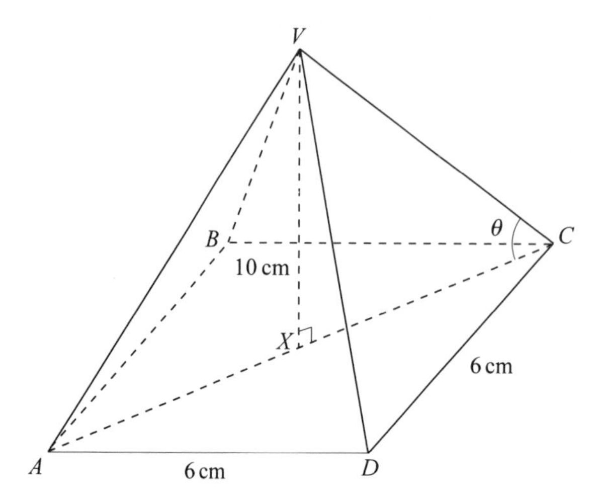

$VABCD$ is a right pyramid (one where the vertex V lies vertically above the centre of the base of the pyramid) on a square base of side 6 cm. The vertex V is vertically above the centre of the base X. The height of the pyramid is 10 cm. Calculate the angle between a sloping edge of the pyramid and its base.

Since the pyramid is on a square base and it is a right pyramid, the angles between each of the sloping edges VA, VB, VC, VD and the base are the same.

Consider the edge VC. The projection of VC on the base $ABCD$ is the line XC, where X is the mid-point of AC. So the angle you need to calculate is the one between VC and XC. That is, $\angle VCX = \theta$.

However, in $\triangle VXC$ the only known side is $VX = 10$ cm and the only known angle is $\angle VXC = 90°$. So you need to calculate either XC or VC.

Consider $\triangle ADC$.

By Pythagoras' theorem:

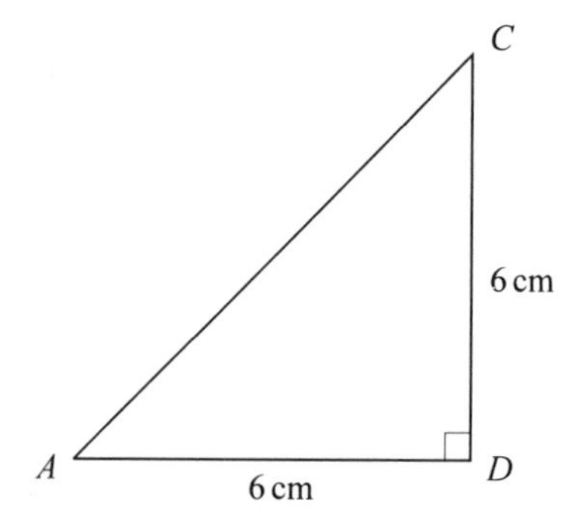

$$AC^2 = 6^2 + 6^2$$

$$= 72$$

$$AC = 8.4852 \text{ cm}$$

So:

$$XC = 4.2426 \text{ cm}$$

Now consider the $\triangle VXC$.

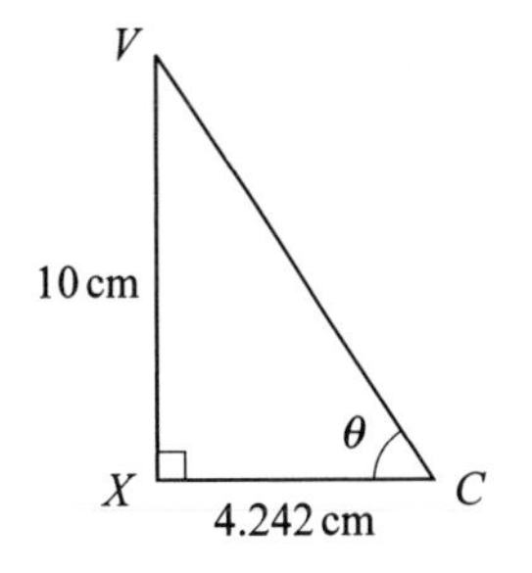

$$\tan\theta = \frac{10}{4.242}$$
$$= 2{:}3570$$

So: $\theta = 67.01° = 67.0°\,(1 \text{ d.p.})$

The angle between two planes

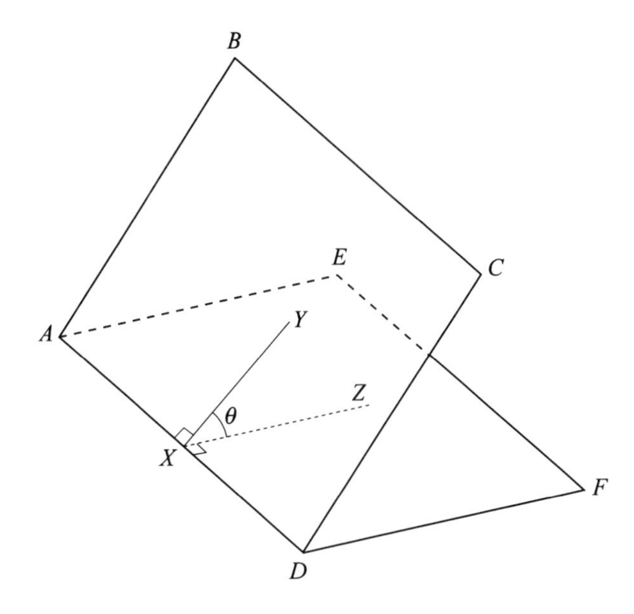

Two planes always meet in a common line. The two planes in this diagram, $ABCD$ and $AEFD$, meet in the line AD.

Take any point X on the common line. From X, draw a perpendicular (XY) to the common line AD which lies in the first plane $ABCD$. Now draw a perpendicular XZ to the common line AD which lies in the second plane $AEFD$. The angle between these two perpendiculars is the angle between the planes $ABCD$ and $AEFD$.

Example 15

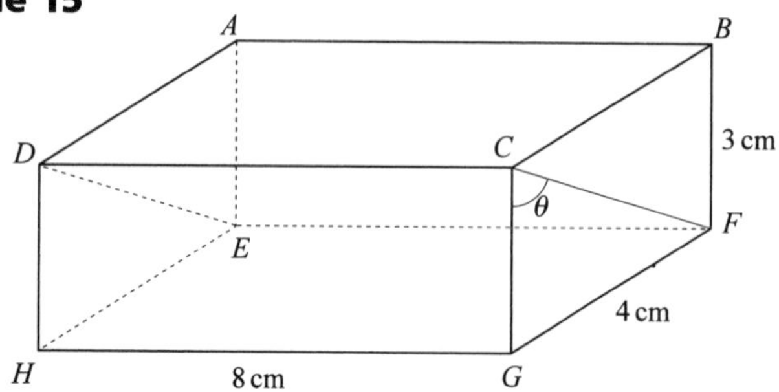

The cuboid $ABCDEFGH$ is 8 cm long, 4 cm wide and 3 cm high. Calculate the angle between the planes $EFCD$ and $DCGH$.

The planes $EFCD$ and $DCGH$ meet in the common line DC. Choose the point C on this common line.

The line FC is in the plane $EFCD$ and is perpendicular to DC.

The line GC is in the plane $DCGH$ and is perpendicular to DC.

So the angle required is that between the lines FC and GC. It is $\angle GCF = \theta$.

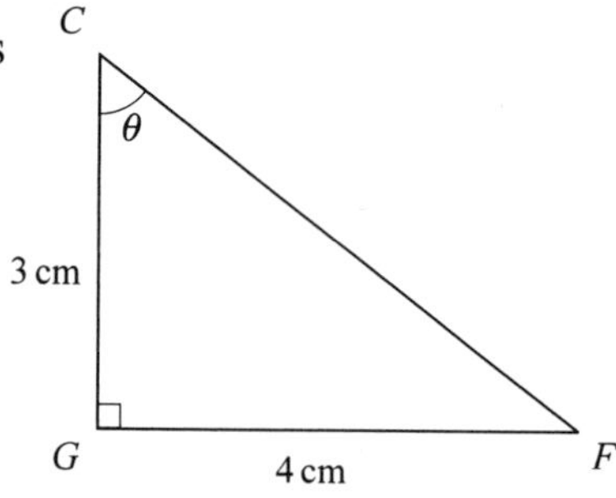

In triangle CGF,

$$\tan\theta = \tfrac{4}{3}$$
$$= 1.3333$$

So: $$\theta = 53.13° = 53.1° \text{ (1 d.p.)}$$

Example 16

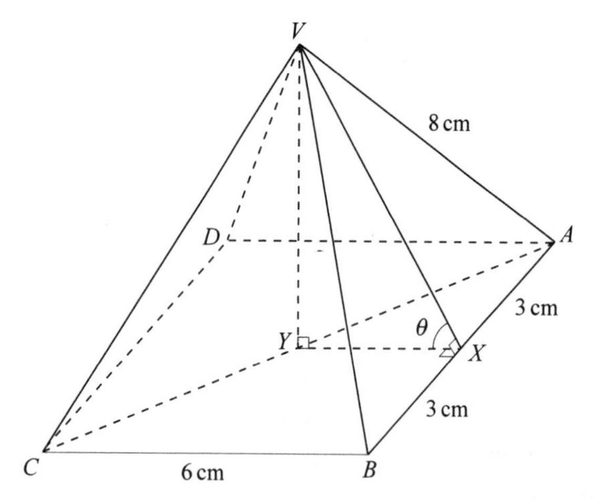

$VABCD$ is a right pyramid on a square base of side 6 cm. The edge VA is of length 8 cm and the vertex V is vertically above the centre of the base Y. Calculate the angle between the plane VAB and the base.

The planes VAB and $ABCD$ meet in the common line AB. Take the point X as the mid-point of AB. The perpendicular to AB at X drawn on the plane VAB passes through V because $\triangle VAB$ is isosceles.

The perpendicular to AB at X drawn on the plane $ABCD$ passes through Y, the mid-point of the base. The angle you want is the one between VX and YX. That is, $\angle VXY = \theta$.

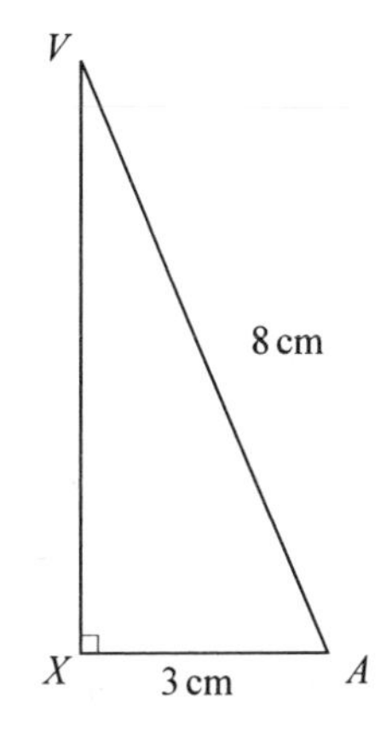

Consider $\triangle VAX$:

By Pythagoras' theorem:

$$VX^2 = 8^2 - 3^2$$
$$= 55$$

$$VX = 7.416 \text{ cm}$$

Now look at $\triangle VYX$:

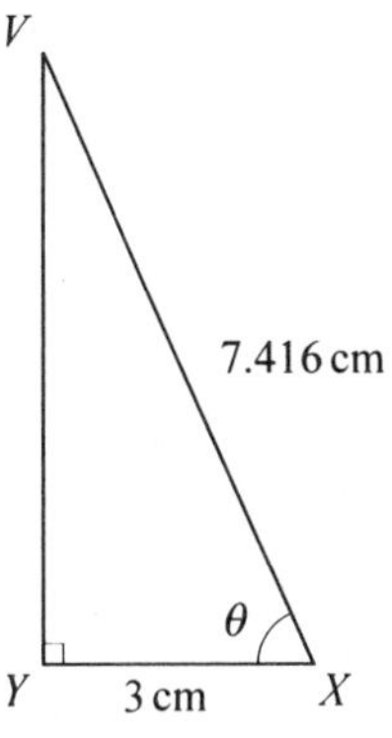

$$\cos\theta = \frac{3}{7.416}$$
$$= 0.4045$$

So: $\theta = 66.13° = 66.1°$ (1 d.p.)

Exercise 7D

1

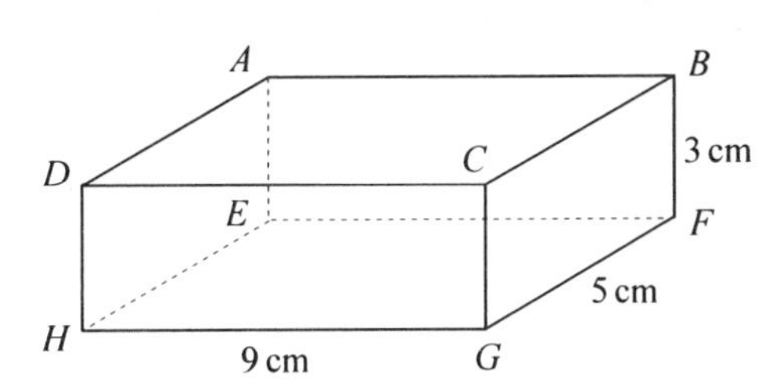

The cuboid $ABCDEFGH$ has $GH = 9$ cm, $FG = 5$ cm and $BF = 3$ cm.

Calculate, to 0.1°, the size of

(a) the angle between BH and the base $EFGH$

(b) the angle between the plane $EFCD$ and the plane $DCGH$.

2

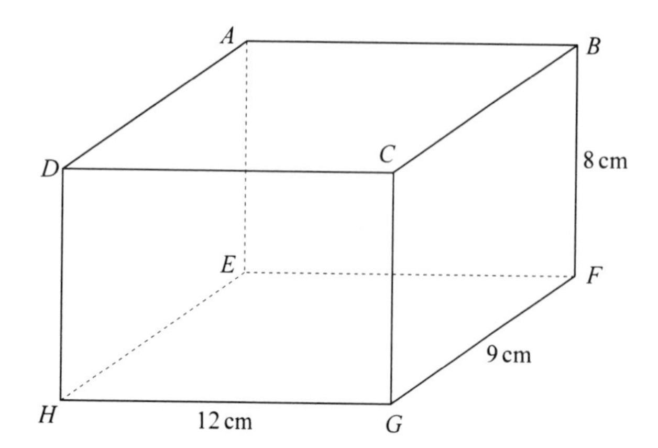

The cuboid $ABCDEFGH$ has $GH = 12\,\text{cm}$, $FG = 9\,\text{cm}$ and $BF = 8\,\text{cm}$.

Calculate, to 0.1°, the size of

(a) the angle between AG and the plane $ABCD$

(b) the angle between the plane $BCHE$ and the plane $EFGH$.

3

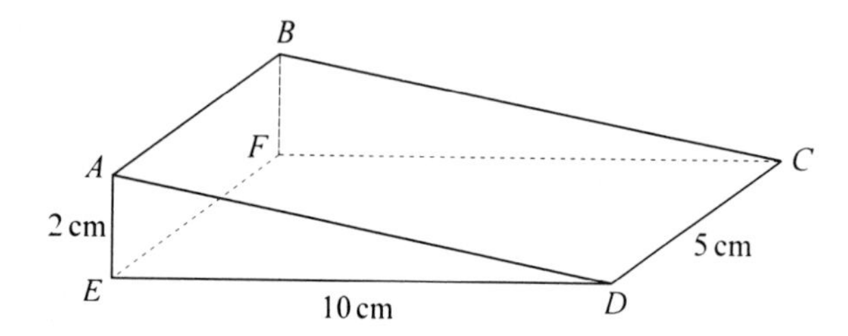

The triangular prism $ABCDEF$ has a rectangular, horizontal base $CDEF$ and a vertical rectangular end $ABFE$.

Given that $DE = 10\,\text{cm}$, $CD = 5\,\text{cm}$ and $AE = 2\,\text{cm}$, calculate, to 0.1°, the size of

(a) the angle between the line AC and the base

(b) the angle between the plane $ABCD$ and the base

(c) the angle between the line BE and the line BD.

4

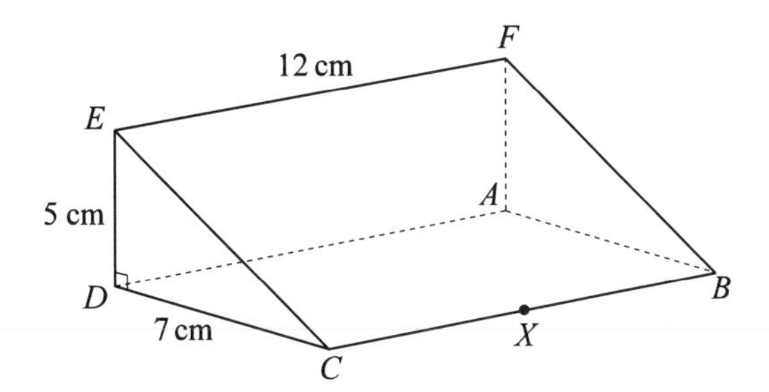

The base $ABCD$ of the triangular prism $ABCDEF$ is a horizontal rectangle and the end $EFAD$ is a vertical rectangle. Given that X is the mid-point of BC, that $EF = 12$ cm, $DE = 5$ cm and $CD = 7$ cm, calculate, to $0.1°$, the size of

(a) the angle between CF and the base $ABCD$

(b) the angle between XF and the base $ABCD$.

5 The right pyramid $VABCD$ has a square base $ABCD$ of length 9 cm. The length of VA is 11 cm. Calculate, to $0.1°$, the size of

(a) the angle between VA and the base

(b) the angle between the plane VAB and the base.

6 The right pyramid $VABCD$ has a square base $ABCD$ of side 7 cm. The height of the pyramid is 10 cm. Calculate, to $0.1°$, the size of

(a) the angle between VA and the base

(b) the angle between the plane VAB and the base.

7

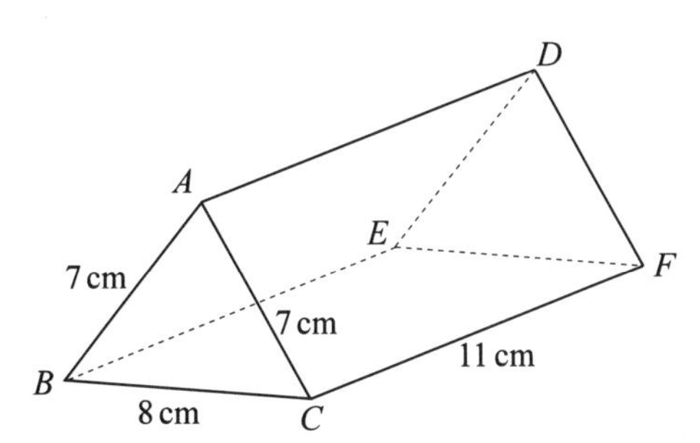

The right triangular prism $ABCDEF$ has a horizontal, rectangular base $BEFC$. The ends are vertical isosceles triangles with $AB = AC = 7$ cm and $BC = 8$ cm. The prism has length 11 cm. Calculate, to $0.1°$, the size of

(a) the angle between the plane DBC and the base

(b) the angle between DC and the base.

8

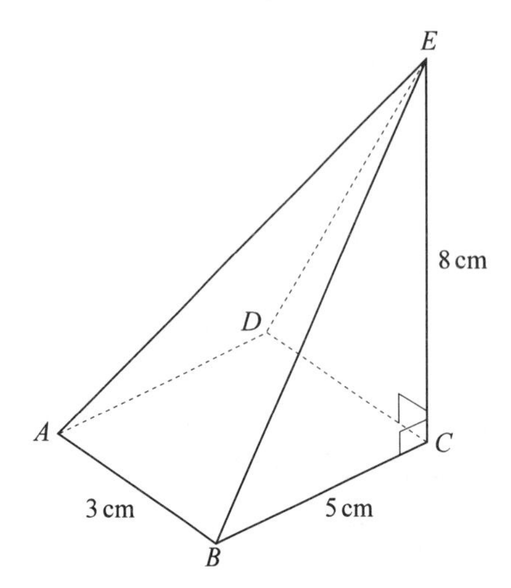

The figure shows a 'lop-sided' pyramid $ABCDE$, which has a horizontal rectangular base $ABCD$ in which $AB = 3$ cm and $BC = 5$ cm, and with the vertex E vertically above C, so that $CE = 8$ cm. Calculate:

(a) the lengths, in cm to 2 decimal places, of BE and AE

(b) the angles, to the nearest tenth of a degree, between

(i) AE and the plane $ABCD$

(ii) AE and the plane BEC. [L]

9

$PQLM$ and $PQRS$ are rectangles, with the plane $PQLM$ horizontal. The lines LR and MS are equal and vertical. Given that $PQ = 16$ m, $QL = 12$ m and $LR = 21$ m, calculate

(a) QM

(b) QS

(c) the angle between the line SQ and the plane $PQLM$

(d) the angle between the plane $PQRS$ and the plane $PQLM$.

[L]

10

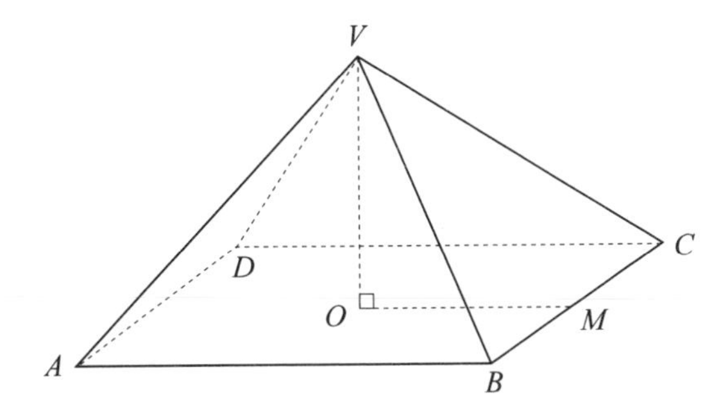

The figure shows a pyramid $VABCD$, with a rectangular base in which $AB = 6$ cm, $BC = 4$ cm. The point O is in the plane $ABCD$, the line VO is perpendicular to the plane $ABCD$, and $VO = 5$ cm. The point M is the midpoint of BC.

(a) Sketch two cross-sections of the pyramid, one through V, O and M, and the other through V, A and C.

(b) Calculate, in cm to 2 decimal places, the lengths of VM, AC and VC.

(c) Calculate, to the nearest tenth of a degree, the angle AVC and the angle between the planes VAD and VBC. [L]

11 Three points A, B and C on horizontal ground are such that $AB = 10$ m, $BC = 21$ m and $\cos \angle ABC = \frac{3}{5}$. The point P lies on BC and $\angle APB = 90°$. Calculate

(a) the length of AC

(b) the length of AP.

A vertical pole VA, of height 19 m is placed at A. Calculate, to the nearest tenth of a degree, the acute angle between

(c) VB and the horizontal

(d) the plane VBC and the horizontal. [L]

12 A pole TA of length 11 m is held in a vertical position with A on horizontal ground by three wires PT, QT and RT of equal length. The points P, Q and R form an equilateral triangle of side 13 m. Calculate:

(a) the length, in metres to 3 significant figures, of each wire

(b) the size of the angle, to 0.1°, between a wire and the horizontal

(c) the size of the angle, to 0.1°, between the planes TPQ and PQR.

13

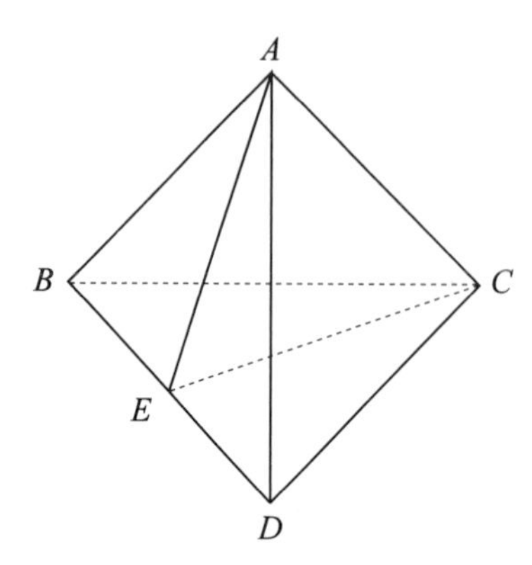

$ABCD$ is a regular tetrahedron with the plane BCD horizontal and E is the mid-point of BD. Prove that $\triangle AEC$ is isosceles.
Hence, or otherwise, calculate in degrees to 3 significant figures,
(a) the angle between the edge AC and the horizontal
(b) the angle between the plane ABD and the horizontal. [L]

7.6 Cartesian coordinates in three dimensions

In Book P1 you were introduced to the ideas of coordinate geometry. Coordinate geometry is the study of the geometry of points, lines, curves and planes using algebraic methods. In Book P1 you looked at lines or curves in two dimensions (2D) only. Now we have a look at some figures in three dimensions (3D).

If two axes at right angles to each other are laid on a plane, then you can identify any point in the plane, using coordinates that show how far the point is from each axis. One of these axes is usually called the x-axis, the other is called the y-axis and the point where the axes meet is called the origin and is labelled O.

A similar set-up can be used in three dimensions by adding a third axis, the z-axis, at right angles to both the x- and y-axes.

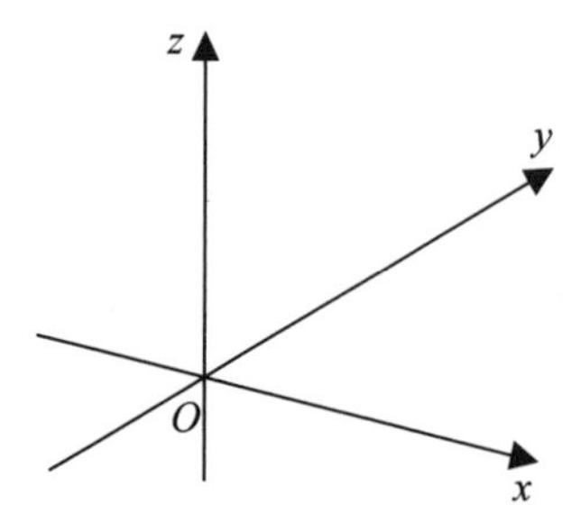

To define a point P in space, you must state the perpendicular distance of the point from the yz-plane, its perpendicular distance from the xz-plane and its perpendicular distance from the xy-plane. This defines the position of the point uniquely.

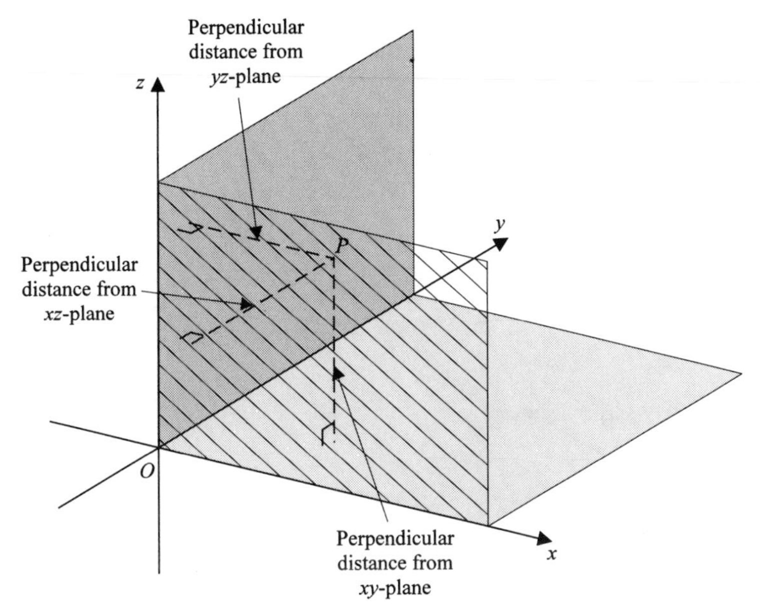

As in two dimensions, part of each axis is positive and the rest is negative. The part of the x-axis from O in the direction of the arrow in the diagram is the positive x-axis and the rest is the negative x-axis; the part of the y-axis from O in the direction of the arrow is the positive y-axis and the rest is the negative y-axis; and the part of the

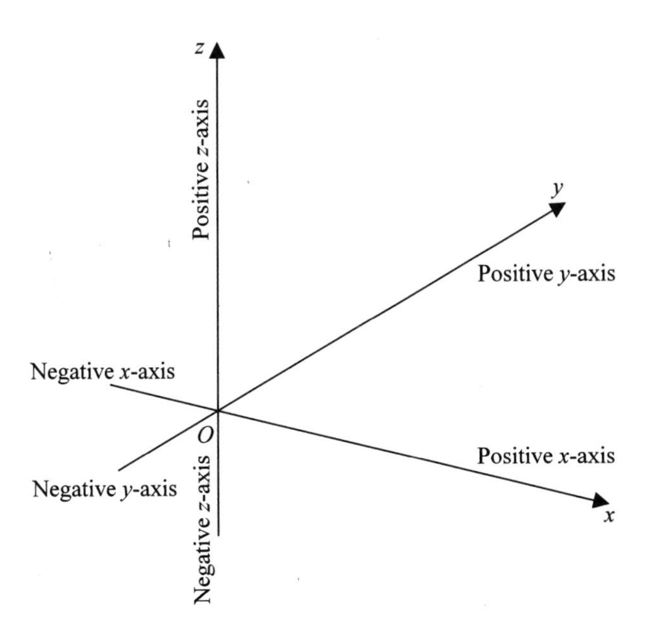

z-axis from O in the direction of the arrow is the positive z-axis and the rest is the negative z-axis.

The perpendicular distance of a point in space from the yz-plane is called the **x-coordinate**, the perpendicular distance of the point from the xz-plane is called the **y-coordinate** and the perpendicular distance of the point from the xy-plane is called the **z-coordinate**. The coordinates of a point are written (x, y, z) where the first number is the x-coordinate, the second number is the y-coordinate and the third number is the z-coordinate. This is the point $P(2, 5, 4)$:

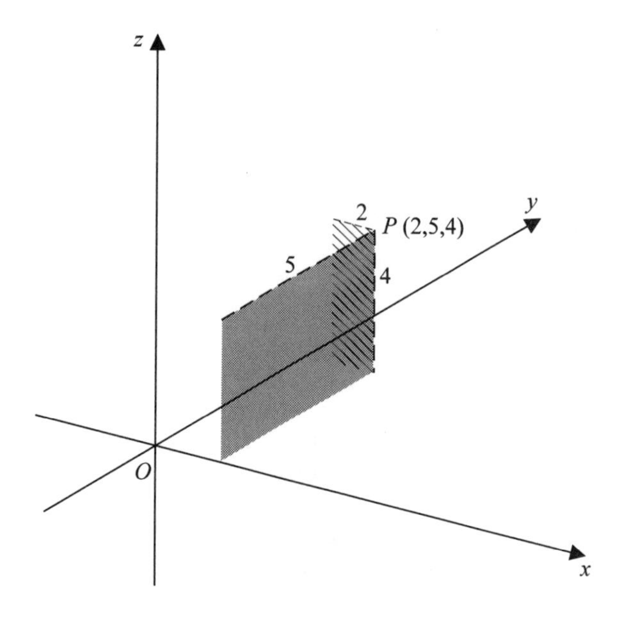

It is usual to define the directions of the coordinate axes like this:

(i) Oy is a 90° rotation *anticlockwise* from Ox in the xy-plane when viewed from the z-positive side of that plane.

(ii) Oz is a 90° rotation *anticlockwise* from Oy in the yz-plane when viewed from the x-positive side of that plane.

(iii) Ox is a 90° rotation *anticlockwise* from Oz in the xz-plane when viewed from the y-positive side of that plane.

Finding the distance between two points

Suppose you want to find the distance AB between the points $A(2, 1, 3)$ and $B(4, 5, 8)$.

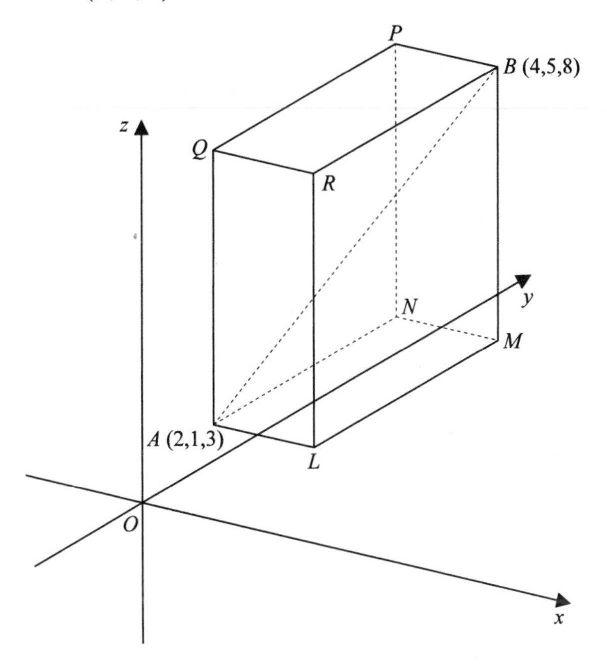

Draw a cuboid that has AB as a diagonal and has each of its faces parallel to the coordinate planes. In the diagram,

(i) the faces $LMBR$ and $ANPQ$ are parallel to the yz-plane,

(ii) the faces $MBPN$ and $LRQA$ are parallel to the xz-plane,

(iii) the faces $LMNA$ and $RBPQ$ are parallel to the xy-plane.

So:

$$AL = NM = 4 - 2 = 2$$

$$LM = AN = 5 - 1 = 4$$

$$MB = AQ = 8 - 3 = 5$$

In $\triangle AMB$:

$$AB^2 = AM^2 + MB^2$$
$$AB^2 = AM^2 + 5^2$$

In $\triangle ALM$:

$$AM^2 = AL^2 + LM^2$$
$$AM^2 = 2^2 + 4^2$$

So: $$AB^2 = (2^2 + 4^2) + 5^2$$
$$= 4 + 16 + 25$$
and $$AB = \sqrt{45}$$

To find a formula that can always be used to find the distance between two points, you can generalise the process by using two points with coordinates (x_1, y_1, z_1) and (x_2, y_2, z_2).

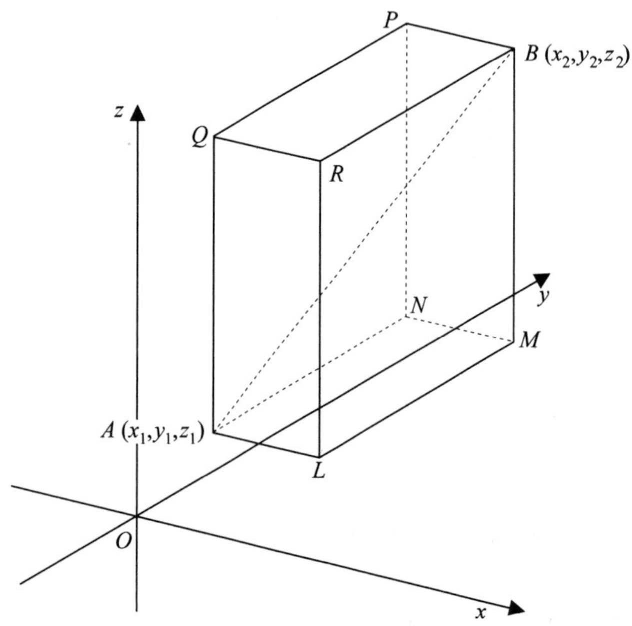

As before, draw the cuboid with AB as diagonal and with each face parallel to the coordinate planes.

So: $$AL = NM = x_2 - x_1$$
$$LM = AN = y_2 - y_1$$
$$MB = AQ = z_2 - z_1$$

In $\triangle AMB$:

$$AB^2 = AM^2 + MB^2$$
$$AB^2 = AM^2 + (z_2 - z_1)^2$$

In $\triangle ALM$:

$$AM^2 = AL^2 + LM^2$$
$$AM^2 = (x_2 - x_1)^2 + (y_2 - y_1)^2$$

So: $$AB^2 = [(x_2 - x_1)^2 + (y_2 - y_1)^2] + (z_2 - z_1)^2$$

or

■ $$\mathbf{AB = \sqrt{[(x_2 - x_1)^2 + (y_2 - y_1)^2 + (z_2 - z_1)^2]}}$$

Example 17

Find the distance between the points $A(1, 3, 5)$ and $B(2, 6, 7)$.

Using the formula:

$$\begin{aligned} AB &= \sqrt{[(x_2 - x_1)^2 + (y_2 - y_1)^2 + (z_2 - z_1)^2]} \\ &= \sqrt{[(2-1)^2 + (6-3)^2 + (7-5)^2]} \\ &= \sqrt{(1^2 + 3^2 + 2^2)} \\ &= \sqrt{(1 + 9 + 4)} \\ &= \sqrt{14} \\ &= 3.74 \text{ (3 s.f.)} \end{aligned}$$

Example 18

Find the distance between the points $A(6, -1, -3)$ and $B(2, 4, -1)$.

Using the formula:

$$\begin{aligned} AB &= \sqrt{[(x_2 - x_1)^2 + (y_2 - y_1)^2 + (z_2 - z_1)^2]} \\ &= \sqrt{[(2-6)^2 + (4+1)^2 + (-1+3)^2]} \\ &= \sqrt{[(-4)^2 + 5^2 + 2^2]} \\ &= \sqrt{(16 + 25 + 4)} \\ &= \sqrt{45} \\ &= 6.71 \text{ (3 s.f.)} \end{aligned}$$

Example 19

The length of the line joining $A(3, 1, 4)$ to $B(1, t, -2)$ is 7. Calculate the two possible values of t.

By the formula:

$$\begin{aligned} AB^2 &= (3-1)^2 + (1-t)^2 + (4+2)^2 \\ &= 2^2 + (1 - 2t + t^2) + 6^2 \\ &= 4 + 1 - 2t + t^2 + 36 \\ &= t^2 - 2t + 41 \end{aligned}$$

But if $AB = 7$ then $AB^2 = 49$

So:

$$\begin{aligned} t^2 - 2t + 41 &= 49 \\ t^2 - 2t - 8 &= 0 \\ (t+2)(t-4) &= 0 \\ t &= -2 \text{ or } 4 \end{aligned}$$

Example 20

The point A has coordinates $(1, 2, 1)$, the point B has coordinates $(-1, -1, 3)$ and the point C has coordinates $(2, 3, -1)$. Calculate, to $0.1°$, the size of the angle ABC.

$$\begin{aligned} AB^2 &= (-1-1)^2 + (-1-2)^2 + (3-1)^2 \\ &= [(-2)^2 + (-3)^2 + (2)^2] \\ &= 4 + 9 + 4 \\ AB^2 &= 17 \end{aligned}$$

$$\begin{aligned} BC^2 &= (2+1)^2 + (3+1)^2 + (-1-3)^2 \\ &= 3^2 + 4^2 + (-4)^2 \\ &= 9 + 16 + 16 \\ BC^2 &= 41 \end{aligned}$$

$$\begin{aligned} AC^2 &= (2-1)^2 + (3-2)^2 + (-1-1)^2 \\ &= 1^2 + 1^2 + (-2)^2 \\ &= 1 + 1 + 4 \\ AC^2 &= 6 \end{aligned}$$

Using the cosine rule:

$$\begin{aligned} \cos \angle ABC &= \frac{41 + 17 - 6}{2 \times \sqrt{41} \times \sqrt{17}} \\ &= \frac{52}{52.801} \\ \cos \angle ABC &= 0.9848 \end{aligned}$$

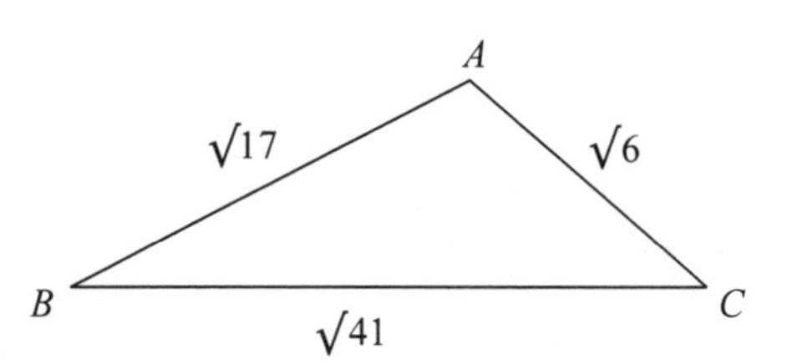

So:

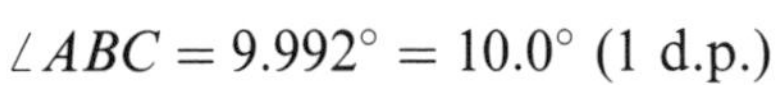

$$\angle ABC = 9.992° = 10.0° \text{ (1 d.p.)}$$

Exercise 7E

1. Find the distance between $A(1, 2, 5)$ and $B(2, 3, 4)$.
2. Find the distance between $A(3, 1, 5)$ and $B(1, 7, 3)$.
3. Find the distance between $A(3, -2, -4)$ and $B(6, 1, 0)$.
4. Find the distance between $A(-1, -3, 5)$ and $B(2, -5, 4)$.
5. Find the distance between $A(-3, 2, -6)$ and $B(-4, -3, 7)$.
6. The point A has coordinates $(2, 1, 7)$ and the point B has coordinates $(3, t, 4)$. The distance AB is 7. Calculate the two possible values of t.

7 In the isosceles triangle ABC, the point A has coordinates $(5, 2, -1)$, the point B has coordinates $(1, 3, -t)$ and the point C has coordinates $(2, 5, 3)$. The sides AB and BC of the triangle are equal in length. Calculate the value of t.

8 In the triangle ABC, A is the point with coordinates $(1, 2, -1)$, B is the point with coordinates $(2, -4, 1)$ and C is the point with coordinates $(1, 3, -2)$. Calculate the size of $\angle ABC$.

9 The point A has coordinates $(3, -2t, -5)$ and B has coordinates $(-4, 2, -7)$. The distance AB is 9. Calculate the two possible values of t.

10 The point A has coordinates $(1, -7, 2)$ and the point B has coordinates $(1, 2t, 4)$. The distance AB is 5. Calculate the two possible values of t.

11 In $\triangle ABC$, A is the point with coordinates $(2, -3, 1)$, B is the point with coordinates $(-5, 4, 3)$ and C is the point with coordinates $(1, -2, -3)$. Calculate the size of $\angle BAC$.

12 The point A has coordinates $(1, 1, 3)$ and the point B has coordinates $(1, 5, -2)$. Show that $\triangle OAB$ is right-angled, where O is the origin.

13 The point A has coordinates $(5, 3, -3)$ and the point B has coordinates $(-t, t, 3t)$.

(a) Find an expression for AB^2.

(b) Find the value of t that makes AB^2 a minimum.

(c) Calculate the minimum value of AB^2.

14 The point A has coordinates $(3, 2t, 1)$ and the point B has coordinates $(-2t, 5, t)$.

(a) Find an expression for AB^2.

(b) Find the value of t that makes AB^2 a minimum.

(c) Calculate the minimum value of AB^2.

15 The point A has coordinates $(3 + t, 2, 6 - t)$ and the point B has coordinates $(5, 1 + t, -3)$.

(a) Find an expression for AB^2.

(b) Find the value of t that makes AB^2 a minimum.

(c) Calculate the minimum value of AB^2.

16 The point A has coordinates $(3, 0, 0)$, the point B has coordinates $(0, 3, 0)$ and the point C has coordinates $(0, 0, 7)$. Find, to 0.1°, the size of the angle between the planes OAB and ABC, where O is the origin.

17 The point A has coordinates $(5, 0, 0)$, the point B has coordinates $(0, 7, 0)$ and the point C has coordinates $(0, 0, 5)$. Find, to $0.1°$, the size of the angle between the planes OAC and ABC, where O is the origin.

18

F (2,11,5) G (6,11,5)
E (2,7,5) H
B C
A (2,7,1) D

In the cube $ABCDEFGH$ the point A has coordinates $(2, 7, 1)$, the point E has coordinates $(2, 7, 5)$, the point F has coordinates $(2, 11, 5)$ and the point G has coordinates $(6, 11, 5)$.

(a) Write down the coordinates of the points B, C, D and H.

(b) Calculate, to 3 significant figures, the length AG.

(c) Calculate, to $0.1°$, the size of $\angle GAC$.

19

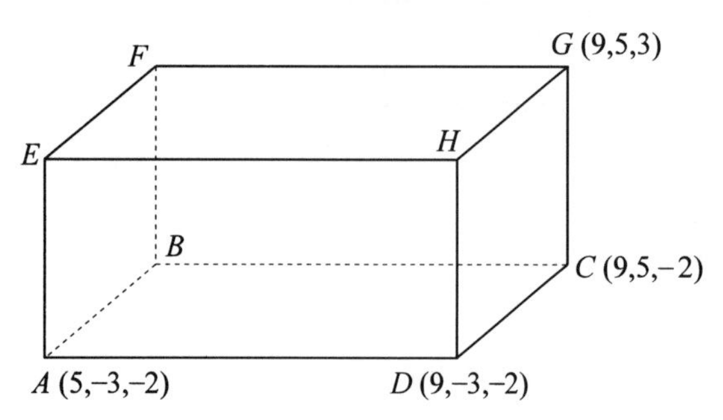

In the cuboid $ABCDEFGH$ the point A has coordinates $(5, -3, -2)$, the point C has coordinates $(9, 5, -2)$, the point D has coordinates $(9, -3, -2)$ and the point G has coordinates $(9, 5, 3)$.

(a) Write down the coordinates of the points B, E, F and H.

X is the mid-point of GH.

(b) Calculate the size of the angle between the planes ABX and $ABCD$.

20 In the triangle ABC, A has coordinates $(2, 1, -3)$, B has coordinates $(1, 0, 4)$ and C has coordinates $(-3, 2, 1)$. Calculate the size of $\angle ACB$.

SUMMARY OF KEY POINTS

1 In any $\triangle ABC$, the sine rule states that

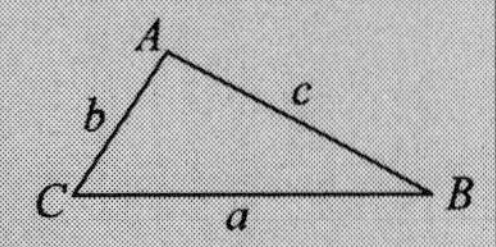

$$\frac{a}{\sin A} = \frac{b}{\sin B} = \frac{c}{\sin C}$$

2 In any $\triangle ABC$ the cosine rule states that

$$a^2 = b^2 + c^2 - 2bc \cos A$$

and that

$$\cos A = \frac{b^2 + c^2 - a^2}{2bc}$$

3 The area of $\triangle ABC$ is given by $\frac{1}{2}ab \sin C$.

4 The angle between a line and a plane is defined to be the angle between the line and the projection of the line on the plane.

4 The angle between two planes is defined to be the angle between the perpendicular to their common line drawn in the first plane and the perpendicular to their common line from the same point drawn in the second plane.

5 The distance between the points (x_1, y_1, z_1) and (x_2, y_2, z_2) is

$$\sqrt{[(x_2 - x_1)^2 + (y_2 - y_1)^2 + (z_2 - z_1)^2]}$$

Differentiation

8

8.1 Composite functions

In chapter 3 of Book P1 you learned how to form and to evaluate composite functions.

Example 1
Given that $f(x) \equiv x^2$ and $g(x) \equiv 3x - 1$, then the composite function fg is given by

$$fg(x) \equiv f[3x - 1] \equiv (3x - 1)^2$$

The composite function gf is given by

$$gf(x) \equiv g[x^2] \equiv 3x^2 - 1$$

You have now reached the stage where you will need to differentiate composite functions. In chapter 8 of Book P1, you learned these important results:

y or $f(x)$	$\frac{dy}{dx}$ or $f'(x)$
x^n (n rational)	nx^{n-1}
e^x	e^x
$\ln x$	$\frac{1}{x}$
$u \pm v$	$\frac{du}{dx} \pm \frac{dv}{dx}$

Consider how you could differentiate $(3x - 1)^2$ using your knowledge from Book P1.

First write $\quad y = (3x - 1)^2$

Expand to give $\quad y = 9x^2 - 6x + 1$

Differentiate: $\quad \frac{dy}{dx} = 18x - 6 = 6(3x - 1)$

This approach is laborious, and it gets worse! Consider, for example, differentiating $(3x - 1)^6$ in this way. Especially when the actual answer, $18(3x - 1)^5$, can be obtained so simply by the method you are about to learn.

8.2 Differentiating composite functions using the chain rule

Suppose that y is a function of t and that t is a function of x. The idea of taking an increment (small change) was introduced in Book P1. Now suppose that a small change δx in the variable x gives rise to small changes δy and δt in the variables y and t respectively.

Also $$\frac{\delta y}{\delta x} = \frac{\delta y}{\delta t} \cdot \frac{\delta t}{\delta x}$$

As δy, δt and $\delta x \to 0$ we may assume that

$$\frac{\delta y}{\delta x} \to \frac{dy}{dx}, \frac{\delta y}{\delta t} \to \frac{dy}{dt} \text{ and } \frac{\delta t}{\delta x} \to \frac{dt}{dx}$$

This leads to the important result

■ $$\frac{\mathbf{d}y}{\mathbf{d}x} = \frac{\mathbf{d}y}{\mathbf{d}t} \cdot \frac{\mathbf{d}t}{\mathbf{d}x}$$

This is called the **chain rule**. You will find that almost every differentiation you undertake will involve some use of the chain rule. You need to know this result, but you will not be expected to prove it.

■ **Knowing how to use the chain rule is of prime importance.**

Example 2

Differentiate (a) $(3x - 1)^2$ (b) $(3x - 1)^6$ with respect to x.

(a) Write $y = (3x - 1)^2$ and $t = 3x - 1$

Then: $$y = t^2 \text{ and } t = 3x - 1$$

Differentiating gives:

$$\frac{dy}{dt} = 2t \quad \text{and} \quad \frac{dt}{dx} = 3$$

Using the chain rule gives

$$\frac{dy}{dx} = \frac{dy}{dt} \cdot \frac{dt}{dx} = 2t \times 3 = 6t$$

But $t = 3x - 1$, so

$$\frac{dy}{dx} = 6(3x - 1)$$

as we found earlier on page 208.

(b) Write $y = (3x - 1)^6$ and $t = 3x - 1$

Then: $y = t^6$ and $t = 3x - 1$

Differentiating: $\frac{dy}{dt} = 6t^5$ and $\frac{dt}{dx} = 3$

Using the chain rule:

$$\frac{dy}{dx} = \frac{dy}{dt} \cdot \frac{dt}{dx} = 6t^5 \times 3 = 18t^5$$

But $t = 3x - 1$; so

$$\frac{dy}{dx} = 18(3x - 1)^5$$

Example 3

Find $\frac{dy}{dx}$, where $y = \ln(x^2 - 3x + 5)$.

Write $t = x^2 - 3x + 5$; then:

$$y = \ln t \quad \text{and} \quad t = x^2 - 3x + 5$$

Differentiating: $\frac{dy}{dt} = \frac{1}{t}$ and $\frac{dt}{dx} = 2x - 3$

Using the chain rule:

$$\frac{dy}{dx} = \frac{dy}{dt} \cdot \frac{dt}{dx} = \frac{1}{t} \times (2x - 3)$$

But $t = x^2 - 3x + 5$; so:

$$\frac{dy}{dx} = \frac{2x - 3}{x^2 - 3x + 5}$$

Example 4

Find $\frac{dy}{dx}$, where $y = e^{\sqrt{x}}$.

Write $t = \sqrt{x} = x^{\frac{1}{2}}$; then:

$$y = e^t$$

Differentiating: $\frac{dt}{dx} = \frac{1}{2}x^{-\frac{1}{2}} = \frac{1}{2\sqrt{x}}$

and $$\frac{dy}{dt} = e^t$$

Using the chain rule:

$$\frac{dy}{dx} = \frac{dy}{dt} \cdot \frac{dt}{dx} = e^t \times \frac{1}{2\sqrt{x}}$$

But $t = \sqrt{x}$; so:

$$\frac{dy}{dx} = \frac{1}{2\sqrt{x}} e^{\sqrt{x}}$$

Example 5

Find the gradient of the curve $y = \frac{1}{(3x+2)^2}$ at the point where $x = \frac{1}{3}$.

Write $t = 3x + 2$ and then $y = t^{-2}$.

Differentiating: $$\frac{dt}{dx} = 3 \text{ and } \frac{dy}{dt} = -2t^{-3}$$

Using the chain rule:

$$\frac{dy}{dx} = \frac{dy}{dt} \cdot \frac{dt}{dx} = -6t^{-3}$$

But $t = 3x + 2$; so:

$$\frac{dy}{dx} = -6(3x+2)^{-3} = \frac{-6}{(3x+2)^3}$$

The gradient of the curve $y = \frac{1}{(3x+2)^2}$ at the point where $x = \frac{1}{3}$ is found by putting $x = \frac{1}{3}$ in the expression for $\frac{dy}{dx}$.

That is, at $x = \frac{1}{3}$,

$$\frac{dy}{dx} = \frac{-6}{(1+2)^3} = \frac{-6}{27} = -\frac{2}{9}$$

Exercise 8A

Differentiate with respect to x:

1 $(x+2)^2$ **2** $(x-5)^7$ **3** $(3-x)^4$ **4** $(2x-1)^3$

5 $(3x-7)^4$ **6** $(x-4)^{-2}$ **7** $(3-x)^{-3}$ **8** $\frac{1}{x-7}$

9 $\dfrac{3}{5-x}$ **10** $(6x^2+1)^4$ **11** $\ln(x-2)$ **12** e^{x^2}

13 e^{3-x} **14** $\ln(6-x^2)$ **15** $\dfrac{1}{x^2-7x+2}$ **16** $\dfrac{-6}{3-4x^3}$

17 $\sqrt{(5+4x)}$ **18** $(1-x^{\frac{1}{2}})^4$ **19** $(x^{\frac{2}{3}}+5)^{-3}$ **20** e^{x^2-3x}

21 $\left(1+\dfrac{1}{x}\right)^4$ **22** $\left(x^2-\dfrac{1}{x}\right)^{-3}$ **23** $(1-x^2)^{\frac{1}{2}}$ **24** $\dfrac{1}{(1+x^2)^2}$

25 $\ln(3x^2-4x+3)$ **26** $\ln(x^{\frac{1}{2}}+x^{-\frac{1}{2}})$ **27** $\ln(e^x+5)$ **28** $e^{\ln x}$

29 $\dfrac{1}{(6x^3-5)^3}$ **30** $\dfrac{1}{\ln x}$

31 Given that $y=\dfrac{1}{(1+\sqrt{x})^3}$, find the value of $\dfrac{dy}{dx}$ at $x=4$.

32 Given that $y=\ln(x^3+4x)$, find the value of $\dfrac{dy}{dx}$ at $x=1$.

33 Given that $y=e^{x^2-x}$, show that the value of x if $\dfrac{dy}{dx}=1$ is 1 and $\dfrac{dy}{dx}=-1$, if $x=0$.

34 Given that $y=(7x^2-1)^{\frac{1}{3}}$, find the value of $\dfrac{dy}{dx}$ at $x=2$.

35 Find $\dfrac{d}{dx}(2x^3-5x)^{-4}$ at $x=-1$.

8.3 Differentiating products

You know already how to differentiate sums and differences of two functions, u and v, of x using the formula

$$\frac{d}{dx}(u \pm v) = \frac{du}{dx} \pm \frac{dv}{dx}$$

Now you will learn how to differentiate uv, where u and v are functions of x.

Consider $y = uv$, where u and v are functions of x.

Suppose that you make a small change, δx, in x and this in turn gives rise to small changes δy, δu and δv in y, u and v respectively.

Then:
$$\begin{aligned} y+\delta y &= (u+\delta u)(v+\delta v) \\ &= uv + u\delta v + v\delta u + \delta u\delta v \end{aligned}$$

or: $$\delta y = uv + u\delta v + v\delta u + \delta u\delta v - y$$

But $y = uv$, so

$$\delta y = u\delta v + v\delta u + \delta u\delta v$$

So: $$\frac{\delta y}{\delta x} = u\frac{\delta v}{\delta x} + v\frac{\delta u}{\delta x} + \frac{\delta u}{\delta x}\delta v$$

As $\delta x \to 0$, $\frac{\delta u}{\delta x}\delta v = \frac{\delta u}{\delta x} \cdot \frac{\delta v}{\delta x} \cdot \delta x \to 0$

and $\frac{\delta y}{\delta x} \to \frac{dy}{dx}$, $\frac{\delta u}{\delta x} \to \frac{du}{dx}$ and $\frac{\delta v}{\delta x} \to \frac{dv}{dx}$

So:

■ $$\mathbf{\frac{dy}{dx} = v\frac{du}{dx} + u\frac{dv}{dx}}$$

This is known as the **product rule**. You should learn it because you will often need to use it. You are not expected to know how to prove this formula in examinations.

Example 6

Given that $y = 3x^3(2x - 5)^4$, find $\frac{dy}{dx}$.

Put $u = 3x^3$; so $\frac{du}{dx} = 9x^2$

Put $v = (2x - 5)^4$; then by the chain rule:

$$\frac{dv}{dx} = 4(2x-5)^3(2) = 8(2x-5)^3$$

$$\begin{aligned}\frac{dy}{dx} &= v\frac{du}{dx} + u\frac{dv}{dx}\\ &= 9x^2(2x-5)^4 + 3x^3 \cdot 8(2x-5)^3\\ &= 3x^2(2x-5)^3[3(2x-5) + 8x]\\ &= 3x^2(2x-5)^3[6x - 15 + 8x]\\ &= 3x^2(2x-5)^3(14x - 15)\end{aligned}$$

Notice that you will usually need to 'tidy up' your answer after applying the product formula.

Example 7

Given that $y = e^{-2x}\sqrt{(x^2+1)}$, find $\frac{dy}{dx}$.

Write $u = e^{-2x}$; then $\frac{du}{dx} = -2e^{-2x}$

Put $v = \sqrt{(x^2+1)} = (x^2+1)^{\frac{1}{2}}$

By the chain rule:
$$\frac{dv}{dx} = \tfrac{1}{2}(x^2+1)^{-\frac{1}{2}}(2x) = \frac{x}{\sqrt{(x^2+1)}}$$

$$\frac{dy}{dx} = v\frac{du}{dx} + u\frac{dv}{dx} = (-2e^{-2x})\sqrt{(x^2+1)} + e^{-2x}\cdot\frac{x}{\sqrt{(x^2+1)}}$$

Take out the common factor $\frac{e^{-2x}}{\sqrt{(x^2+1)}}$ to give

$$\frac{dy}{dx} = \frac{e^{-2x}}{\sqrt{(x^2+1)}}[x - 2\sqrt{(x^2+1)}\sqrt{(x^2+1)}] = \frac{e^{-2x}}{\sqrt{(x^2+1)}}(x - 2x^2 - 2)$$

8.4 Differentiating quotients

You can write the quotient $\frac{u}{v}$, where u and v are functions of x, as:

$$y = \frac{u}{v} = uv^{-1}$$

By the product rule then,

$$\frac{dy}{dx} = v^{-1}\frac{du}{dx} + u\frac{d}{dx}(v^{-1})$$

As v is a function of x, you can write $t = v^{-1}$.

Then:
$$\frac{dt}{dv} = -v^{-2}$$

By the chain rule,

$$\frac{dt}{dx} = \frac{dt}{dv}\cdot\frac{dv}{dx} = -v^{-2}\frac{dv}{dx}$$

So:
$$\frac{dy}{dx} = v^{-1}\frac{du}{dx} - uv^{-2}\frac{dv}{dx} = \frac{1}{v}\frac{du}{dx} - \frac{u}{v^2}\frac{dv}{dx}$$

■
$$\mathbf{\frac{dy}{dx} = \frac{v\frac{du}{dx} - u\frac{dv}{dx}}{v^2}}$$

This result is known as the **quotient rule**. Here are some examples of how to use it.

Example 8

Given that $y = \frac{x^2 - 1}{x^2 + 1}$, find $\frac{dy}{dx}$.

Put $u = x^2 - 1$, then: $\frac{du}{dx} = 2x$

Put $v = x^2 + 1$, then: $\frac{dv}{dx} = 2x$

Apply the quotient formula:

$$\frac{dy}{dx} = \frac{v\frac{du}{dx} - u\frac{dv}{dx}}{v^2}$$

$$\begin{aligned}\frac{dy}{dx} &= \frac{2x(x^2 + 1) - 2x(x^2 - 1)}{(x^2 + 1)^2} \\ &= \frac{2x^3 + 2x - 2x^3 + 2x}{(x^2 + 1)^2}\end{aligned}$$

So: $$\frac{dy}{dx} = \frac{4x}{(x^2 + 1)^2}$$

Alternatively, you could write $y = (x^2 - 1)(x^2 + 1)^{-1}$ and use the product rule. Then:

$$u = x^2 - 1 \qquad \frac{du}{dx} = 2x$$

$v = (x^2 + 1)^{-1}$ and, by the chain rule, $\frac{dv}{dx} = -2x(x^2 + 1)^{-2}$

$$\begin{aligned}\frac{dy}{dx} &= v\frac{du}{dx} + u\frac{dv}{dx}\\ &= 2x(x^2+1)^{-1} + (x^2-1)[-2x(x^2+1)^{-2}]\\ &= \frac{2x}{x^2+1} - \frac{2x(x^2-1)}{(x^2+1)^2}\\ &= \frac{2x(x^2+1) - 2x(x^2-1)}{(x^2+1)^2}\\ &= \frac{2x^3+2x-2x^3+2x}{(x^2+1)^2}\\ &= \frac{4x}{(x^2+1)^2}\end{aligned}$$

As you can see, the quotient rule formula reduces the amount of algebraic processing you have to do and this is why the first method is recommended.

Example 9

Given that $y = \frac{\sqrt{(3-4x)}}{\ln(2x-3)}$, find $\frac{dy}{dx}$.

Put $u = \sqrt{(3-4x)} = (3-4x)^{\frac{1}{2}}$; then: $\frac{du}{dx} = \frac{1}{2}(3-4x)^{-\frac{1}{2}}(-4)$
$= -2(3-4x)^{-\frac{1}{2}}$

Put $v = \ln(2x-3)$; then: $\frac{dv}{dx} = \frac{2}{2x-3}$

using the chain rule in each case.

So:

$$\begin{aligned}\frac{dy}{dx} &= \frac{v\frac{du}{dx} - u\frac{dv}{dx}}{v^2}\\ &= \frac{-2(3-4x)^{-\frac{1}{2}}\ln(2x-3) - \frac{2(3-4x)^{\frac{1}{2}}}{2x-3}}{[\ln(2x-3)]^2}\\ &= \frac{\frac{-2\ln(2x-3)}{\sqrt{(3-4x)}} - \frac{2\sqrt{(3-4x)}}{2x-3}}{[\ln(2x-3)]^2}\\ &= \frac{-2(2x-3)\ln(2x-3) - 2(3-4x)}{(2x-3)\sqrt{(3-4x)}[\ln(2x-3)]^2}\end{aligned}$$

Exercise 8B

Differentiate each of the following with respect to x:

1 $x^3(x-1)^2$ **2** $x^4(2x+1)^3$ **3** x^2e^x

4 $(2x-1)\ln x$ **5** $e^{2x}\ln x$ **6** $(x+1)\ln(3x-1)$

7 $x(2x-1)^5$ **8** $(2x-3)^3(x^2+1)^2$ **9** $x(x^2+1)^4$

10 $(2x-3)\sqrt{(x^2-1)}$ **11** $(x-1)^2\sqrt{x}$

12 $(x-2)^{\frac{2}{3}}(x+2)^{\frac{3}{4}}$ **13** $(2x^2-1)\sqrt{(4x-3)}$

14 $x^{\frac{1}{2}}\ln(x^2-1)$ **15** $(x-1)^3e^{x^2-2x}$ **16** $\dfrac{x}{x-1}$

17 $\dfrac{x^2}{1-x}$ **18** $\dfrac{x^3}{2-x^2}$ **19** $\dfrac{e^x}{2x+1}$

20 $\dfrac{\ln(1-x^2)}{x^3}$ **21** $\dfrac{x^3-1}{x^3+1}$ **22** $\dfrac{2x}{\sqrt{(x-1)}}$

23 $\dfrac{2x-1}{x^2-1}$ **24** $\dfrac{\sqrt{x}}{x^4+1}$ **25** $\dfrac{3e^x-1}{3e^x+1}$

26 $\dfrac{e^{x^2}}{\ln(2x+1)}$ **27** $\dfrac{\ln(x^2+1)}{x^2-1}$ **28** $\dfrac{x^2(x-\frac{1}{x})}{x+\frac{1}{x}}$

29 $\dfrac{\sqrt{(4x^3-1)}}{\ln(2x-3)}$ **30** $\dfrac{\ln(x^2+1)}{e^x+1}$

31 Given that $y=\dfrac{x}{x^2+1}$, find the values of x for which $\dfrac{dy}{dx}=0$.

32 Given that $y=(1+x^2)^{\frac{1}{2}}$, find an expression for $\dfrac{dy}{dx}$. Hence find the value of $\dfrac{dy}{dx}$ when $x=0$ and when $x=\frac{3}{4}$.

33 Given that $y=x^3(x-3)^2$, find the values of x and y when $\dfrac{dy}{dx}=0$.

34 Given that $y=\dfrac{x^{\frac{1}{2}}-1}{x^{\frac{1}{2}}+1}$, find the value of $\dfrac{dy}{dx}$ when $x=9$.

35 Given that $f(x)=x\ln(x^3-4)$, find the value of $f'(2)$.

8.5 Related rates of change

You learned in Book P1 that the derivative is used to measure rates of change. By using the chain rule you can find rates of change that are related to other rates of change, as the following examples illustrate. Sometimes you may also need to use the product or quotient rule.

Example 10

The radius of a spherical balloon is increasing at the rate of $0.2\,\text{m}\,\text{s}^{-1}$. Find the rate of increase of (a) the volume (b) the surface area of the balloon at the instant when the radius is 1.6 m.

(a) The volume $V\,\text{m}^3$ of a sphere, radius r m, is given by the formula $V = \frac{4}{3}\pi r^3$.

We know that $\dfrac{dr}{dt} = 0.2$ and also, by differentiating, that:

$$\frac{dV}{dr} = 4\pi r^2$$

Use the chain rule:

$$\frac{dV}{dt} = \frac{dV}{dr} \cdot \frac{dr}{dt}$$

When $r = 1.6$, the rate of change of V with respect to t is

$$\begin{aligned}\frac{dV}{dt} &= 4\pi(1.6)^2(0.2)\ \text{m}^3\,\text{s}^{-1} \\ &= 6.43\ \text{m}^3\,\text{s}^{-1}\end{aligned}$$

(b) The surface area $A\,\text{m}^2$ of a sphere, radius r m, is given by the formula $A = 4\pi r^2$.

We know that $\dfrac{dr}{dt} = 0.2$ and also, by differentiating, that:

$$\frac{dA}{dr} = 8\pi r$$

Use the chain rule:

$$\frac{dA}{dt} = \frac{dA}{dr} \cdot \frac{dr}{dt}$$

When $r = 1.6$, the rate of change of A with respect to t is

$$\begin{aligned}\frac{dA}{dt} &= 8\pi(1.6)(0.2)\ \text{m}^2\,\text{s}^{-1} \\ &= 8.04\ \text{m}^2\,\text{s}^{-1}\end{aligned}$$

Example 11

Given that $P = x(x^2+4)^{\frac{1}{2}}$, find $\frac{dP}{dt}$ when $x = 2$ and $\frac{dx}{dt} = 3$.

First you need to find $\frac{dP}{dx}$ using the product rule, so write:

$$u = x \quad \text{and} \quad v = (x^2+4)^{\frac{1}{2}}$$

Then:

$$\frac{du}{dx} = 1 \quad \text{and} \quad \frac{dv}{dx} = \tfrac{1}{2}(x^2+4)^{-\frac{1}{2}}(2x)$$
$$= x(x^2+4)^{-\frac{1}{2}}$$

$$\frac{dP}{dx} = v\frac{du}{dx} + u\frac{dv}{dx}$$
$$= (x^2+4)^{\frac{1}{2}} + x^2(x^2+4)^{-\frac{1}{2}}$$

At $x = 2$,
$$\frac{dP}{dx} = (2^2+4)^{\frac{1}{2}} + 2^2(2^2+4)^{-\frac{1}{2}}$$
$$= 8^{\frac{1}{2}} + \frac{4}{8^{\frac{1}{2}}}$$
$$= \frac{8+4}{8^{\frac{1}{2}}} = \frac{12}{2 \times 2^{\frac{1}{2}}} = \frac{6}{2^{\frac{1}{2}}} \times \frac{2^{\frac{1}{2}}}{2^{\frac{1}{2}}} = 3 \times 2^{\frac{1}{2}}$$

$$\frac{dP}{dt} = \frac{dP}{dx} \cdot \frac{dx}{dt} \quad \text{(chain rule)}$$
$$= 3 \times 2^{\frac{1}{2}} \times 3$$
$$= 9 \times 2^{\frac{1}{2}} \text{ or } 9\sqrt{2}$$

8.6 Second derivatives

For the function f, given by $y = f(x)$, the first derivative is $f'(x)$ or $\frac{dy}{dx}$ and we write

$$\frac{dy}{dx} = f'(x)$$

The **second derivative** of y with respect to x is obtained by differentiating $f'(x)$ and is denoted by $f''(x)$; we write:

$$\frac{d}{dx}\left(\frac{dy}{dx}\right) = f''(x)$$

$\frac{d}{dx}\left(\frac{dy}{dx}\right)$ is written in shorthand form as $\frac{d^2y}{dx^2}$ and read as 'dee two y by dee x squared'. $\frac{d^2y}{dx^2}$ is called the **second derivative of y with respect to x.**

This notation can be extended; for example:

$$\frac{d}{dx}\left(\frac{d^2y}{dx^2}\right) = \frac{d^3y}{dx^3}$$

and this is called the **third derivative** of y with respect to x.

The nth derivative of y with respect to x is written as $\frac{d^ny}{dx^n}$ or as $f^{(n)}(x)$. You will do more work on higher derivatives in Books P3 and P4.

Example 12

Express $\frac{7x+1}{(1-2x)(1+x)}$ in partial fractions and hence find the values of $\frac{dy}{dx}$ and $\frac{d^2y}{dx^2}$, where $y = \frac{7x+1}{(1-2x)(1+x)}$, at $x = 0$.

Using the work from chapter 1, you can write

$$\frac{7x+1}{(1-2x)(1+x)} \equiv \frac{A}{1-2x} + \frac{B}{1+x}$$

So $7x + 1 \equiv A(1+x) + B(1-2x)$ is true for all values of x.

$x = -1$: $\quad -7 + 1 = 0 + B(1+2) \Rightarrow B = -2$

$x = \frac{1}{2}$: $\quad \frac{7}{2} + 1 = A(1 + \frac{1}{2}) + 0 \Rightarrow A = 3$

Hence:

$$y = \frac{7x+1}{(1-2x)(1+x)} \equiv \frac{3}{(1-2x)} - \frac{2}{1+x}$$
$$\equiv 3(1-2x)^{-1} - 2(1+x)^{-1}$$

Using the chain rule of differentiation:

$$\frac{dy}{dx} = 3(-2)(-1)(1-2x)^{-2} - 2(-1)(1+x)^{-2}$$
$$= \frac{6}{(1-2x)^2} + \frac{2}{(1+x)^2}$$

At $x = 0$: $$\frac{dy}{dx} = \frac{6}{1} + \frac{2}{1} = 8$$

$$\frac{d^2y}{dx^2} = \frac{d}{dx}\left(\frac{dy}{dx}\right) = \frac{d}{dx}\left[\frac{6}{(1-2x)^2} + \frac{2}{(1+x)^2}\right]$$

$$\frac{d^2y}{dx^2} = \frac{d}{dx}[6(1-2x)^{-2} + 2(1+x)^{-2}]$$

Using the chain rule:

$$\frac{d^2y}{dx^2} = 6(-2)(-2)(1-2x)^{-3} + 2(-2)(1+x)^{-3}$$
$$= \frac{24}{(1-2x)^3} - \frac{4}{(1+x)^3}$$

At $x = 0$: $$\frac{d^2y}{dx^2} = \frac{24}{1^3} - \frac{4}{1^3} = 20$$

8.7 Points of inflexion

For any point P on the curve $y = f(x)$ at which $\frac{d^2y}{dx^2} = 0$ **and** $\frac{d^3y}{dx^3} \neq 0$, the point P is called a **point of inflexion**. At points of inflexion, the value of $\frac{dy}{dx}$ is stationary: it is neither increasing nor decreasing.

Here are two illustrations of points of inflexion.

(i) The curve $y = x^3$ at the origin.

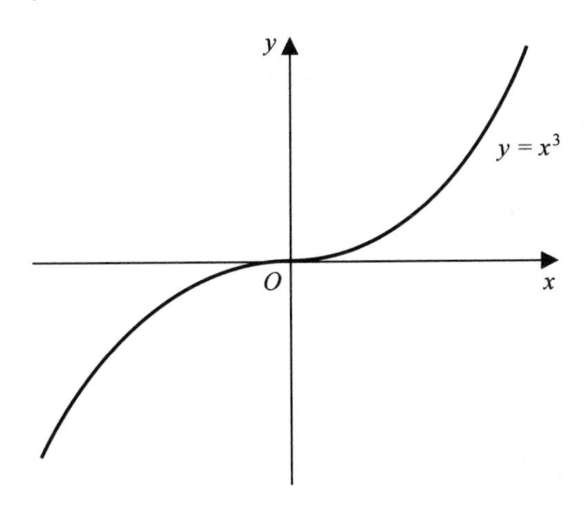

At O, $\frac{dy}{dx} = 0$ and $\frac{d^2y}{dx^2} = 0$

But $\frac{d^3y}{dx^3} = 6 \neq 0$, so O is a point of inflexion.

(ii) The curve $y = x^3 - x^2$ has $\frac{d^2y}{dx^2} = 6x - 2$, so $\frac{d^2y}{dx^2} = 0$ at $x = \frac{1}{3}$.

At $x = \frac{1}{3}$, $y = -\frac{2}{27}$

Also $\frac{d^3y}{dx^3} = 6 \neq 0$

At $P(\frac{1}{3}, -\frac{2}{27})$ there is a point of inflexion.

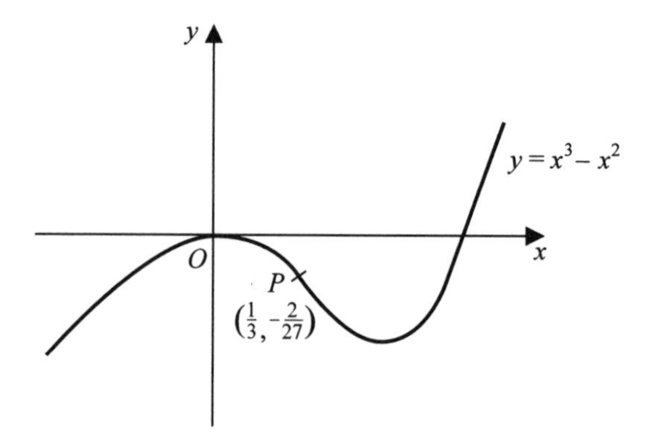

8.8 Finding the nature of stationary points

The second derivative is very useful when you want to find out what kind of stationary point you have identified on a curve: maximum, minimum or point of inflexion. Suppose we have $y = f(x)$ and $f'(x_1)$, $f'(x_2)$and $f'(x_3)$ are all zero. If the corresponding values of y are y_1, y_2 and y_3 respectively, then (x_1, y_1), (x_2, y_2) and (x_3, y_3) are stationary points of $y = f(x)$.

If $f''(x_1) < 0$, then y_1 is a maximum value of $f(x)$.

If $f''(x_2) > 0$, then y_2 is a minimum value of $f(x)$.

If $f''(x) = 0$ and $f'''(x) \neq 0$, then y_3 is an inflexion point.

Example 13

Find the turning points and the points of inflexion on the curve $y = x^2e^x$.

Using the product rule: $u = x^2 \qquad v = e^x$

$$\frac{du}{dx} = 2x \qquad \frac{dv}{dx} = e^x$$

$$\frac{dy}{dx} = v\frac{du}{dx} + u\frac{dv}{dx} = 2xe^x + x^2e^x$$
$$= xe^x(x+2)$$

For turning points $\frac{dy}{dx} = 0$.

$$xe^x(x+2) = 0 \Rightarrow x = 0 \quad \text{or} \quad x = -2$$

Remember that $e^x > 0$ for all x because the graph of $y = e^x$ lies above the x-axis. So the turning points are $O(0, 0)$ and $A(-2, 4e^{-2})$.

$$\frac{dy}{dx} = e^x(x^2 + 2x)$$

For the points of inflexion, you need to solve the equation $\frac{d^2y}{dx^2} = 0$.

Using the product rule again:

$$u = x^2 + 2x \qquad v = e^x$$

$$\frac{du}{dx} = 2x + 2 \qquad \frac{dv}{dx} = e^x$$

$$\frac{d^2y}{dx^2} = v\frac{du}{dx} + u\frac{dv}{dx} = (2x+2)e^x + (x^2+2x)e^x$$
$$= (x^2 + 4x + 2)e^x$$

$e^x > 0 \Rightarrow$ possible points of inflexion occur at

$$x^2 + 4x + 2 = 0$$

$$x = \frac{-4 \pm \sqrt{(16-8)}}{2} = -2 \pm \sqrt{2}$$

So there may be points of inflexion at the two points

$$x = -2 - \sqrt{2}, \quad y \approx 0.38$$

$$x = -2 + \sqrt{2}, \quad y \approx 0.19$$

$$\frac{d^3y}{dx^3} = \frac{d}{dx}[(x^2 + 4x + 2)e^x]$$
$$= e^x\frac{d}{dx}(x^2 + 4x + 2) + (x^2 + 4x + 2)\frac{d}{dx}(e^x)$$

(by the product rule)

So: $$\frac{d^3y}{dx^3} = e^x(2x+4) + (x^2+4x+2)e^x$$

$$= e^x(x^2+6x+6)$$

At $x = -2+\sqrt{2}$ and at $x = -2-\sqrt{2}$ you can show that $\frac{d^3y}{dx^3} \neq 0$ and therefore the points $(-2-\sqrt{2}, 0.38)$ and $(-2+\sqrt{2}, 0.19)$ are points of inflexion on the curve $y = x^2e^x$.

8.9 Parallel and perpendicular lines

In Book P1 you were shown how to find the gradient of a straight line and how to find the equation of a straight line in the forms $y = mx + c$ and $y - y_1 = m(x - x_1)$. The gradient of a line is defined as change in y divided by change in x over some specific interval.

Over the interval P to Q shown, the y increase is QR and the x increase is PR.

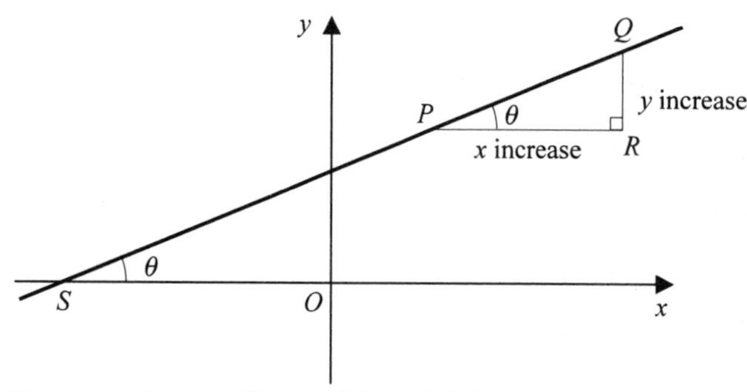

In the diagram, the gradient of line PQ is

$$\frac{y \text{ increase}}{x \text{ increase}} = \frac{QR}{PR} = \tan \angle QPR$$

Extend line QP until it meets the x-axis at the point S. Then $\angle PSO = \angle QPR = \theta$, say, because PR and the x-axis are parallel.

The gradient of the line PQ then is $\tan\theta$, where θ is the angle made by the extended line QP and the *positive* direction of the x-axis.

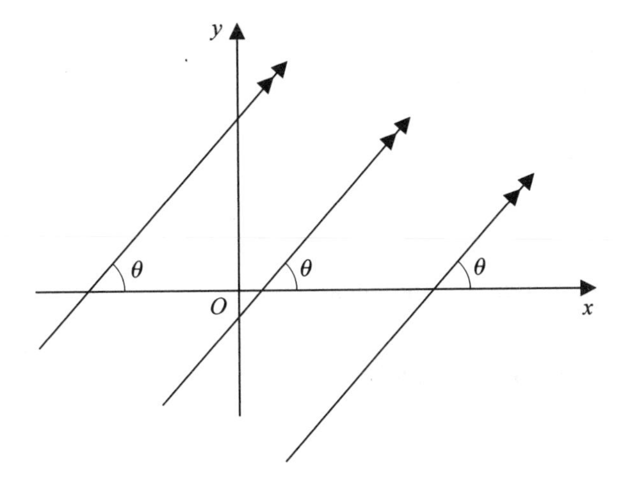

Lines that are parallel have the same gradient, $\tan\theta$, as shown.

■ **That is, two lines with gradients m_1 and m_2 are parallel if, and only if, $m_1 = m_2$.**

Consider now two lines, one of gradient m, which are perpendicular and intersect at the point A.

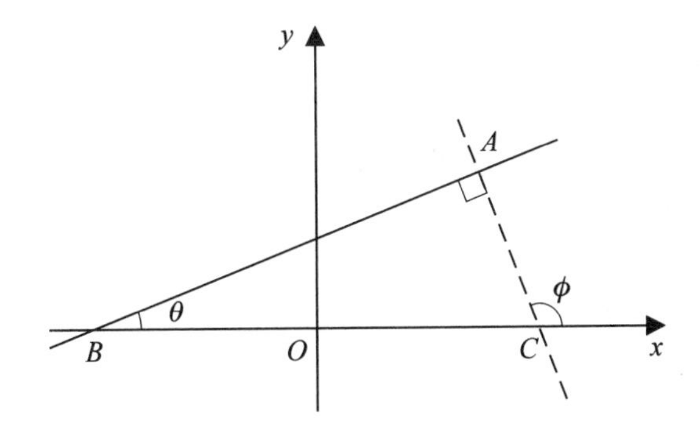

The line with gradient m meets the x-axis at B, $\angle ABC = \theta$ and the perpendicular line meets the x-axis at C. Now $\tan\theta = m$. By definition the gradient of AC is $\tan\phi$, where ϕ is the angle between AC and the positive direction of the x-axis. But by geometry, $\phi = 90° + \theta$ (angle sum of triangle is 180°).

So the gradient of AC is

$$\tan(90° + \theta) = \frac{\sin(90° + \theta)}{\cos(90° + \theta)}$$

$$= \frac{\sin 90° \cos\theta + \cos 90° \sin\theta}{\cos 90° \cos\theta - \sin 90° \sin\theta}$$

But $\sin 90° = 1$ and $\cos 90° = 0$.

So: Gradient of $AC = -\dfrac{\cos\theta}{\sin\theta}$

$$= -\frac{1}{\tan\theta}$$

The gradient of AC is $-\dfrac{1}{m}$.

Lines that are perpendicular have gradients which are the negative reciprocals of each other.

- **That is, two lines with gradients m_1 and m_2 are perpendicular if, and only if, $m_1 m_2 = -1$.**

Example 14

Find the equation of the line through the point $(3, -2)$ which is (a) parallel to (b) perpendicular to the line $y = 2x - 4$.

The line $y = 2x - 4$ has gradient 2 and any line parallel to it has gradient 2.

(a) We require a line through $(3, -2)$ with gradient 2. The equation is

$$y + 2 = 2(x - 3)$$

because a line of gradient m, passing through the point (x_1, y_1) has equation $y - y_1 = m(x - x_1)$.

(This was shown in Book P1, chapter 4.)

(b) Any line perpendicular to the line $y = 2x - 4$ has gradient $-\frac{1}{2}$ (since $(-\frac{1}{2}) \times 2 = -1$). The required equation is

$$y + 2 = -\tfrac{1}{2}(x - 3)$$

Example 15

Show that the lines with equations

$$x + 3y + 1 = 0$$
$$3x - y - 7 = 0$$

meet at right angles.

The gradient of $x + 3y + 1 = 0$ is found by rearranging the equation in the form

$$y = -\tfrac{1}{3}x - \tfrac{1}{3}$$

If you compare this with $y = mx + c$, you see that the gradient is $-\frac{1}{3}$. Similarly, $3x - y - 7 = 0$ can be rearranged as

$$y = 3x - 7$$

and the gradient is 3.

The lines have gradients $-\frac{1}{3}$ and 3. The product of these is

$$-\tfrac{1}{3} \times 3 = -1$$

So the lines are perpendicular.

8.10 Tangents and normals to curves

The gradient at any point on the curve $f(x)$ is

$$\frac{dy}{dx} = f'(x)$$

At $P(a, b)$ $$\frac{dy}{dx} = f'(a)$$

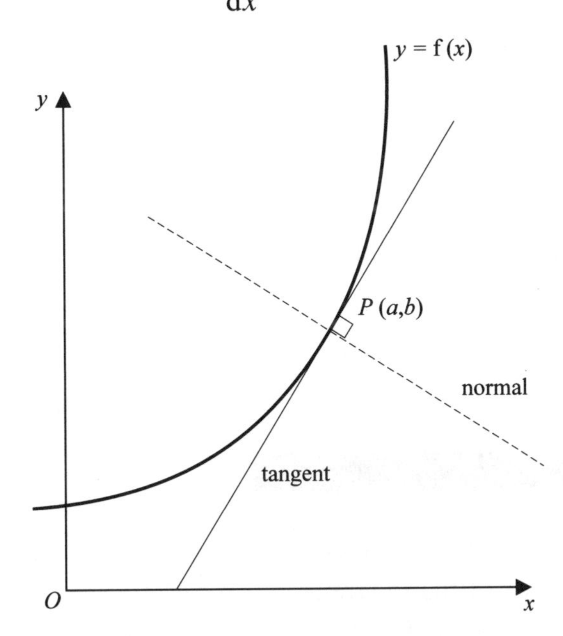

As you learned in Book P1, the gradient of the tangent to the curve at P is the same as the gradient of the curve at P.

- **The equation of the tangent at P to the curve is**

$$y - b = f'(a)(x - a)$$

By definition the normal to the curve at P is perpendicular to the tangent. The gradient of the normal is therefore $-\frac{1}{f'(a)}$, as the product of the gradients of lines at right angles is -1.

- **The equation of the normal is**

$$y - b = -\frac{1}{f'(a)}(x - a)$$

Example 16

Find the equations of the tangent and normal to the curve $y = x(x - 1)^{\frac{1}{2}}$ at the point $(5, 10)$.

First you need to find $\frac{dy}{dx}$ using the product rule:

$$u = x \qquad v = (x-1)^{\frac{1}{2}}$$

$$\frac{du}{dx} = 1 \qquad \frac{dv}{dx} = \tfrac{1}{2}(x-1)^{-\frac{1}{2}}$$

So:

$$\frac{dy}{dx} = v\frac{du}{dx} + u\frac{dv}{dx}$$

$$= (x-1)^{\frac{1}{2}} + \tfrac{x}{2}(x-1)^{-\frac{1}{2}}$$

At $(5, 10)$:

$$\frac{dy}{dx} = 2 + \tfrac{5}{2} \times \tfrac{1}{2} = \tfrac{13}{4}$$

The equation of the tangent at $(5, 10)$ to $y = x(x-1)^{\frac{1}{2}}$ is

$$y - 10 = \tfrac{13}{4}(x - 5)$$

that is,

$$13x - 4y - 25 = 0$$

The normal has gradient $-\frac{4}{13}$. The equation of the normal is

$$y - 10 = -\tfrac{4}{13}(x - 5)$$

that is,

$$4x + 13y - 150 = 0$$

Exercise 8C

1 The sides of a square are increasing at a constant rate of 0.1 cm s^{-1}. Find the rate at which the area of the square is increasing when each side of the square is of length 10 cm.

2 A circular oil-slick is increasing in area at a constant rate of $3\text{ m}^2\text{ s}^{-1}$. Find the rate of increase of the radius of the slick at the instant when the area is 1200 m^2.

3 You are told that $A = 14x^2$ and that x is increasing at 0.5 cm s^{-1}. Find the rate of change of A at the instant when $x = 8$ cm.

4 Each side of a contracting cube is decreasing at a rate of 0.06 cm s^{-1}. Find the rate of decrease of (a) the volume (b) the outer surface area of the cube when the sides of the cube are each 4 cm.

5 The volume of an expanding sphere is increasing at a rate of $24\,\text{cm}^3\,\text{s}^{-1}$. Find the rate of increase of (a) the radius (b) the surface area when the radius of the sphere is 20 cm.

6 Given that $y = (3t - 1)^2$ and $t = 4x^{\frac{1}{4}}$, find the value of $\frac{\mathrm{d}y}{\mathrm{d}x}$ when $x = 81$.

7 (a) Given that

$$y = \frac{\ln x}{x}, \ x > 0$$

find the value of x for which $\frac{\mathrm{d}y}{\mathrm{d}x} = 0$.

(b)

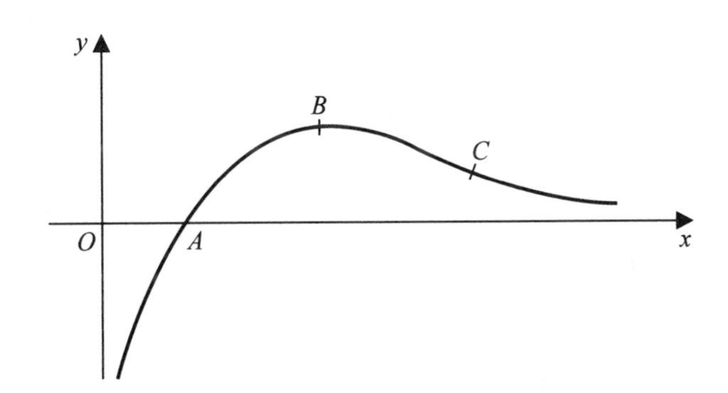

The diagram shows the curve $y = \frac{\ln x}{x}$. The curve crosses the x-axis at A; at B, $\frac{\mathrm{d}y}{\mathrm{d}x} = 0$; at C, $\frac{\mathrm{d}^2y}{\mathrm{d}x^2} = 0$. Determine the coordinates of A, B and C.

8 Show that the curve $y = \frac{\mathrm{e}^{-x}}{x}$ has a maximum value of $-\mathrm{e}$ at $x = -1$.

9 Given that

$$y = \frac{2 - 8x}{x^2 + 2x}$$

find the values of x and of y when $\frac{\mathrm{d}y}{\mathrm{d}x} = 0$. Investigate the nature of these stationary values of y.

10 When the depth of liquid in a container is x cm, the volume of liquid is $x(x^2 + 25)$ cm^3. Liquid is added to the container at a constant rate of 2 cm^3 s^{-1}. Find the rate of change of the depth of liquid at the instant when $x = 11$.

11 Express $\frac{2x}{x^2 - 4}$ in partial fractions. Given that $f(x) \equiv \frac{2x}{x^2 - 4}$,

find $f'(4)$ and $f''(0)$.

12 Given that

$$y = \frac{1}{x^2 + 4}$$

find the value of $\frac{d^2y}{dx^2}$ at $x = 1$.

13 Given that

$$y = \frac{e^x}{x + 1}$$

find, in terms of e, the value of $\frac{dy}{dx}$ and $\frac{d^2y}{dx^2}$ at $x = 1$.

14 Find the equation of the straight line:

(a) parallel to the line $y = 4x - 5$, passing through $(2, 3)$

(b) parallel to the line $y = 4 - 6x$, passing through $(-1, 3)$

(c) parallel to the line $2y + 3x = 7$, passing through $(2, -5)$.

15 Find the equation of the straight line:

(a) perpendicular to the line $y = 3x + 5$, passing through $(1, 7)$

(b) perpendicular to the line $y = 2 - 5x$, passing through $(3, -5)$

(c) perpendicular to the line $3y + 2x = 7$, passing through $(1, -\frac{1}{2})$.

16 (a) Show that $\triangle ABC$, where A is $(0, 2)$, B is $(8, 6)$ and C is $(2, 8)$, contains a right angle.

(b) $\triangle DEF$, where D is $(-2, 0)$, E is $(\frac{1}{2}, y)$ and F is $(3\frac{1}{2}, -3\frac{1}{2})$ is right-angled at E. Determine the value of y.

(c) Quadrilateral $ABCD$, where A is at $(4, 5)$ and C is at $(3, -2)$, is a square. Find the coordinates of B and D and the area of the square.

17 Find the equations of the tangent and the normal at the point (3, 4) on the curve

$$y = (x^2 + 7)^{\frac{1}{2}}$$

18

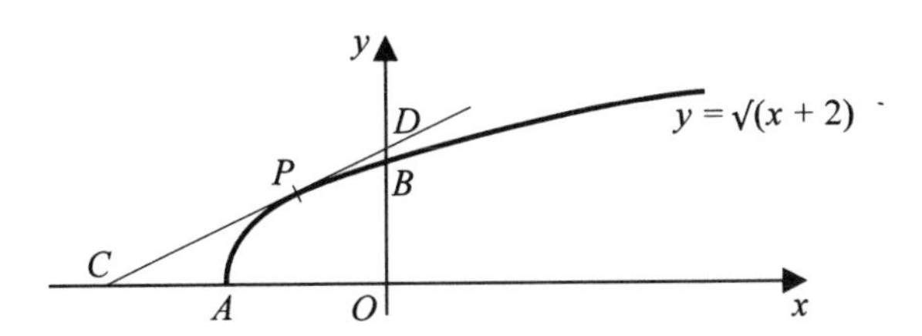

The curve $y = \sqrt{(x + 2)}$ meets the x-axis at A and the y-axis at B, as shown.

(a) Find the coordinates of A and B.

At the point P on the curve, the x-coordinate is -1. The tangent to the curve at P meets the x-axis at C and the y-axis at D.

(b) Find the equation of the tangent.

(c) Determine the length of CD.

The normal at P to the curve meets the y-axis at E.

(d) Determine the distance BE.

19 Show that the line $x + y = 0$ is the normal to the curve $y = xe^x$ at the origin O.

20 Find the equations of the tangent and normal to the curve $y = x \ln x$ at the point P, whose x-coordinate is e.
The tangent and normal meet the x-axis at Q and R respectively. Find the length of QR and the area of $\triangle PQR$.

21 Given that $y = 4e^{3x} - 5e^{-3x}$, show that

$$\frac{d^2y}{dx^2} = 9y$$

22 On the curve $y = (3x + 1)^{\frac{1}{2}}$, the points P and Q have x-coordinates of 1 and 8 respectively. Find the equations of the normals to the curve at P and Q.

23 Prove that the curve $y = 4x^5 + \lambda x^3$ has two turning points for $\lambda < 0$ and none for $\lambda \geqslant 0$. Given that $\lambda = -\frac{5}{3}$, find the coordinates of the turning points and distinguish between them.

24 The curve $y = e^x(px^2 + qx + r)$ is such that the tangents at $x = 1$ and $x = 3$ are parallel to the x-axis. The point $(0, 9)$ is on the curve. Find the values of p, q and r.

25 (a) Find the equation of the tangent and the normal at $(\frac{1}{2}, 1)$ on the curve $y = \dfrac{x}{1-x}$.

(b) Find the coordinates of the point where the normal meets the curve again.

26 For the curve $y = \dfrac{x}{(x+3)^2}$ find:

(a) the coordinates of the stationary point

(b) the coordinates of the point of inflexion.

27 For the curve $y = \dfrac{2x}{1+x^2}$,

(a) show that

$$\frac{dy}{dx} = \frac{2(1-x^2)}{(1+x^2)^2}$$

(b) find the coordinates of the stationary points and distinguish between them.

28 The curve $y = x^2 \ln x$ is defined for positive values of x.

(a) Determine the coordinates of the stationary point.

(b) Find the equation of the tangent at the point (e, e^2).

29 Given that

$$y = \frac{2x^2 - x + 1}{(x+1)(x^2+1)}$$

express y in terms of partial fractions.

Hence determine the value of $\dfrac{dy}{dx}$ at $x = 1$.

30 Find the values of A, B and C for which

$$f(x) \equiv \frac{3x^2 - 2x + 1}{x^2(1-x)} \equiv \frac{A}{1-x} + \frac{B}{x^2} + \frac{C}{x}$$

Hence find $f'(2)$ and $f''(2)$.

8.11 Differentiating trigonometric functions

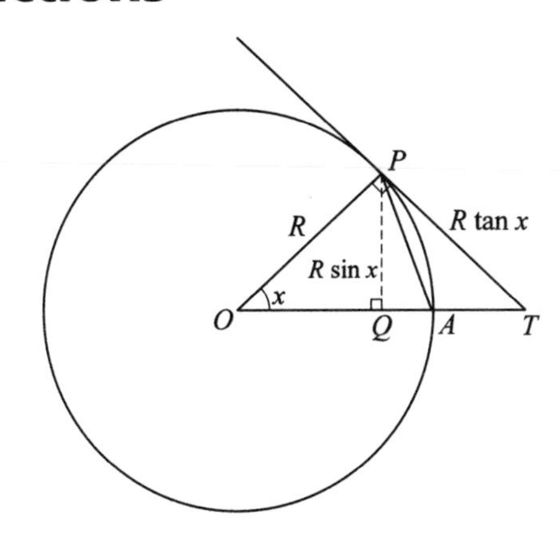

This circle, centre O and radius R, has the angle $POA = x$ radians, where $x < \frac{\pi}{2}$. The tangent to the circle at P meets OA produced at the point T. The radius OP is perpendicular to the tangent PT. The line PQ is perpendicular to OA. In $\triangle POQ$, $PQ = R\sin x$; in $\triangle OPT$, $PT = R\tan x$. Think about the areas of parts of the diagram. You can see that

$$\text{area } \triangle OAP < \text{area sector } OAP < \text{area } \triangle OPT$$

$$\tfrac{1}{2}R^2 \sin x < \tfrac{1}{2}R^2 x < \tfrac{1}{2}R^2 \tan x$$

So:

$$\sin x < x < \frac{\sin x}{\cos x}$$

Divide by $\sin x$, which is positive, as $x < \frac{\pi}{2}$:

$$1 < \frac{x}{\sin x} < \frac{1}{\cos x}$$

As $x \to 0$, $\cos\ x \to 1$ and so $\dfrac{x}{\sin x} \to 1$

or you could say

$$\lim_{x \to 0} \left(\frac{x}{\sin x}\right) = 1$$

This limit is of vital importance in finding the derivative of $\sin x$ because if you consider

$$\begin{aligned} y &= \sin x \\ y + \delta y &= \sin(x + \delta x) \\ &= \sin x \cos \delta x + \cos x \sin \delta x \end{aligned}$$

[using the identity for $\sin(A + B)$]

When δx is small enough, $\cos \delta x \approx 1$ and since

$$\lim_{\delta x \to 0} \left(\frac{\sin \delta x}{\delta x} \right) = 1$$

$\sin \delta x \approx \delta x$.

So when δx is sufficiently small,

$$y + \delta y \approx \sin x + \delta x \cos x$$

$$\delta y \approx \sin x + \delta x \cos x - y$$

But $y = \sin x$

So: $$\delta y \approx \delta x \cos x \text{ and } \frac{\delta y}{\delta x} \approx \cos x$$

So: $$\lim_{\delta x \to 0} \left(\frac{\delta y}{\delta x} \right) = \frac{\mathrm{d}y}{\mathrm{d}x} = \cos x$$

■ $\boldsymbol{\dfrac{\mathrm{d}}{\mathrm{d}x}(\sin x) = \cos x}$

Now $\cos x \equiv \sin\left(\frac{\pi}{2} - x\right)$. Write

$$y = \sin\left(\frac{\pi}{2} - x\right)$$

Put $t = \frac{\pi}{2} - x$; then $y = \sin t$ and differentiating gives

$$\frac{\mathrm{d}t}{\mathrm{d}x} = -1 \quad \text{and} \quad \frac{\mathrm{d}y}{\mathrm{d}t} = \cos t$$

By the chain rule:

$$\frac{\mathrm{d}y}{\mathrm{d}x} = \frac{\mathrm{d}y}{\mathrm{d}t} \cdot \frac{\mathrm{d}t}{\mathrm{d}x} = -\cos t = -\cos\left(\frac{\pi}{2} - x\right)$$

But $\cos\left(\frac{\pi}{2} - x\right) \equiv \sin x$ and so $\frac{\mathrm{d}y}{\mathrm{d}x} = -\sin x$

■ $\boldsymbol{\dfrac{\mathrm{d}}{\mathrm{d}x}(\cos x) = -\sin x}$

In your examination, you need to know the derivatives of $\sin x$ and $\cos x$ but you will not be expected to prove the results just established. The derivation of these results has assumed that **the angle x is always measured in radians**. In all future work connected with differentiation and integration of trigonometric functions you must always take x to be measured in radians. Expressions such as $\sin x$, $\cos^2 \theta$, $\tan 3y$ all imply that x, θ and y are in radians and you should *never* use degrees in this level of work, except in practical trigonometry and, perhaps, in solving trigonometric equations, when you are instructed to give the answer in degrees.

You should memorise the following important results obtained by using the chain rule:

- $\frac{\mathrm{d}}{\mathrm{d}x}(\sin nx) = n \cos nx$
- $\frac{\mathrm{d}}{\mathrm{d}x}(\cos nx) = -n \sin nx$
- $\frac{\mathrm{d}}{\mathrm{d}x}(\sin^n x) = n \sin^{n-1} x \cos x$
- $\frac{\mathrm{d}}{\mathrm{d}x}(\cos^n x) = -n \cos^{n-1} x \sin x$

The derivative of $\tan x$ is found by using the quotient rule because

$$y = \tan x = \frac{\sin x}{\cos x}$$

Take $\qquad u = \sin x \qquad v = \cos x$

Then: $\qquad \frac{\mathrm{d}u}{\mathrm{d}x} = \cos x \qquad \frac{\mathrm{d}v}{\mathrm{d}x} = -\sin x$

$$\frac{\mathrm{d}y}{\mathrm{d}x} = \frac{v\frac{\mathrm{d}u}{\mathrm{d}x} - u\frac{\mathrm{d}v}{\mathrm{d}x}}{v^2}$$

$$= \frac{\cos^2 x - (-\sin^2 x)}{\cos^2 x} = \frac{\cos^2 x + \sin^2 x}{\cos^2 x}$$

But $\cos^2 x + \sin^2 x \equiv 1$ and so:

$$\frac{\mathrm{d}}{\mathrm{d}x}(\tan x) = \frac{1}{\cos^2 x} = \sec^2 x$$

Similarly you can show that

$$\frac{\mathrm{d}}{\mathrm{d}x}(\cot x) = -\mathrm{cosec}^2 x$$

The derivative of $\sec x$ is found by using the chain rule:

$$y = \sec x = \frac{1}{\cos x} = (\cos x)^{-1}$$

Put $t = \cos x$ and then $y = t^{-1}$

$$\frac{\mathrm{d}t}{\mathrm{d}x} = -\sin x$$

$$\frac{\mathrm{d}y}{\mathrm{d}t} = -t^{-2} = \frac{-1}{t^2} = \frac{-1}{\cos^2 x}$$

So: $$\frac{dy}{dx} = \frac{dy}{dt} \cdot \frac{dt}{dx} = \frac{\sin x}{\cos^2 x}$$

$$= \frac{1}{\cos x} \cdot \frac{\sin x}{\cos x}$$

$$= \sec x \tan x$$

Similarly you can show that

$$\frac{d}{dx}(\operatorname{cosec} x) = -\operatorname{cosec} x \cot x$$

You should learn all these results:

- $\dfrac{\mathbf{d}}{\mathbf{d}x}(\sec x) = \sec x \tan x$
- $\dfrac{\mathbf{d}}{\mathbf{d}x}(\operatorname{cosec} x) = -\operatorname{cosec} x \cot x$
- $\dfrac{\mathbf{d}}{\mathbf{d}x}(\tan x) = \sec^2 x$
- $\dfrac{\mathbf{d}}{\mathbf{d}x}(\cot x) = -\operatorname{cosec}^2 x$

Example 17

Differentiate with respect to x:

(a) $\cos^4 x$ (b) $\tan(2x - 3)$ (c) $x^2 \sin 3x$

(a) Write $y = \cos^4 x$ and $\cos x = t$

Then: $$y = t^4 \quad \text{and} \quad t = \cos x$$

$$\frac{dy}{dt} = 4t^3 \text{ and } \frac{dt}{dx} = -\sin x$$

Using the chain rule:

$$\frac{dy}{dx} = \frac{dy}{dt} \cdot \frac{dt}{dx} = -4t^3 \sin x$$

So: $$\frac{dy}{dx} = -4\cos^3 x \sin x$$

(b) Write $y = \tan(2x - 3)$ and $2x - 3 = t$

Then: $$y = \tan t \quad \text{and} \quad t = 2x - 3$$

$$\frac{dy}{dt} = \sec^2 t \text{ and } \frac{dt}{dx} = 2$$

Using the chain rule:

$$\frac{dy}{dx} = \frac{dy}{dt} \cdot \frac{dt}{dx} = \sec^2 t \times 2$$

So: $$\frac{dy}{dx} = 2\sec^2(2x - 3)$$

(c) Take $y = x^2 \sin 3x$ and write

$$u = x^2, \qquad v = \sin 3x$$

Then: $$\frac{du}{dx} = 2x, \quad \frac{dv}{dx} = 3\cos 3x$$

Using the product rule:

$$\frac{dy}{dx} = v\frac{du}{dx} + u\frac{dv}{dx}$$

So: $$\frac{dy}{dx} = 2x\sin 3x + 3x^2 \cos 3x$$

Example 18

Given that $y = \sin x + \cos 2x$, $0 \leqslant x \leqslant \pi$, find (a) the stationary values of y (b) the nature of these stationary values.

(a) By differentiation: $\dfrac{dy}{dx} = \cos x - 2\sin 2x$

But $$\sin 2x \equiv 2\sin x \cos x$$

So: $$\frac{dy}{dx} = \cos x - 4\sin x \cos x = \cos x(1 - 4\sin x)$$

At stationary values of y, $\dfrac{dy}{dx} = 0$

So: $$\cos x = 0 \quad \text{or} \quad \sin x = \tfrac{1}{4}$$

$$x = \frac{\pi}{2} \quad \text{or} \quad 0.253 \quad \text{or} \quad 2.889$$

Stationary values are

$$y = 0 \quad \text{or} \quad 1.125 \quad \text{or} \quad 1.125$$

The coordinates of the stationary points are

$$\left(\frac{\pi}{2}, 0\right), \quad (0.253, 1.125), \quad (2.889, 1.125)$$

(b) By differentiation of $\frac{dy}{dx} = \cos x - 2\sin 2x$, you have

$$\frac{d^2y}{dx^2} = -\sin x - 4\cos 2x$$

At $x = 0.253$, $\frac{d^2y}{dx^2} = -3.75 < 0$

At $x = \frac{\pi}{2}$, $\frac{d^2y}{dx^2} = 3 > 0$

At $x = 2.889$, $\frac{d^2y}{dx^2} = -3.75 < 0$

At $(0.253, 1.125)$, $\frac{dy}{dx} = 0$, $\frac{d^2y}{dx^2} < 0$

This stationary point is a *maximum*.

At $\left(\frac{\pi}{2}, 0\right)$, $\frac{dy}{dx} = 0$, $\frac{d^2y}{dx^2} > 0$

This stationary point is a *minimum*.

At $(2.889, 1.125)$, $\frac{dy}{dx} = 0$, $\frac{d^2y}{dx^2} < 0$

This stationary point is a *maximum*.

Exercise 8D

Differentiate with respect to x:

1 $\sin 3x$ **2** $\sin \frac{1}{2}x$ **3** $\cos 4x$ **4** $\cos \frac{2}{3}x$

5 $\tan 2x$ **6** $\tan \frac{x}{4}$ **7** $\sec 5x$ **8** $\operatorname{cosec} \frac{1}{2}x$

9 $\cot 6x$ **10** $\sec \frac{x}{2}$ **11** $\cot \frac{3x}{2}$ **12** $\operatorname{cosec} \frac{2x}{3}$

Find $\frac{dy}{dx}$ in each of these questions:

13 $y = \sin^2 x$ **14** $y = \sin^3 x$ **15** $y = \sqrt{(\sin x)}$

16 $y = \cos^4 x$ **17** $y = \cos^5 x$ **18** $y = (\cos x)^{\frac{1}{3}}$

19 $y = \tan^2 x$ **20** $y = \sqrt{(\tan x)}$ **21** $y = \frac{1}{\tan x}$

22 $y = \frac{1}{\sin^2 x}$ **23** $y = \frac{4}{\cos^4 x}$ **24** $y = \operatorname{cosec}^2 x$

Differentiate with respect to x:

25 $\sin^2 2x$ **26** $\cos^2 3x$ **27** $\tan^3 2x$

28 $\sec^2 2x$ **29** $\sin(3x + 5)$ **30** $\cos^2(2x - 4)$

31 $\tan^2(1 - 2x)$ **32** $\cot^2 3x$ **33** $\operatorname{cosec}^2 4x$

34 $\sin^2 x + \cos^2 x$ **35** $(\cot x - \operatorname{cosec} x)^2$ **36** $(\sin x - \cos x)^2$

Differentiate with respect to x:

37 $\sin^2 x \cos x$ **38** $\sin x \cos^3 x$ **39** $\tan x \sec x$

40 $\sec x \operatorname{cosec} x$ **41** $x \sin^2 x$ **42** $x^2 \cos x$

43 $x^2 \tan 2x$ **44** $\dfrac{x}{\sin x}$ **45** $\dfrac{\sin x}{x}$

46 $\dfrac{x^2}{\cos x}$ **47** $\dfrac{\cos 3x}{x^3}$ **48** $\tan 2x \sec 3x$

49 $\sin^2 x \cos^3 x$ **50** $e^x \sin x$ **51** $\dfrac{e^{2x}}{\cos^2 x}$

52 $\sin^2 x \ln x$ **53** $\dfrac{e^{2x}}{\sin x + \cos x}$ **54** $\dfrac{\ln(2x - 5)}{\cos^2 3x}$

55 Given that $y = \sin 3x$, show that $\dfrac{d^2y}{dx^2} = -9y$.

56 Given that $y = A\cos 2x + B\sin 2x$, where A and B are constants, show that

$$\frac{d^2y}{dx^2} + 4y = 0$$

57 Find the value of $\sin\theta$ when $7\sec\theta - 3\tan\theta$ is a minimum.

58 Find the equation of the tangent and normal to the curve $y = \sin x$, $0 < x < \frac{\pi}{2}$, at the point where $\sin x = \frac{3}{5}$.

59 Find the value of $\dfrac{d}{dx}[\sec^2 x + \tan^3 x]$ at $x = \frac{3\pi}{4}$.

60 Differentiate $\dfrac{1 + \cos 2x}{1 - \cos 2x}$.

61 Show that $\dfrac{d}{dx}\left[\dfrac{1 + \cot x}{1 - \cot x}\right] = \dfrac{2}{\sin 2x - 1}$.

62 Given that $y = (A + x)\cos x$, where A is a constant, show that $\dfrac{d^2y}{dx^2} + y$ is independent of A.

63 The tangent to the curve $y = \tan 2x$ at the point where $x = \frac{\pi}{8}$ meets the y-axis at the point Y. Find the distance OY, where O is the origin.

64 The normal to the curve $y = \sec^2 x$ at the point $P(\frac{\pi}{4}, 2)$ meets the line $y = x$ at the point Q. Find PQ^2.

65 Find the equation of the normal to the curve $y = e^x(\cos x + \sin x)$ at the point (0, 1).

66 The volume of a cone with base radius $a \cos\theta$ and height $a \sin\theta$ is V, where

$$V = \tfrac{1}{3}\pi a^3 \cos^2\theta \sin\theta$$

Given that θ can vary and a is a constant, find the maximum value of V in terms of a.

67 Find the coordinates of the turning points of the curve $y = e^{2x} \cos x$ in the interval $0 \leqslant x \leqslant 2\pi$ and distinguish between them.

68 Show that the function f given by

$$f(x) = \sin x - x \cos x$$

is increasing throughout the interval $0 \leqslant x \leqslant \frac{\pi}{2}$.

8.12 Differentiating relations given implicitly

Up to this point you have learned how to find $\frac{dy}{dx}$ from a function given **explicitly** as $y = f(x)$.

For example: $\quad y = x^3 \quad \frac{dy}{dx} = 3x^2$

But if $y = x^3$, then $x = y^{\frac{1}{3}}$ so that differentiating with respect to y gives

$$\frac{dx}{dy} = \tfrac{1}{3}y^{-\frac{2}{3}}$$

That is,

$$\frac{dx}{dy} = \frac{1}{3y^{\frac{2}{3}}} = \frac{1}{3x^2}$$

since $x^2 = y^{\frac{2}{3}}$.

In this case then,

$$\frac{dy}{dx} = \frac{1}{\frac{dx}{dy}}$$

We can generalise this result by considering a small change δx in x giving rise to a small change δy in y.

Notice that $\frac{\delta y}{\delta x} \cdot \frac{\delta x}{\delta y} = 1$

So:

$$\frac{\delta y}{\delta x} = \frac{1}{\frac{\delta x}{\delta y}}$$

You also know that

$$\lim_{\delta x \to 0}\left(\frac{\delta y}{\delta x}\right) = \frac{dy}{dx} \quad \text{and} \quad \lim_{\delta y \to 0}\left(\frac{\delta x}{\delta y}\right) = \frac{dx}{dy}$$

So it is generally true that:

■ $$\frac{\mathbf{d}y}{\mathbf{d}x} = \frac{1}{\frac{\mathbf{d}x}{\mathbf{d}y}}$$

Often you are not given y as a function of x. Instead, relations between the two variables x and y are given **implicitly**. For example:

$$x^2 + y^2 = 16x$$

$$\sin(x + y) = \cos y$$

These relations can be differentiated directly by using the chain rule and the product or quotient rules if required.

If, for example, you want to differentiate y^3 with respect to x, you can use the chain rule, first differentiating y^3 with respect to y to get $3y^2$, then differentiating y with respect to x, to get $\frac{dy}{dx}$.

Thus: $$\frac{d}{dx}(y^3) = 3y^2\frac{dy}{dx} \quad \text{by the chain rule}$$

Similarly: $$\frac{d}{dx}(y^n) = ny^{n-1}\frac{dy}{dx} \quad \text{by the chain rule}$$

If you want to differentiate xy with respect to x, you can use the product rule and write

$$\begin{aligned}\frac{d}{dx}(xy) &= \frac{d}{dx}(x)\ y + x\ \frac{d}{dx}(y)\\ &= 1\ y + x\ \frac{dy}{dx}\\ &= y + x\frac{dy}{dx}\end{aligned}$$

Example 19

Find $\frac{dy}{dx}$ in terms of x and y for (a) $x^2 + y^2 = 16x$

(b) $\sin(x + y) = \cos y$.

(a) Differentiating with respect to x:

$$2x + 2y\ \frac{dy}{dx} = 16$$

So:

$$y\ \frac{dy}{dx} = 8 - x$$

$$\frac{dy}{dx} = \frac{8 - x}{y}$$

(b) Using the chain rule:

$$\begin{aligned}\frac{d}{dx}[\sin(x + y)] &= \cos(x + y)\left[\frac{d}{dx}(x + y)\right]\\ &= \cos(x + y)\left[1 + \frac{dy}{dx}\right]\end{aligned}$$

$$\frac{d}{dx}(\cos y) = -\sin y\left(\frac{dy}{dx}\right)$$

So when you differentiate the relation $\sin(x + y) = \cos y$ with respect to x you get:

$$\left(1 + \frac{dy}{dx}\right)\cos(x + y) = -\sin y\frac{dy}{dx}$$

That is:

$$\frac{dy}{dx}[\cos(x + y) + \sin y] = -\cos(x + y)$$

$$\frac{dy}{dx} = \frac{-\cos(x + y)}{\cos(x + y) + \sin y}$$

Example 20
Find the equation of the normal at the point $(2, 1)$ on the curve $y^2 + 3xy = 2x^2 - 1$.

First differentiate with respect to x to obtain:

$$2y\frac{\mathrm{d}y}{\mathrm{d}x} + 3\left(y + x\frac{\mathrm{d}y}{\mathrm{d}x}\right) = 4x$$

Taking $x = 2$, $y = 1$ gives

$$2\frac{\mathrm{d}y}{\mathrm{d}x} + 3\left(1 + 2\frac{\mathrm{d}y}{\mathrm{d}x}\right) = 8$$

So:
$$2\frac{\mathrm{d}y}{\mathrm{d}x} + 3 + 6\frac{\mathrm{d}y}{\mathrm{d}x} = 8 \Rightarrow \frac{\mathrm{d}y}{\mathrm{d}x} = \frac{5}{8}$$

So the gradient of the tangent to the curve at $(2, 1)$ is $\frac{5}{8}$.

$\therefore$ Gradient of normal at $(2, 1)$ is $-\frac{8}{5}$.

The equation of the normal is

$$y - 1 = -\tfrac{8}{5}(x - 2)$$

That is,
$$5y + 8x = 21$$

Example 21
Given that $y = a^x$, $a > 0$, find $\frac{\mathrm{d}y}{\mathrm{d}x}$.

Taking logs to the base e gives

$$\ln y = \ln a^x = x \ln a$$

Differentiating with respect to x, using the chain rule,

$$\frac{1}{y}\frac{\mathrm{d}y}{\mathrm{d}x} = \ln a$$

$\Rightarrow$
$$\frac{\mathrm{d}y}{\mathrm{d}x} = y \ln a = a^x \ln a$$

- **This is an important result, which you should memorise:**

$$\boldsymbol{\frac{\mathrm{d}}{\mathrm{d}x}(a^x) = a^x \ln a}$$

8.13 Differentiating functions given parametrically

You will often come across relationships between the variables x and y where x and y are given in terms of another variable t, where t is called a **parameter**. For example, the curve $x^2 + y^2 = a^2$, which is a circle centre the origin and radius a, can also be represented by the equations

$$x = a\cos t, \qquad y = a\sin t, \qquad 0 \leqslant t < 2\pi$$

where t is a parameter.

You can move from the **parametric equations** to the (x, y) equation (the cartesian equation), by eliminating t (if this is possible and simple). In this example, you know that

$$\cos t = \frac{x}{a} \qquad \sin t = \frac{y}{a}$$

and $$\cos^2 t + \sin^2 t = 1$$

So: $$\left(\frac{x}{a}\right)^2 + \left(\frac{y}{a}\right)^2 = 1$$

giving $$x^2 + y^2 = a^2$$

Parametric equations are often used to reduce the amount of working needed to solve problems, particularly in coordinate geometry.

By using the chain rule, expressions for $\frac{dy}{dx}$ can be easily obtained:

$$\frac{dy}{dx} = \frac{dy}{dt} \cdot \frac{dt}{dx} = \frac{\frac{dy}{dt}}{\frac{dx}{dt}}$$

because $\frac{dt}{dx} = \frac{1}{\frac{dx}{dt}}$, as shown in section 8.12.

Example 22

Find the gradient at the point P where $t = -1$, on the curve given parametrically by

$$x = t^2 - t, \quad y = t^3 - t^2$$

Hence find the equation of the tangent and normal at P.

Differentiating the x and y relations with respect to t,

$$\frac{dx}{dt} = 2t - 1 \qquad \frac{dy}{dt} = 3t^2 - 2t$$

$$\frac{dy}{dx} = \frac{\frac{dy}{dt}}{\frac{dx}{dt}} = \frac{3t^2 - 2t}{2t - 1}$$

At $t = -1$, $x = 2$, $y = -2$, and

$$\frac{dy}{dx} = \frac{5}{-3} = -\tfrac{5}{3}$$

The equation of the tangent at P is

$$y + 2 = -\tfrac{5}{3}(x - 2)$$

The normal has gradient $\frac{3}{5}$ and its equation is

$$y + 2 = \tfrac{3}{5}(x - 2)$$

Example 23

Find the equation of the tangent at the point P where $t = \frac{\pi}{4}$, on the curve

$$x = 2\cos t - \cos 2t, \quad y = 2\sin t - \sin 2t, \quad 0 \leqslant t < \pi$$

Differentiating both x and y with respect to t:

$$\frac{dx}{dt} = -2\sin t + 2\sin 2t, \quad \frac{dy}{dt} = 2\cos t - 2\cos 2t$$

Using the chain rule:

$$\frac{dy}{dx} = \frac{\frac{dy}{dt}}{\frac{dx}{dt}} = \frac{2\cos t - 2\cos 2t}{2\sin 2t - 2\sin t} = \frac{\cos t - \cos 2t}{\sin 2t - \sin t}$$

At $t = \frac{\pi}{4}$, $x = \sqrt{2}$, $y = \sqrt{2} - 1$, $\dfrac{dy}{dx} = \dfrac{\frac{1}{\sqrt{2}}}{1 - \frac{1}{\sqrt{2}}} = \dfrac{1}{\sqrt{2} - 1}$

$$\frac{dy}{dx} = \frac{1}{\sqrt{2} - 1} \times \frac{\sqrt{2} + 1}{\sqrt{2} + 1} = \frac{\sqrt{2} + 1}{2 - 1} = \sqrt{2} + 1$$

The equation of the tangent at P is:

$$y - (\sqrt{2} - 1) = (\sqrt{2} + 1)(x - \sqrt{2})$$

$$y - \sqrt{2} + 1 = x(\sqrt{2} + 1) - 2 - \sqrt{2}$$

That is, $\quad x(\sqrt{2} + 1) - y = 3$

Exercise 8E

Find $\frac{dy}{dx}$ for each of these relations:

1 $y^2 = 2x + 1$ **2** $x^2 + y^2 = 4$ **3** $xy^2 = 16$
4 $x^2 + 2xy + 3y^2 = 6$ **5** $\sin x \cos y = 1$ **6** $e^{xy} = 4$
7 $\sin(x + y) = \frac{1}{2}$ **8** $\ln y \ln x = 3$ **9** $e^x \ln y = x$
10 $y(x + y) = 12$

Find $\frac{dy}{dx}$ for each of the following where t, θ and u are parameters; a,b and c are constants.

11 $x = 3t^2,\ y = 2t$ **12** $x = t^2,\ y = t^3$

13 $x = ct,\ y = \frac{c}{t^2}$ **14** $x = a\cos t,\ y = b\sin t$

15 $x = a\sec t,\ y = a\tan t$ **16** $x = e^t \cos t,\ y = e^t \sin t$
17 $x = t\cos t,\ y = t\sin t$
18 $x = a(\theta - \cos\theta),\ y = a(1 + \sin\theta)$
19 $x = e^u,\ y = e^u - e^{-u}$ **20** $x = \sin^2\theta,\ y = \cos\theta\sin\theta$

21 Find the equations of the tangents to the curve $y = 2^x$ at the points P and Q where $x = 2$ and $x = 5$ respectively. These tangents meet at the point R. Find the x-coordinate of R.

22 Find the equation of the normal to the curve $x = 3t - t^3,\ y = t^2$ at the point where $t = 2$.

23 A curve is given by the equations

$$x = t^2, \quad y = t^5$$

where t is a parameter.
Find the equations of the tangent and normal to the curve at the point P where $t = -1$.
Find the cartesian equation of the curve.

24 Find the equation of the tangent to the curve

$$x = (1 - \tfrac{1}{3}t^3), \quad y = 4t^2$$

at the point P, where $t = -2$.

25 The tangents at the points P and Q in the first quadrant, where $x = 1$ and $x = 4$, respectively to the curve $y^2 = 4x$ meet at the point R. Find the coordinates of R.

26 Find the equation of the tangent at (2, 1) the curve

$$y(x+y)^2 = 3(x^3 - 5)$$

27 Find $\frac{dy}{dx}$ in terms of t for the curve

$$x = \frac{t}{1-t}, \quad y = \frac{t^2}{1-t}$$

Deduce the equation of the normal at the point where $t = \frac{1}{2}$.

28 For the curve $xy(x+y) = 84$, find $\frac{dy}{dx}$ at (3, 4).

29 Differentiate with respect to x:

(a) 10^x (b) 2^{x^2} (c) 5^{-x}

30 For the curve $2\cos y \sin x = 1$, find the equation of the normal at $(\frac{\pi}{4}, \frac{\pi}{4})$.

SUMMARY OF KEY POINTS

1 If y is a function of t and t is a function of x, then

$$\frac{dy}{dx} = \frac{dy}{dt} \cdot \frac{dt}{dx}$$

This is called the chain rule.

2 If $y = uv$, where u and v are functions of x, then

$$\frac{dy}{dx} = v\frac{du}{dx} + u\frac{dv}{dx}$$

This is called the product rule.

3 If $y = \frac{u}{v}$, where u and v are functions of x, then

$$\frac{dy}{dx} = \frac{v\frac{du}{dx} - u\frac{dv}{dx}}{v^2}$$

This is called the quotient rule.

4 The chain rule is used to find related rates of change; for example:

$$\frac{dV}{dt} = \frac{dV}{dr} \cdot \frac{dr}{dt}$$

5 The second derivative of y with respect to x is written as $\frac{d^2y}{dx^2}$ and $\frac{d}{dx}\left(\frac{dy}{dx}\right) = \frac{d^2y}{dx^2}$. The notation $y = f(x)$,

$\frac{dy}{dx} = f'(x)$, $\frac{d^2y}{dx^2} = f''(x)$ is also commonly used.

6 Points of inflexion occur at points on the curve $y = f(x)$, where $\frac{d^2y}{dx^2} = 0$ and $\frac{d^3y}{dx^3} \neq 0$.

7 For the stationary points (x_1, y_1), (x_2, y_2) and (x_3, y_3) on the curve $y = f(x)$,

$f'(x_1) = 0$	$f'(x_2) = 0$	$f'(x_3) = 0$	
$f''(x_1) < 0$	$f''(x_2) > 0$	$f''(x_3) = 0$	$f'''(x_3) \neq 0$
y_1 is a maximum value of $f(x)$	y_2 is a minimum value of $f(x)$	At (x_3, y_3) there is a point of inflexion	

8 The tangent at (x_1, y_1) on the curve $y = f(x)$ has gradient $f'(x_1)$ and equation

$$y - y_1 = f'(x_1)[x - x_1]$$

The normal at (x_1, y_1) on the curve $y = f(x)$ has gradient $-\frac{1}{f'(x_1)}$ and equation

$$y - y_1 = \frac{-1}{f'(x_1)}(x - x_1)$$

9 You should know that $\lim_{x \to 0}\left(\frac{\sin x}{x}\right) = 1$.

10 Memorise these standard formulae for derivatives:
(a, b, k, n are constants)

$y = f(x)$	$\frac{dy}{dx} = f'(x)$
x^n	nx^{n-1}
$(ax + b)^n$	$na(ax + b)^{n-1}$
e^x	e^x
e^{ax+b}	ae^{ax+b}
a^x	$a^x(\ln a)$

$\ln x$	$\frac{1}{x}$
$\ln(ax+b)$	$\frac{a}{ax+b}$
$\sin x$	$\cos x$
$\cos x$	$-\sin x$
$\tan x$	$\sec^2 x$
$\cot x$	$-\text{cosec}^2 x$
$\sec x$	$\sec x \tan x$
$\text{cosec}\, x$	$-\text{cosec}\, x \cot x$
$\sin kx$	$k \cos kx$
$\sin^n x$	$n \sin^{n-1} x \cos x$
$\sin^n kx$	$nk \sin^{n-1} kx \cos kx$
$u \pm v$	$\frac{du}{dx} \pm \frac{dv}{dx}$
$x = g(t),\ y = f(t)$	$\frac{dy}{dx} = \frac{dy}{dt} \cdot \frac{dt}{dx} = \frac{\frac{dy}{dt}}{\frac{dx}{dt}} = \frac{f'(t)}{g'(t)}$
uv	$v\frac{du}{dx} + u\frac{dv}{dx}$
$\frac{u}{v}$	$\frac{v\frac{du}{dx} - u\frac{dv}{dx}}{v^2}$
y^n	$ny^{n-1}\frac{dy}{dx}$
xy	$y + x\frac{dy}{dx}$

Review exercise 2

1 Differentiate with respect to x

(a) $\dfrac{\sin x}{e^x}$ (b) $\ln(1 + \tan^2 x)$ [L]

2 When a metal cube is heated, the length of each edge increases at the rate of $0.03\,\text{cm}\,\text{s}^{-1}$. Find the rate of increase, in $\text{cm}^2\,\text{s}^{-1}$, of the total surface area of the cube, when the length of each side is 8 cm. [L]

3 (i) Given that $x + \dfrac{1}{x} = 3$,

(a) expand $\left(x + \dfrac{1}{x}\right)^3$ and use your expression to show that

$$x^3 + \frac{1}{x^3} = 18$$

(b) expand $\left(x + \dfrac{1}{x}\right)^5$

Use your expansion and the previous result to find the value of

$$x^5 + \frac{1}{x^5}$$

(ii) Given that

$$(1 + 3x)^{15} \equiv 1 + Ax + Bx^2 + Cx^3 + \dots,$$

find the values of the constants A, B and C. [L]

4 Given that α is acute and $\tan\alpha = \frac{3}{4}$, prove that

$$3\sin(\theta + \alpha) + 4\cos(\theta + \alpha) \equiv 5\cos\theta \qquad [L]$$

5

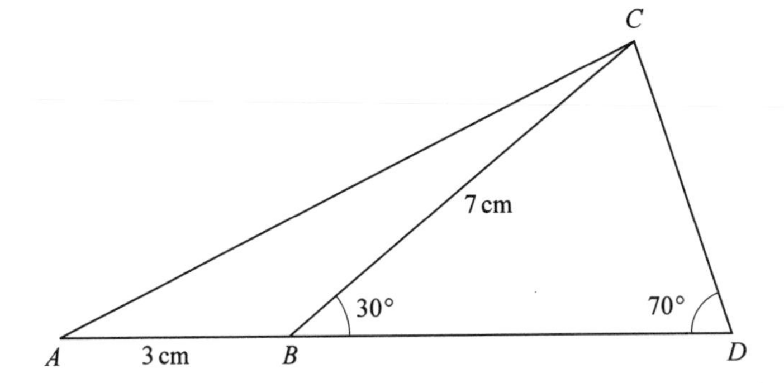

The figure shows $\triangle ABC$ in which $AB = 3$ cm, $BC = 7$ cm. AB is produced to D so that $\angle CBD = 30°$ and $\angle BDC = 70°$.
Calculate, to 3 significant figures,

(a) the length, in cm, of AC

(b) the length, in cm, of BD

(c) the area, in cm^2, of $\triangle ACD$ [L]

6 Find the equation of the normal at the point P with parameter t on the curve with parametric equations

$$x = t^2, \quad y = 2t$$

Show that, if this normal meets the x-axis at G, and S is the point $(1, 0)$, then $SP = SG$.
Find also the equation of the tangent at P, and show that, if the tangent meets the y-axis at Z, then SZ is parallel to the normal at P. [L]

7 Find the coordinates of the turning points on the curve

$$y^3 + 3xy^2 - x^3 = 3 \qquad [L]$$

8 Use the binomial expansion to express $x^4(1 - x)^4$ as a polynomial in x.
Hence, or otherwise, verify that

$$x^4(1 - x)^4 \equiv (1 + x^2)(x^6 - 4x^5 + 5x^4 - 4x^2 + 4) - 4 \qquad [L]$$

9

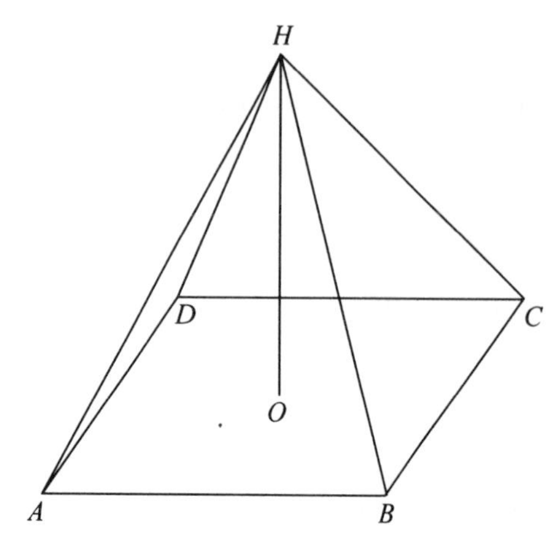

The figure shows a pyramid $HABCD$ standing on horizontal ground. The points A, B, C and D are the corners of its square base. The length of a side of the square is 12 m and its diagonals intersect at O. Each sloping edge makes an angle of $28°$ with the ground. Calculate

(a) the height, OH, in m to 3 significant figures,

(b) the size, to the nearest degree, of the angle which the plane HCB makes with the ground.

The point E lies on AD and is such that $AE : ED = 1 : 4$.

(c) Calculate the size, to the nearest degree, of $\angle OEH$. [L]

10 Differentiate with respect to x

(a) $e^{2x} \cos x$ (b) $\dfrac{x^2}{2x+1}$ [L]

11 Given that $\sin x = 0.6$ and $\cos x = -0.8$, evaluate $\cos(x + 270°)$ and $\cos(x + 540°)$. [L]

12 In the binomial expansion of $\left(1 + \dfrac{x}{n}\right)^n$ in ascending powers of x, the coefficient of x^2 is $\frac{7}{16}$. Given that n is a positive integer,

(a) find the value of n

(b) evaluate the coefficient of x^3 in the expansion. [L]

13 The radius of a circular ink blot is increasing at the rate of 0.3 cm s^{-1}. Find, in $\text{cm}^2 \text{ s}^{-1}$ to 2 significant figures, the rate at which the area of the blot is increasing at the instant when the radius of the blot is 0.8 cm. [L]

14 A vertical wall, 2.7 m high, runs parallel to the wall of a house and is at a horizontal distance of 6.4 m from the house. An extending ladder is placed to rest on the top B of the wall with one end C against the house and the other end, A, resting on horizontal ground, as shown in the figure.

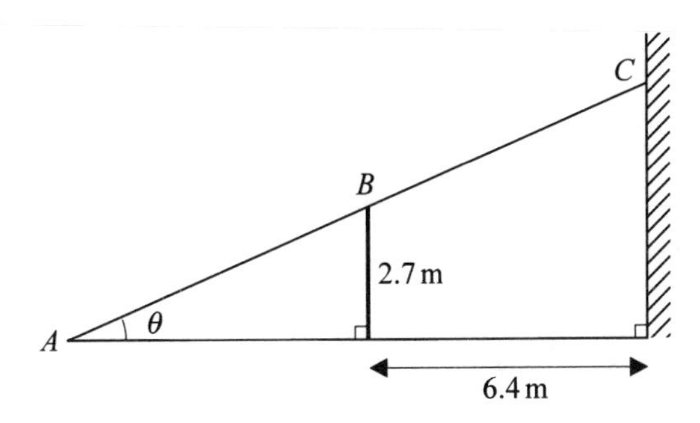

The points A, B and C are in a vertical plane at right angles to the wall and the ladder makes an angle θ, where $0 < \theta < \frac{\pi}{2}$, with the horizontal. Show that the length, y metres, of the ladder is given by

$$y = \frac{2.7}{\sin\theta} + \frac{6.4}{\cos\theta}$$

As θ varies, find the value of $\tan\theta$ for which y is a minimum. Hence find the minimum value of y. [L]

15 Three points A, B and C are on horizontal ground. The distance AB is 60 m. The bearing of B from A is 040°. The distance BC is 90 m. The bearing of C from B is 120°. D is the point on CB produced which is nearest to A. Calculate

(a) the distance AD, in m to 3 significant figures

(b) the distance AC, in m to 3 significant figures

(c) the bearing of C from A, to the nearest degree. [L]

16 In the binomial expansion of $\left(1 + \frac{x}{k}\right)^n$, where k is a constant and n is a positive integer, the coefficients of x and x^2 are equal.

(a) Show that $2k = n - 1$.

For the case when $n = 7$

(b) deduce the value of k.

(c) Hence find the first three terms in the expansion in ascending powers of x. [L]

17 Solve the equation $\cos\theta + \sin\frac{1}{2}\theta = 0$ where $-2\pi < \theta \leqslant 2\pi$, giving your answers in radians. [L]

18

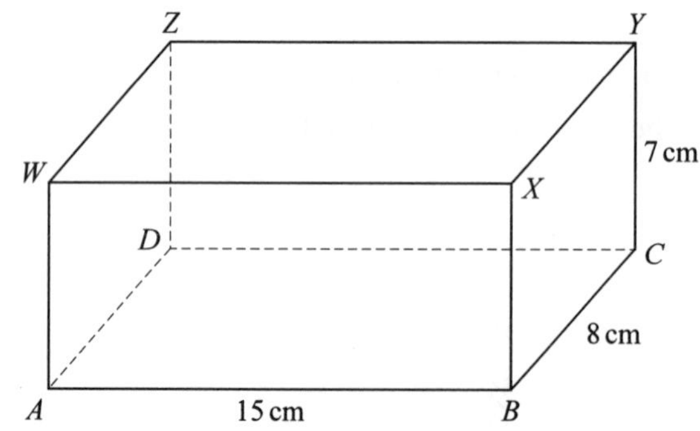

The figure shows a cuboid $ABCDWXYZ$ in which $AB = 15$ cm, $BC = 8$ cm and $CY = 7$ cm. CY is vertical. Calculate

(a) the length, in cm, of AC

(b) the length, in cm to 3 significant figures, of AY

(c) the size of the angle between AY and the horizontal, giving your answer to the nearest degree

(d) the area, in cm^2 to 3 significant figures, of triangle YAB.

Given that M is the mid-point of AB, calculate

(e) the length, in cm to 3 significant figures, of the perpendicular from M to AY. [L]

19 A line through $P(-2, 4)$ is perpendicular to the line whose equation is $x - 2y = 5$. Given that these two lines intersect at the point Q, calculate

(a) the coordinates of Q

(b) the distance PQ. [L]

20 Find the distance AB where A is the point $(2, -7, 4)$ and B is the point $(-3, -5, 6)$. [L]

21 Find $\frac{dy}{dx}$ when

(a) $y = \sin 3x \cos x + \cos 3x \sin x$

(b) $y = \frac{e^{3x}}{x^2}$ [L]

22 A curve has parametric equations

$$x = 3e^{2t} - t, \quad y = e^{3t} - 2t$$

Find, in terms of t,

(a) $\frac{dx}{dt}$ (b) $\frac{dy}{dt}$ (c) $\frac{dy}{dx}$

The gradient of the normal to the curve at the point P is $-\frac{1}{2}$.

(d) Find the value of t at P, giving your answer in the form $t = \ln k$, where k is a constant. [L]

23

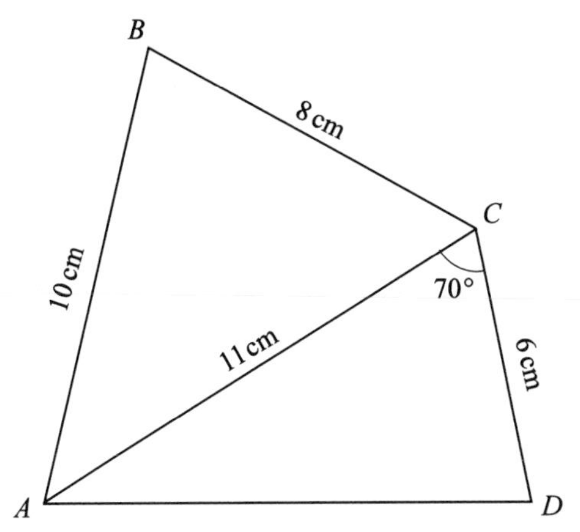

In the figure, $ABCD$ is a plane quadrilateral in which $AB = 10\,\text{cm}$, $BC = 8\,\text{cm}$, $CD = 6\,\text{cm}$, $AC = 11\,\text{cm}$ and $\angle ACD = 70°$. Calculate

(a) the size of angle ABC

(b) the length of AD

(c) the area of the quadrilateral

(d) the length of the side of a square which has the same area as the quadrilateral. [L]

24 A symmetrical pyramid stands on its base, which is a regular pentagon of side 20 cm. The height of the pyramid is 50 cm. Calculate, to 3 significant figures,

(a) the angle, in degrees, between a sloping edge and the base

(b) the length, in cm, of a sloping edge

(c) the angle, in degrees, between a sloping face and the base

(d) the area, in cm^2, of a sloping face. [L]

25 Given that $(1.01)^{30} = 1.3478$ to 5 significant figures, show that the magnitude of the error in using the first three terms of the binomial expansion of $(1+x)^{30}$ to estimate the value of $(1.01)^{30}$ is less than 0.01. Find also the numerical value of the coefficient of x^5 and x^{25} in this expansion. [L]

26 (a) Solve, for $0 \leqslant x < 360$, the equation

$$2\cos(x+50)° = \sin(x+40)°$$

giving your answers to one decimal place.

(b) Solve, for $0 \leqslant x < 2\pi$, the equation

$$\cos 2x = 2\sin^2 x$$

giving your answer in terms of π.

(c)

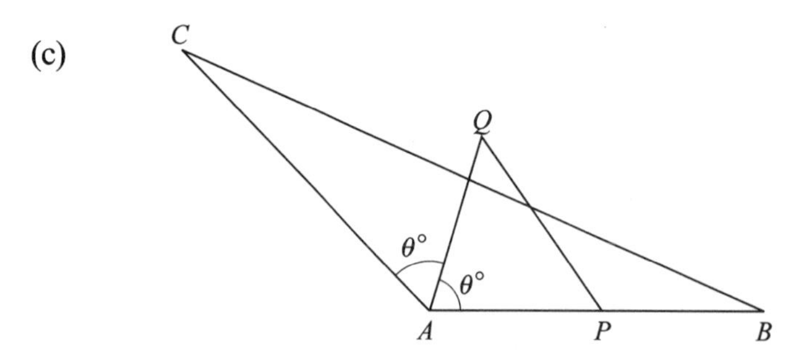

In the figure $AQ = AP = PB = a$ cm, $AC = 2a$ cm and $\angle BAQ = \angle CAQ = \theta°$. Given that

$$(\text{area } \triangle APQ) = \tfrac{1}{3}(\text{area } \triangle ABC)$$

find θ to one decimal place. [L]

27 The distance between $(7, -3, -5)$ and $(-2, -4, t)$, where t is a constant, is $\sqrt{278}$. Find the values of t.

28 A curve has equation $x^3 + 2x^2y^2 + y^3 = 0$.

(a) Show that $\frac{dy}{dx} = -1$ at $(-1, -1)$.

(b) Find an equation of the tangent to the curve at $(-1, -1)$. [L]

29 Given that $(1 + 2x)^{22} = 1 + Ax + Bx^2 + Cx^3 + \ldots$, find the values of the integers A, B and C. [L]

30 Differentiate with respect to x

(a) $\frac{\sin 2x}{x}$ (b) x^2e^{3x} (c) $\cos(3x^2)$ [L]

31 The points A, B and C lie on horizontal ground and are such that $AB = 19$ m, $BC = 16$ m and $CA = 21$ m.

(a) Calculate the size of $\angle ACB$.

A vertical mast AH of height 11 m is placed at A.

(b) Calculate the size of the acute angle between the planes HBC and ABC, giving your answer to the nearest 0.1°. [L]

32 A petrol tanker is damaged in a road accident, and petrol leaks onto a flat section of motorway. The leaking petrol begins to spread in a circle of thickness 2 mm. Petrol is leaking from the tanker at a rate of $0.0084\,\text{m}^3\,\text{s}^{-1}$. Find the rate at which the radius of the circle of petrol is increasing at the instant when the radius of the circle is 3 m, giving your answer in $\text{m}\,\text{s}^{-1}$ to 2 decimal places. [L]

33 The curve C has parametric equations

$$x = at, \quad y = \frac{a}{t}, \quad t \in \mathbb{R}, t \neq 0,$$

where t is a parameter and a is a positive constant.

(a) Find $\frac{dy}{dx}$ in terms of t.

The point P on C has parameter $t = 2$.

(b) Show that an equation of the normal to C at P is

$$2y = 8x - 15a$$

This normal meets C again at the point Q.

(c) Find the value of t at Q. [L]

34 (i) Given that $0 < \theta \leqslant 360$, find all the values of θ, to 1 decimal place, such that

(a) $\cos(\theta + 30)^\circ = 0.6$

(b) $\sin\theta^\circ + 2\cos\theta^\circ = 0$

(ii) A pyramid $ABCDE$ has a rectangular horizontal base $ABCD$ and its vertical height is 6 cm. Each sloping edge is inclined at 40° to the horizontal. Given that $AB = 6x$ cm and $BC = 2x$ cm, calculate, to 3 significant figures,

(a) the value of x

(b) the acute angle between the plane EAB and the horizontal.

[L]

35 The power output of a generator, P watts, is given by

$$P = \frac{E^2R}{(R+r)^2}$$

where E volts is the constant electromotive force of the generator, r ohms is its constant internal resistance, and R ohms is the variable load resistance.

Find $\frac{dP}{dR}$.

Hence find R in terms of r when the power output is a maximum. [L]

36 In the binomial expansion of $(1 + kx)^n$, where k is a constant and n is a positive integer, the coefficients of x and x^2 are equal.

(a) Show that $k(n - 1) = 2$.

Given that $nk = 2\frac{1}{3}$, find

(b) the value of k

(c) the value of n. [L]

37

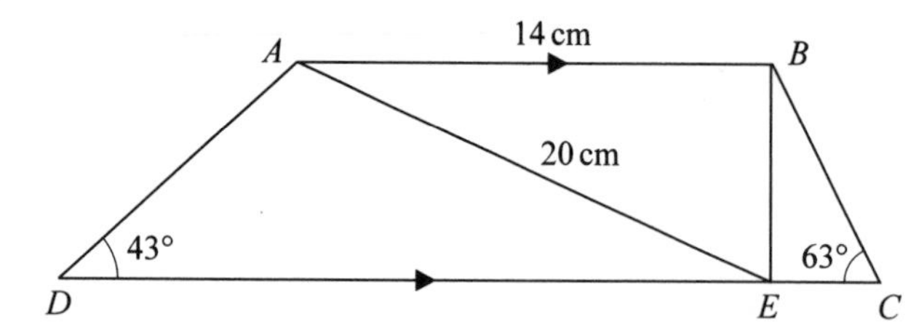

The figure shows a trapezium $ABCD$ in which AB is parallel to DC. The perpendicular from B to DC meets DC at E. Given that $AB = 14$ cm, $AE = 20$ cm, $\angle BCE = 63°$ and $\angle ADC = 43°$, calculate, to 3 significant figures,

(a) the length, in cm, of AD

(b) the area, in cm^2, of the trapezium $ABCD$. [L]

38 Airfield C is situated 50 km from airfield A on a bearing 048°. Airfield B is situated 100 km from airfield A on a bearing of 300°. Calculate

(a) the distance, in kilometres to 3 significant figures, from B to C

(b) the bearing, to the nearest degree, of C from B

(c) the area of $\triangle ABC$ giving your answer to the nearest 10 km^2. [L]

39 Find, in terms of π, the complete set of values of θ in the interval $0 \leqslant \theta \leqslant 2\pi$ for which the roots of the equation

$$x^2 + 2x\sin\theta + 3\cos^2\theta = 0$$

are real.

Show that the roots of the equation

$$x^2 + (5\cos 2\theta + 1)x + 9\cos^4\theta = 0$$

are the squares of the roots of the equation

$$x^2 + 2x\sin\theta + 3\cos^2\theta = 0$$

[L]

40 (a) Differentiate with respect to x

(i) $\ln(x^2)$ (ii) $x^2 \sin 3x$

(b) Find the gradient of the curve with equation

$$5x^2 + 5y^2 - 6xy = 13$$

at the point $(1, 2)$. [L]

41 Differentiate $e^{2x} \cos x$ with respect to x.

The curve C has equation $y = e^{2x} \cos x$.

(a) Show that the turning points on C occur where $\tan x = 2$.

(b) Find an equation of the tangent to C at the point where $x = 0$. [L]

42 Show that an equation of the normal to the curve with parametric equations $x = ct$, $y = \frac{c}{t}$, $t \neq 0$, at the point $\left(cp, \frac{c}{p}\right)$ is

$$y - \frac{c}{p} = xp^2 - cp^3$$

[L]

43 The edges of a cube are of length x cm. Given that the volume of the cube is being increased at a rate of $p\,\text{cm}^3\,\text{s}^{-1}$, where p is a constant, calculate, in terms of p, in $\text{cm}^2\,\text{s}^{-1}$, the rate at which the surface area of the cube is increasing when $x = 5$. [L]

44 Given that $y = \left(x + \frac{1}{x}\right)^3 + \left(x - \frac{1}{x}\right)^3$, where $x \in \mathbb{R}$, $x \neq 0$,

(a) prove that $y = 2x^3 + 6x^{-1}$

(b) find the values of y for which $\frac{dy}{dx} = 0$. [L]

45 (a) Differentiate the following functions with respect to x, simplifying your answers where possible:

(i) $\frac{\sqrt{(1 + x^3)}}{x^2}$ (ii) $\ln\left(\frac{2 + \cos x}{3 - \sin x}\right)$

(b) If $y = e^{3x} \sin 4x$ show that

$$\frac{d^2y}{dx^2} - 6\frac{dy}{dx} + 25y = 0$$

[L]

46

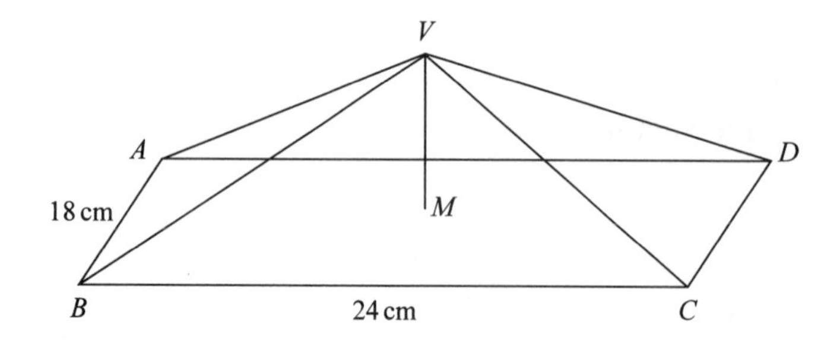

The horizontal base $ABCD$ of a pyramid $VABCD$, shown in the figure, is a rectangle with $AB = 18$ cm and $BC = 24$ cm. The diagonals AC and BD meet at M and VM is vertical.
Given that $VB = 17$ cm, calculate

(a) the length, in cm, of VM

(b) the size, to the nearest degree, of the angle which the plane VAB makes with the base of the pyramid

(c) the area, in cm^2 to 3 significant figures, of $\triangle VAB$.

(d) Hence, or otherwise, calculate the perpendicular distance, in cm to 3 significant figures, from A to VB. [L]

47 The points $A(5, 3, 1)$, $B(1, 4, -1)$ and $C(-1, 2, t)$ are such that $AB = BC$. Find, to 3 significant figures, the possible values of the constant t.

48 (a) If A is the acute angle such that $\sin A = \frac{3}{5}$ and B is the obtuse angle such that $\sin B = \frac{5}{13}$, find without using a calculator the values of $\cos(A + B)$ and $\tan(A - B)$.

(b) Find the solutions of the equation

$$\tan\theta + 3\cot\theta = 5\sec\theta$$

for which $0 < \theta < 2\pi$. [L]

49 Expand $(a + b)^6$, simplifying each coefficient.
By taking $a = 1$ and $b = -0.003$ and showing all your working, find the value of $(0.997)^6$, giving your answer to 8 decimal places.
Evaluate $(0.003)^6$, giving your answer in the standard form $p \times 10^{-q}$, where $1 \leqslant p < 10$ and q is an integer.
State the number of decimal places required to express the exact value of $(0.997)^6$ when it is written as a decimal number.

[L]

50 (a) If $y = \tan nx$, express both $\dfrac{\mathrm{d}y}{\mathrm{d}x}$ and $\dfrac{\mathrm{d}^2y}{\mathrm{d}x^2}$ in terms of n and y only and prove that

$$\frac{\mathrm{d}^2y}{\mathrm{d}x^2} = 2ny\frac{\mathrm{d}y}{\mathrm{d}x}$$

(b) Find the stationary values of the function $y = x^2\mathrm{e}^{-x}$. [L]

51

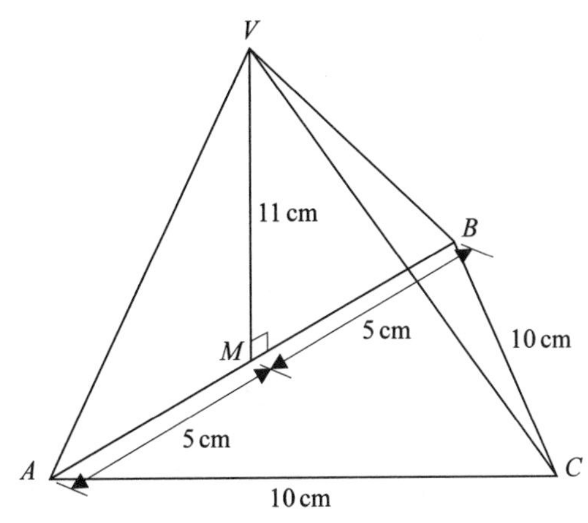

The base ABC of the pyramid $VABC$ shown in the figure is a horizontal equilateral triangle and V is vertically above M, the mid-point of the side AB of the base. Given that $AB = BC = CA = 10$ cm and $VM = 11$ cm, calculate

(a) the length, in cm correct to 3 significant figures, of VA

(b) the length, in cm, of VC

(c) the size of the angle, to the nearest degree, between VC and the horizontal

(d) the volume, in cm^3 correct to 3 significant figures, of the pyramid $VABC$. [L]

52 (i) Given that $\left(x + \dfrac{1}{x}\right) = p$,

(a) expand $\left(x + \dfrac{1}{x}\right)^2$ and use your expansion to show that

$x^2 + \dfrac{1}{x^2} = p^2 - 2$

(b) expand $\left(x + \dfrac{1}{x}\right)^4$ and hence find, in terms of p, the value of $x^4 + \dfrac{1}{x^4}$.

(ii) $(1 + qx)^7 \equiv 1 + \frac{7}{5}px + (4 - p)x^2 + \ldots + q^7x^7$, where p and q are constants. Find the possible values of p and q. [L]

53 In each part of this question give all the answers, to 1 decimal place, in the interval $0 \leqslant x < 360$.

(a) Find the values of x for which

$$\cos x^\circ = -0.85$$

(b) Find the value of x for which

$$\cos x^\circ = 0.8 \quad \textit{and} \quad \tan x^\circ = -0.75$$

(c) Write the equation

$$6\sin^2 x^\circ + 5\sin x^\circ + 3 = 6\cos^2 x^\circ$$

as a quadratic equation in $\sin x^\circ$ only.
Hence find the values of x for which

$$6\sin^2 x^\circ + 5\sin x^\circ + 3 = 6\cos^2 x^\circ$$

[L]

54

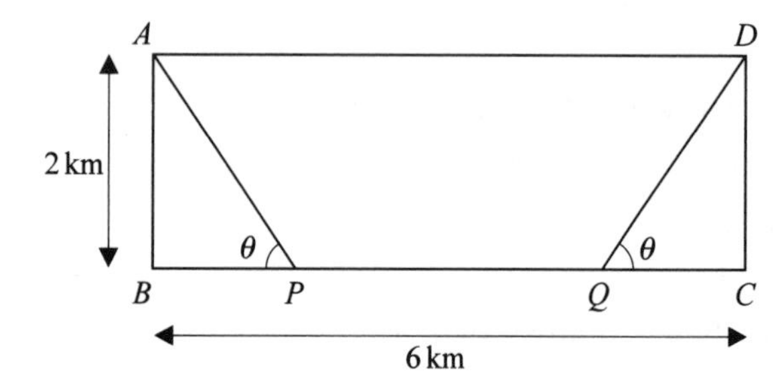

The diagram shows a house at A, a school at D and a straight canal BC, where $ABCD$ is a rectangle with $AB = 2$ km and $BC = 6$ km.
During the winter, when the canal freezes over, a boy travels from A to D by walking to a point P on the canal, skating along the canal to a point Q and then walking from Q to D. The points P and Q being chosen so that the angles APB and DQC are both equal to θ.
Given that the boy walks at a constant speed of $4\,\text{km h}^{-1}$ and skates at a constant speed of $8\,\text{km h}^{-1}$, show that the time, T minutes, taken for the boy to go from A to D along this route is given by

$$T = 15\left(3 + \frac{4}{\sin\theta} - \frac{2}{\tan\theta}\right)$$

Show that, as θ varies, the minimum time for the journey is approximately 97 minutes. [L]

55 Show that

$$(\cos X + \cos Y)^2 + (\sin X + \sin Y)^2 \equiv 4\cos^2\left(\frac{X - Y}{2}\right)$$

Without using a calculator, show that $\cos 15^\circ = \dfrac{\sqrt{2} + \sqrt{6}}{4}$

and that $\cos 15^\circ \cos 75^\circ = \frac{1}{4}$. [L]

56 (i) Differentiate with respect to x

(a) $\dfrac{\sec x + \tan x}{\sec x - \tan x}$ (b) $e^{x \sin x}$

(ii) If the length of the perimeter of a sector of a circle is l, find the largest possible area of the sector. [L]

57 A curve is defined with parameter t by the equations

$$x = at^2, \quad y = 2at$$

The tangent and normal at the point P with parameter t_1 cut the x-axis at T and N respectively. Prove that

$$\frac{PT}{PN} = |t_1|$$ [L]

58 The points O, P and Q have coordinates $(0, 0)$, $(4, 3)$ and (a, b) respectively. Given that OQ is perpendicular to PQ,

(a) show that $a^2 + b^2 = 4a + 3b$.

Given also that $a = 1$,

(b) find, to 2 decimal places, the possible values of b. [L]

59 The volume, $V\,\text{cm}^3$, of water in a container is given by the expression

$$V = 12h^2$$

where h cm is the depth of the water. Water is pouring into the container at a steady rate of $90\,\text{cm}^3\,\text{s}^{-1}$. Find the rate, in $\text{cm}\,\text{s}^{-1}$, at which the depth of water is increasing when $h = 3$. [L]

60 The lengths of the sides of a triangle are 4 cm, 5 cm and 6 cm. The size of the largest angle of the triangle is θ.

(a) Calculate the value of $\cos\theta$.

(b) Hence, or otherwise, show that $\sin\theta = \dfrac{a\sqrt{7}}{b}$, where a and b are integers. [L]

61 To make the sea trip from port A to port B, a ship has first to sail 10.7 km on a bearing 042° to a point C and then 6.2 km on a bearing 293° to the port B. Calculate

(a) the distance AB, in km to 3 significant figures

(b) the bearing of B from A, giving your answer to the nearest degree

(c) the time, in minutes to the nearest minute, taken by a train which travels in a straight line from A to B at an average speed of 80 km h^{-1}

(d) how far east B is of A, giving your answer in km to 3 significant figures. [L]

62 

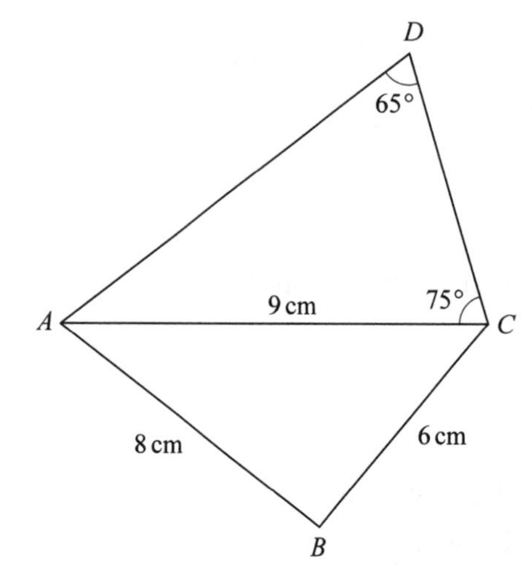

In the figure, $ABCD$ is a quadrilateral in which $AB = 8$ cm, $BC = 6$ cm, $AC = 9$ cm, $\angle DCA = 75°$ and $\angle ADC = 65°$.
Calculate

(a) the length, in cm to 3 significant figures, of AD

(b) the size, to 0.1°, of $\angle BAC$

(c) the length, in cm to 3 significant figures, of BD.

Given that BD meets AC at the point O, calculate

(d) the size, to 0.1°, of $\angle AOD$. [L]

63 Expand $(1 + ax)^8$ in ascending powers of x up to and including the term in x^2.
The coefficients of x and x^2 in the expansion of

$$(1 + bx)(1 + ax)^8$$

are 0 and -36 respectively.
Find the values of a and b, given that $a > 0$ and $b < 0$. [L]

64 (i) Solve, for $0 \leqslant x < 360$, giving your answers to 1 decimal place, where appropriate, the equations

(a) $\tan 2x° = -1$

(b) $2\cos^2 x° = \sin x° \cos x°$

(c) $3\sin x° + 3 = \cos^2 x°$

(ii) A cuboid has a square horizontal base of side 2 cm and a height of 3 cm. Calculate, in degrees to 3 significant figures, the angle that a diagonal of the cuboid makes with

(a) the base

(b) a vertical face of the cuboid. [L]

65 (a) Find, in degrees, all positive angles not greater than 360° which satisfy the equation $\sin 3\theta = \sin^2 \theta$.

(b) Find, in radians for $-2\pi \leqslant x \leqslant 2\pi$ the solutions of the equation

$$\cos\theta + \sin\tfrac{1}{2}\theta = 0$$ [L]

66

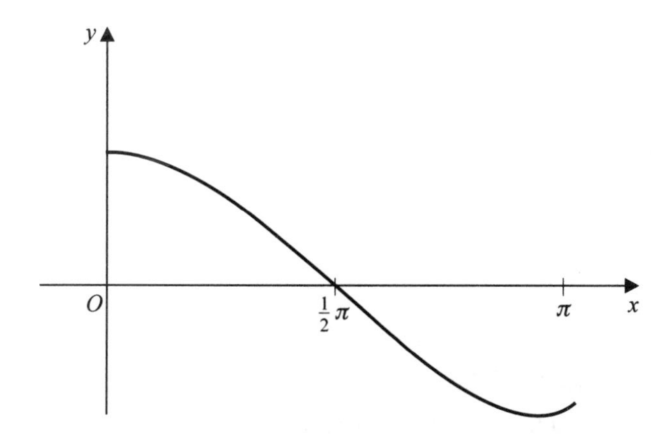

The figure shows a sketch of the curve with equation

$$y = \frac{\cos x}{5 - \sin x}, \quad 0 \leqslant x \leqslant \pi$$

(a) By considering the values of $\sin x$ and $\cos x$ for which

$\frac{dy}{dx} = 0$, show that

$$-\tfrac{1}{12}\sqrt{6} \leqslant y \leqslant \tfrac{1}{12}\sqrt{6}$$

(b) Show that the normal to the curve at the point where $x = \frac{1}{2}\pi$ meets the y-axis at the point $(0, -2\pi)$. [L]

67 (a) If $y = \sin^2 nx \cos 2nx$, show that $\frac{dy}{dx}$ can be expressed in the form $n(\sin ax - \sin bx)$, where a and b are multiples of n.

(b) If $xy^2 = (x-1)^2$, prove that

$$y\frac{d^2y}{dx^2} + \left(\frac{dy}{dx}\right)^2 = \frac{1}{x^3}$$

[L]

68 The tangent to the curve with equation $y = e^{-2x}$, at the point whose x-coordinate is 1, crosses the x-axis at P and the y-axis at Q. Show that the area of $\triangle POQ$, where O is the origin, is $\frac{9}{4e^2}$.

[L]

69

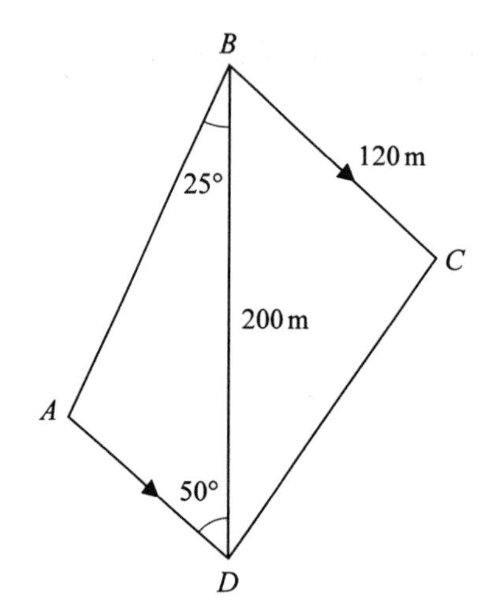

The figure represents a field $ABCD$ in which BC is parallel to AD, $\angle ADB = 50°$, $\angle ABD = 25°$, $BD = 200$ m and $BC = 120$ m.

(a) Calculate, in m, to 3 significant figures, the lengths of the sides AD and CD.

(b) Calculate, in m^2, to 3 significant figures, the area of the field. [L]

70 Given that

$$(1 + kx)^8 = 1 + 12x + px^2 + qx^3 + \dots, \quad \text{for all } x \in \mathbb{R},$$

(a) find the value of k, the value of p and the value of q.

(b) Using your values of k, p and q find the numerical coefficient of the x^3 term in the expansion of

$$(1 - x)(1 + kx)^8$$

[*L*]

71

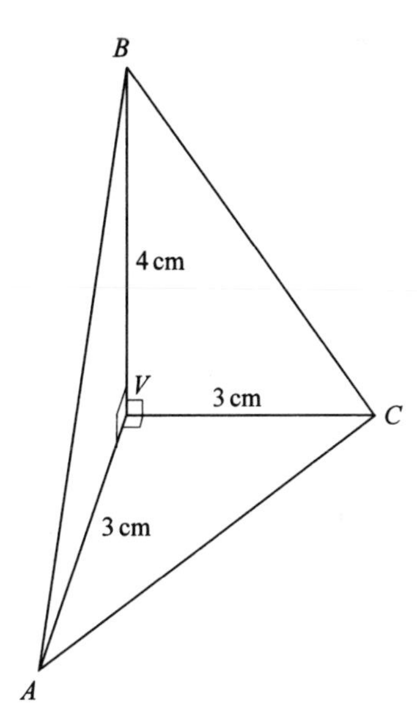

In the figure, the edges VA, VB and VC of the tetrahedron $VABC$ are mutually perpendicular. Given that $VA = VC = 3\,\text{cm}$ and $VB = 4\,\text{cm}$, calculate

(a) the volume, in cm^3, of the tetrahedron

(b) the length, in cm to 2 decimal places, of AC

(c) the angle, in degrees to the nearest tenth of a degree, between the planes BAC and VAC

(d) the area, in cm^2 to 2 decimal places, of triangle ABC

(e) the perpendicular distance, in cm to one decimal place, from V to the plane ABC. [L]

72 (a) Using the identities

$$\sin(A + B) \equiv \sin A \cos B + \cos A \sin B$$

and $\cos(A + B) \equiv \cos A \cos B - \sin A \sin B$

show that $\sin 2A \equiv 2 \sin A \cos A$

and $\cos 2A \equiv 2\cos^2 A - 1$

(b) Solve, for $0 \leqslant x \leqslant 2\pi$, giving your answers in radians, the equations

(i) $\sin 2x \sin x = \cos x$

(ii) $\sin\left(x + \frac{\pi}{3}\right) + \sin\left(x - \frac{\pi}{3}\right) = 1$

(c) In $\triangle ABC$, $AB = 3\,\text{cm}$, $AC = 5\,\text{cm}$, $\angle ABC = 2\theta^\circ$ and $\angle ACB = \theta^\circ$. Calculate, to 3 significant figures, the value of θ. [L]

73 Given that the coefficient of x in the expansion of $(1 + ax)^5$ is equal to the coefficient of x^4 in the expansion of $\left(9 + \frac{x}{3}\right)^6$, calculate the value of a. [L]

74 (a) Solve, for $0 \leqslant \theta < 360$, the equations

(i) $\tan(\theta° - 30°) = 1$

(ii) $(2\cos\theta° - 1)(\sin 2\theta° + 1) = 0$

(iii) $\cos^2\theta° - 2\sin^2\theta° = 1$

(b) Prove that $\cos^4\theta + 2\sin^2\theta - \sin^4\theta \equiv 1$.

(c) Given that $\sin\theta = x - 1$ and $\cos\theta = 2y$ for all values of θ, find an equation, not containing θ, relating x and y. [L]

75 (a) Differentiate with respect to x

(i) $e^{2x}\cos\pi x$ (ii) $\ln\dfrac{x^2+1}{2x+1}$ (iii) $\sqrt{(1+4x^2)}$

(b) Show that, for $x > 0$, the function $\dfrac{\ln x}{x}$ has a maximum at $x = e$ and no other turning point. [L]

76 A is the point with coordinates $(2, 3, -5)$, B is the point with coordinates $(-1, 4, 7)$ and C is the point with coordinates $(2, 3, -4)$. Calculate the size of $\angle BCA$.

77 In the binomial expansion of $(1 - px)^5$ in ascending powers of x, the coefficient of x^3 is 80.

(a) Find the value of p.

(b) Evaluate the coefficient of x^4 in the expansion. [L]

78 (a) *Without the use of a calculator*, find the values of

(i) $\sin 40°\cos 10° - \cos 40°\sin 10°$

(ii) $\dfrac{1}{\sqrt{2}}\cos 15° - \dfrac{1}{\sqrt{2}}\sin 15°$

(iii) $\dfrac{1 - \tan 15°}{1 + \tan 15°}$

(b) Find, to one decimal place, the values of x, $0 \leqslant x \leqslant 360$, which satisfy the equation

$$2\sin x° = \cos(x° - 60°)$$ [L]

79 Show that $\frac{\mathrm{d}}{\mathrm{d}x}\left(\frac{\cos x}{\sin x}\right) = -\frac{1}{\sin^2 x}$

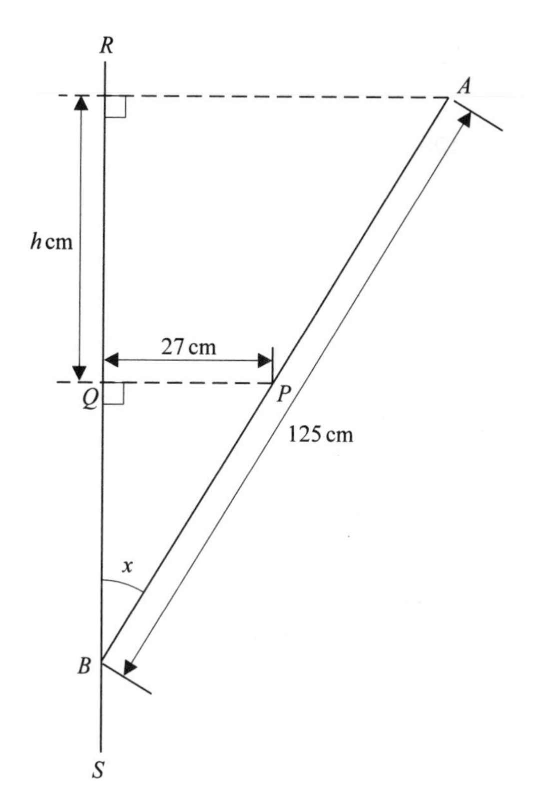

In the diagram, the straight rod AB, of length 125 cm, is supported at an angle x to the vertical by a peg P and a vertical wall RS. The rod is in a plane perpendicular to the wall and the horizontal distance PQ of the peg P from the wall is 27 cm. Show that the end A of the rod is at a vertical height h cm above P where

$$h = 125 \cos x - 27 \frac{\cos x}{\sin x}$$

(a) Find the value of $\sin x$ for which $\frac{\mathrm{d}h}{\mathrm{d}x} = 0$.

(b) Show that as x varies h attains a maximum value and find this value. [L]

80 Three landmarks P, Q and R are on the same horizontal level. Landmark Q is 3 km and on a bearing of 328° from P, landmark R is 6 km and on a bearing of 191° from Q. Calculate the distance and the bearing of R from P, giving your answers in km to one decimal place and in degrees to the nearest degree.

[L]

81 A curve is given parametrically by the equations

$$x = 1 + t^2, \quad y = 2t - 1$$

Show that an equation of the tangent to the curve at the point with parameter t is

$$ty = x + t^2 - t - 1$$

Verify that the tangent at $A(2, 1)$ passes through the point $C(6, 5)$.
Show that the line $5y = x + 19$ passes through C and is also a tangent to the curve.
Find also the coordinates of the point of contact of this line with the given curve. [L]

82 Evaluate $\dfrac{dy}{dx}$ when $y = 1$, given that

(a) $y(x + y) = 3$

(b) $x = \dfrac{1}{(4 - t)^2}$, $y = \dfrac{t}{4 - t}$, $0 < t < 4$. [L]

83 Find the first three terms in the expansion in ascending powers of x of $(1 - 2x)^7$.
Given that x is so small that x^3 and higher powers of x may be neglected, show that

$$(1 - x)(1 - 2x)^7 \equiv 1 + ax + bx^2$$

where a and b are constants to be found. [L]

84

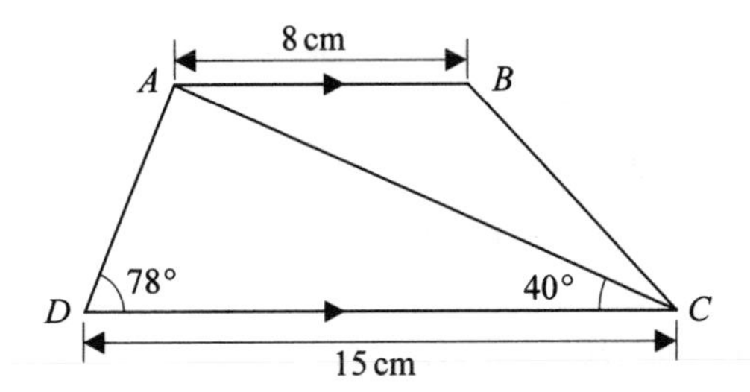

In the figure, AB is parallel to DC, $AB = 8\,\text{cm}$, $DC = 15\,\text{cm}$, $\angle ADC = 78°$ and $\angle DCA = 40°$. Calculate, giving your answers to 3 significant figures,

(a) the length, in cm, of AC

(b) the area, in cm^2, of $ABCD$.

A circle, centre D, radius DC is drawn to cut DA produced at E. Calculate, taking $\pi = 3.14$,

(c) the arc length CE, in cm, giving your answer to 3 significant figures. [L]

85 (a) Given that $180 < x < 540$, find all the value of x such that

(i) $\cos x° = \frac{1}{2}$

(ii) $\tan x° = -\tan 35°$

(b) A cuboid $ABCDEFGH$ has a square horizontal base $ABCD$, where AB is of length $2a$. The vertical edges EA, FB, GC and HD are each of length a.

(i) Calculate the length of GA.

(ii) Calculate, to the nearest half degree, the acute angle between GA and the horizontal.

(iii) Calculate, to the nearest degree, the acute angle between the plane GBD and the horizontal. [L]

86 Obtain the equation of the normal to the curve

$$y = x + \frac{4}{x}$$

at the point $(1, 5)$. Find also the equations of the two tangents to the curve which are parallel to the normal and show that the perpendicular distance between them is $\frac{8}{5}\sqrt{15}$. [L]

87 (a) Given that $x = \sin\theta° - 2\cos\theta°$
and $y = (\sqrt{3})\sin\theta°$, where $0 \leqslant \theta < 180$,
show that $x^2 + y^2 = 2(\sin\theta° - \cos\theta°)^2 + 2$
Deduce the minimum value of $(x^2 + y^2)$ and the value of θ for which it occurs.

(b) The tetrahedron $ABCD$ is such that AB, AC and AD are mutually perpendicular, $AB = AC = 3$ cm and $AD = 4$ cm.
Calculate, to the nearest 0.1°,
(i) the angle between DB and the plane ABC
(ii) the angle between the planes ABC and DBC. [L]

88 In the binomial expansion of $(1 + px)^6$ in ascending powers of x, the coefficient of x^2 is 135. Given that p is a positive integer,
(a) find the value of p
(b) evaluate the coefficient of x^3 in the expansion. [L]

89

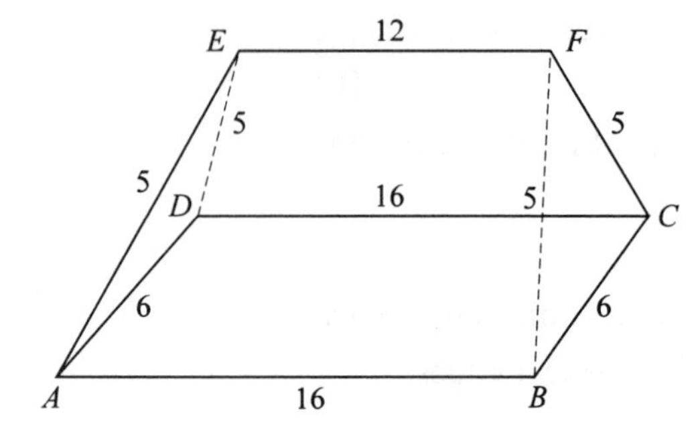

A tent is erected, as shown in the figure. The base $ABCD$ is rectangular and horizontal and the top edge EF is also horizontal.
The lengths, in metres, of the edges are

$$AE = BF = CF = DE = 5, \quad AB = CD = 16,$$
$$AD = BC = 6, \quad EF = 12$$

(a) Calculate the size of $\angle ADE$, giving your answer to the nearest degree.
(b) Show that the vertical height of EF above the base $ABCD$ is $2\sqrt{3}$ m.

Calculate, to the nearest degree, the size of the acute angle between
(c) the face ADE and the horizontal
(d) the edge AE and the horizontal. [L]

90 Water dripping from a ceiling makes a circular stain on a carpet. The area of the stain increases at the rate of $0.5\,\text{cm}^2\,\text{s}^{-1}$. Find the rate of change of the radius of the circle, in $\text{cm}\,\text{s}^{-1}$ to 2 significant figures, when the area is $20\,\text{cm}^2$. [L]

91 Find the gradient of the curve

$$y = \frac{2\sin x}{\sin x + \cos x}$$

at the point $P(\frac{\pi}{4}, 1)$.
Show that P is a point of inflexion on the curve. [L]

92 (a) Differentiate with respect to x

(i) $(5x+2)^{\frac{1}{2}}$ (ii) $\sin^3 x$

(b) The power, W watts, consumed by an electrical appliance is given by

$$W = \frac{2R}{(R+9)^2}$$

where R is the resistance of the appliance in ohms. Use differentiation to estimate the increase in W, to one significant figure, when R is increased by 0.1, in the two cases (i) $R = 3$, (ii) $R = 9$. [L]

93 (a) Solve, in radians, the equation

$$\sin\left(x + \frac{\pi}{6}\right) = 2\cos x$$

giving all solutions in the range $0 \leqslant x < 2\pi$.

(b) (i) Use the formula

$$\tan(A - B) = \frac{\tan A - \tan B}{1 + \tan A \tan B}$$

to show that

$$\tan\left(\theta - \frac{\pi}{4}\right) = \frac{\tan\theta - 1}{\tan\theta + 1}$$

(ii) Hence, or otherwise, solve, in radians to 3 significant figures, the equation

$$\tan\left(\theta - \frac{\pi}{4}\right) = 6\tan\theta$$

giving all solutions in the range $-\pi < \theta < \pi$. [L]

94 It is given that for a right circular cone, base radius r and vertical height h, the volume is $\frac{1}{3}\pi r^2 h$ and the curved surface area is $\pi r\sqrt{(r^2 + h^2)}$.

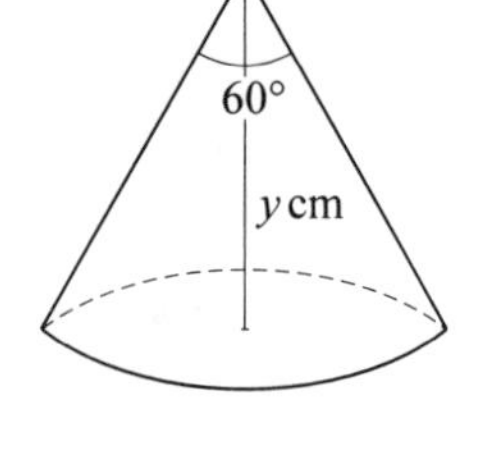

Sand falls on to a horizontal floor at a constant rate of $0.15\,\text{cm}^3\,\text{s}^{-1}$. The sand falls in a heap in the shape of a right circular cone with vertical angle 60°. At time t s after the sand starts to fall, the height of the cone is y cm. By considering the volume of sand at this instant, as shown in the figure,

(a) show that $y^3 = \dfrac{27t}{20\pi}$.

(b) Find, in $\text{cm}\,\text{s}^{-1}$ to 3 decimal places, the rate of change of the height of the conical pile after 60 s.

(c) Find, in $\text{cm}^2\,\text{s}^{-1}$ to 3 decimal places, the rate of change of the curved surface area of the conical pile after 60 s. [L]

95 A curve has parametric equations

$$x = 1 + \sqrt{(32)}\cos\theta, \quad y = 5 + \sqrt{(32)}\sin\theta, \quad 0 \leqslant \theta < 2\pi$$

Show that the tangent to the curve at the point with parameter θ is given by

$$(y - 5)\sin\theta + (x - 1)\cos\theta = \sqrt{(32)}$$

Find the two values of θ such that this tangent passes through the point $A(1, -3)$. Hence, or otherwise, find the equations of the two tangents to the curve from the point A. [L]

96 Using the formula

$$\cos(A + B) \equiv \cos A\cos B - \sin A\sin B$$

(a) show that

$$\cos(A - B) - \cos(A + B) \equiv 2\sin A\sin B$$

(b) Hence show that

$$\cos 2x - \cos 4x \equiv 2\sin 3x\sin x$$

(c) Find all solutions in the range $0 \leqslant x \leqslant \pi$ of the equation

$$\cos 2x - \cos 4x = \sin x$$

giving all your solutions in multiples of π radians. [L]

97

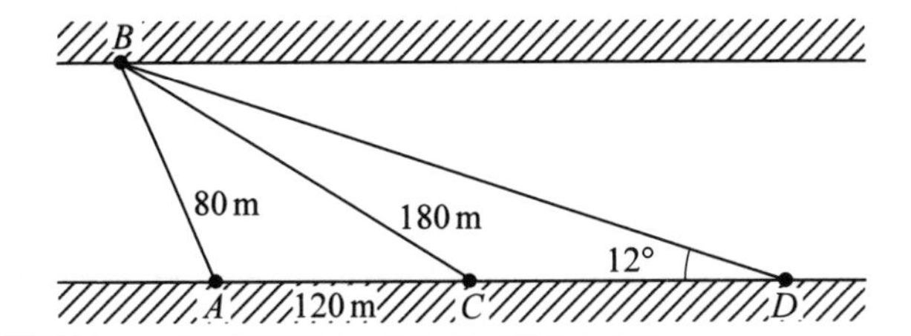

The figure shows a road with parallel straight sides. A man standing at the point A on one side of the road is 80 m from the point B on the other side of the road. He walks 120 m along the road to the point C which is 180 m from B. He then walks to the point D which is such that BD makes an angle of 12° with the side of the road. Calculate

(a) the size, in degrees to 2 decimal places, of $\angle BCA$

(b) the distance AD, in m to 3 significant figures

(c) the size, in degrees to 2 decimal places, of $\angle ABC$

(d) the shortest distance, in m to 3 significant figures, from B to the other side of the road. [L]

98 (a) Differentiate

(i) $\dfrac{(1+x)^2}{1+x^2}$

(ii) $\ln\sqrt{(1+\sin^2 x)}$

simplifying the answers where possible.

(b) If $(1+x)(2+y) = x^2 + y^2$, find $\dfrac{dy}{dx}$ in terms of x and y.

Find the gradient of the curve $(1+x)(2+y) = x^2 + y^2$ at each of the two points where the curve cuts the y-axis.

Show that there are two points at which the tangents to this curve are parallel to the y-axis. [L]

Integration

9

9.1 Integrating standard functions

In Book P1 you learned these standard integrals:

$$\int x^n \, dx = \frac{1}{n+1} x^{n+1} + C, \; n \neq -1$$

$$\int e^{kx} \, dx = \frac{1}{k} e^{kx} + C$$

$$\int \frac{1}{x} \, dx = \ln |x| + C$$

As integration is the reverse process to differentiation, you can employ the results obtained in chapter 8 to produce some more integrals. Here a, b are constants and C is the arbitrary constant of integration.

- $\int \sin x \, dx = -\cos x + C$
- $\int \sin(ax + b) \, dx = -\frac{1}{a} \cos(ax + b) + C$
- $\int \cos x \, dx = \sin x + C$
- $\int \cos(ax + b) \, dx = \frac{1}{a} \sin(ax + b) + C$
- $\int (ax + b)^n \, dx = \frac{1}{a(n+1)} (ax + b)^{n+1} + C, \; n \neq -1$
- $\int \frac{1}{ax + b} \, dx = \frac{1}{a} \ln|ax + b| + C$
- $\int e^{ax+b} \, dx = \frac{1}{a} e^{ax+b} + C$

Often, memory plays a vital role in exercises involving integration. This is why **you should memorise these integrals and the results of chapter 8**.

Example 1

Find (a) $\int \operatorname{cosec}^2 x \, dx$ (b) $\int \sec 2x \tan 2x \, dx$

(a) In chapter 8, this result was found:

$$\frac{d}{dx}(\cot x) = -\operatorname{cosec}^2 x$$

By reversing this result, you get:

$$\int \operatorname{cosec}^2 x \, dx = -\cot x + C$$

(b) In chapter 8 also, the result

$$\frac{d}{dx}(\sec x) = \sec x \tan x$$

was established, using the chain rule.

So: $$\frac{d}{dx}(\sec 2x) = 2 \sec 2x \tan 2x$$

By reversing this result you get

$$\int \sec 2x \tan 2x \, dx = \tfrac{1}{2} \sec 2x + C$$

Example 2

Evaluate $\int_0^{\frac{\pi}{3}} (\cos 3x - 2 \sin x) \, dx$

You know that

$$\int \cos 3x \, dx = \tfrac{1}{3} \sin 3x \quad \text{and} \quad \int \sin x \, dx = -\cos x$$

So: $$\int (\cos 3x - 2 \sin x) \, dx = \tfrac{1}{3} \sin 3x + 2 \cos x + C$$

$$\left[\tfrac{1}{3} \sin 3x + 2 \cos x\right]_0^{\frac{\pi}{3}} = \tfrac{1}{3} \sin \pi + 2 \cos \tfrac{\pi}{3} - (\tfrac{1}{3} \sin 0 + 2 \cos 0)$$
$$= 0 + 2(\tfrac{1}{2}) - 0 - 2(1)$$
$$= -1$$

Example 3

Evaluate $\int_1^6 \frac{1}{3x + 2} \, dx$

We have

$$\int \frac{1}{3x+2}\,\mathrm{d}x = \tfrac{1}{3}\ln|3x+2| + C$$

from the list of standard integrals.

$$\begin{aligned}\int_1^6 \frac{1}{3x+2}\,\mathrm{d}x &= \left[\tfrac{1}{3}\ln|3x+2|\right]_1^6 \\ &= \tfrac{1}{3}\ln 20 - \tfrac{1}{3}\ln 5 \\ &= \tfrac{1}{3}(\ln 20 - \ln 5) \\ &= \tfrac{1}{3}\ln\frac{20}{5} = \tfrac{1}{3}\ln 4\end{aligned}$$

Exercise 9A

Integrate with respect to x:

1 $\cos 4x$

2 $\sin 3x$

3 $\sin\tfrac{1}{2}x$

4 $\cos\dfrac{3x}{2}$

5 $\sec^2 x$

6 $\operatorname{cosec}^2 3x$

7 $\dfrac{1}{\cos^2 2x}$

8 e^{3x-2}

9 $\dfrac{1}{2x-5}$

10 $(4x-3)^3$

11 $\cos(5x+4)$

12 $\sin(3-4x)$

13 $(3-2x)^2$

14 $\dfrac{1}{(3-2x)^2}$

15 $(\mathrm{e}^x - \mathrm{e}^{-x})^2$

16 $\dfrac{1}{\sin^2\frac{1}{2}x}$

17 $\sec 3x\tan 3x$

18 $\operatorname{cosec} 2x\cot 2x$

19 $\sec^2 x - x^2$

20 $2x + \sin 2x$

Evaluate:

21 $\displaystyle\int_{-\frac{\pi}{3}}^{\frac{\pi}{2}} \sin x\,\mathrm{d}x$

22 $\displaystyle\int_{-\frac{\pi}{6}}^{\frac{\pi}{2}} \cos x\,\mathrm{d}x$

23 $\displaystyle\int_{\frac{\pi}{3}}^{\frac{2\pi}{3}} \sin 2x\,\mathrm{d}x$

24 $\displaystyle\int_0^{\frac{\pi}{4}} \sec^2 x\,\mathrm{d}x$

25 $\displaystyle\int_0^{\frac{\pi}{2}} (x + \sin x)\,\mathrm{d}x$

26 $\displaystyle\int_{\frac{\pi}{2}}^{\frac{\pi}{3}} \operatorname{cosec}^2 \tfrac{1}{2}x\,\mathrm{d}x$

27 $\displaystyle\int_0^2 \frac{1}{3x+4}\,\mathrm{d}x$

28 $\displaystyle\int_1^2 (2x-1)^3\,\mathrm{d}x$

29 $\displaystyle\int_{-1}^1 \mathrm{e}^{2x-1}\,\mathrm{d}x$

30 $\displaystyle\int_0^{\frac{2\pi}{3}} \sec\tfrac{1}{2}x\,\tan\tfrac{1}{2}x\,\mathrm{d}x$

9.2 Integration using identities

Until now you have been finding integrals by just thinking of integration as the reverse of differentiation. But often this approach does not work immediately because the function to be integrated is not integrable as it stands. In this case it is often possible to replace the function by equivalent ones using, for example, trigonometric identities, which can then be immediately integrated by using the formula

$$\int(u \pm v)\,\mathrm{d}x = \int u\,\mathrm{d}x \pm \int v\,\mathrm{d}x.$$

The following examples illustrate this.

Example 4

Find $\int \tan^2 x\,\mathrm{d}x$.

Consider the identity

$$\sec^2 x \equiv \tan^2 x + 1$$

which can be rearranged as

$$\tan^2 x \equiv \sec^2 x - 1$$

That is,

$$\int \tan^2 x\,\mathrm{d}x = \int (\sec^2 x - 1)\,\mathrm{d}x$$
$$= \int \sec^2 x\,\mathrm{d}x - \int 1\,\mathrm{d}x$$

So: $\int \tan^2 x\,\mathrm{d}x = \tan x - x + C$

Note: We know that $\int \sec^2 x\,\mathrm{d}x = \tan x + C$ because

$$\frac{\mathrm{d}}{\mathrm{d}x}(\tan x) = \sec^2 x$$

Example 5

Use the identity $\cos 2x \equiv 2\cos^2 x - 1$ to find $\int \cos^2 x\,\mathrm{d}x$.

If you rearrange the identity,

$$\cos^2 x \equiv \frac{1}{2} + \frac{1}{2}\cos 2x$$

So:
$$\int \cos^2 x \, dx = \int (\tfrac{1}{2} + \tfrac{1}{2}\cos 2x) \, dx$$
$$= \int \tfrac{1}{2} \, dx + \int \tfrac{1}{2}\cos 2x \, dx$$
$$= \tfrac{1}{2}x + \tfrac{1}{4}\sin 2x + C$$

Another way to deal with an expression that you cannot immediately integrate is to split it into partial fractions that *can* be integrated.

Example 6

Find $\displaystyle\int \frac{7-5x}{(2x-1)(x+1)} \, dx$.

First you split the expression $\dfrac{7-5x}{(2x-1)(x+1)}$ into partial fractions by writing:

$$\frac{7-5x}{(2x-1)(x+1)} \equiv \frac{A}{2x-1} + \frac{B}{x+1}$$

So:
$$7 - 5x \equiv A(x+1) + B(2x-1)$$

Let $x = -1$, then $B = -4$

Let $x = \frac{1}{2}$, then $A = 3$

Then
$$\frac{7-5x}{(2x-1)(x+1)} \equiv \frac{3}{2x-1} - \frac{4}{x+1}$$

To find the required integral, write:

$$\int \frac{7-5x}{(2x-1)(x+1)} \, dx = \int \frac{3}{2x-1} \, dx - \int \frac{4}{x+1} \, dx$$
$$= \tfrac{3}{2}\ln|2x-1| - 4\ln|x+1| + C$$

Exercise 9B

Integrate with respect to x:

1 $\sin^2 x$ **2** $\cot^2 x$ **3** $\tan^2 2x$

4 $(1 + \cos x)^2$ **5** $(1 - 2\sin x)^2$ **6** $(\cos x + \sec x)^2$

7 $\dfrac{1}{x^2 - 4}$ **8** $\dfrac{4-x}{(x-2)(x-3)}$ **9** $\dfrac{5-2x}{(x-1)(2x+1)}$

10 $\dfrac{x}{(2x+1)(3x+1)}$ **11** $\dfrac{-1}{(2x+1)(3x+1)}$

12 $\dfrac{12}{(3-2x)(3+2x)}$

13 By finding A, B and C so that

$$\frac{x^2}{x^2-1} \equiv A + \frac{B}{x-1} + \frac{C}{x+1}$$

find $\displaystyle\int \frac{x^2}{x^2-1}\,dx$.

14 Show that $\sin^2 x + 3\cos^2 x \equiv 2 + \cos 2x$.

Hence evaluate $\displaystyle\int_{\frac{\pi}{12}}^{\frac{\pi}{4}} (\sin^2 x + 3\cos^2 x)\,dx$.

15 Show that $\dfrac{4\cos 2x}{\sin^2 2x} \equiv \operatorname{cosec}^2 x - \sec^2 x$.

Hence evaluate $\displaystyle\int_{\frac{\pi}{6}}^{\frac{\pi}{3}} \frac{4\cos 2x}{\sin^2 2x}\,dx$.

16 Evaluate $\displaystyle\int_0^{\frac{\pi}{6}} (\sin 3x + \cos 2x)\,dx$.

17 Evaluate (a) $\displaystyle\int_0^{\pi} \sin^2 \tfrac{1}{4}x\,dx$ (b) $\displaystyle\int_0^{\pi} \cos^2 \tfrac{1}{4}x\,dx$.

9.3 Integration using substitutions

The integral $\int f(x)f'(x)\,dx$ is often called **the integral of a function f and its derivative f′**. By a simple substitution of $f(x) = u$, say, you can tranform the integral into an integral which is simpler, with the variable u replacing the variable x.

$$u = f(x) \Rightarrow \frac{du}{dx} = f'(x)$$

So:

$$\int f(x)f'(x)\,dx = \int u\frac{du}{dx}\,dx$$

$$= \int u\,du$$

Integrating gives

$$\int f(x)f'(x)\,dx = \tfrac{1}{2}u^2 + C$$

That is:

■ $\displaystyle\int \mathbf{f}(x)\mathbf{f}'(x)\,\mathbf{d}x = \tfrac{1}{2}[\mathbf{f}(x)]^2 + C$

Also you should note the general results

■ $\displaystyle\int [\mathbf{f}(x)]^n\mathbf{f}'(x)\,\mathbf{d}x = \frac{1}{n+1}[\mathbf{f}(x)]^{n+1} + C,\ n \neq -1$

■ $\displaystyle\int \frac{\mathbf{f}'(x)}{\mathbf{f}(x)}\,\mathbf{d}x = \ln|\mathbf{f}(x)| + C$

Once you are thoroughly practised in the techniques of integration, you will find that, in some simple cases, you can recognise a function and its derivative just by looking at the expression to be integrated. If you can do this, you can move straight to the answer without employing a change of variable. You are advised, however, in the early stages of this work to use the substitution method.

Example 7

Use the substitution $u = \cos x$ to find $\int \tan x\,\mathrm{d}x$.

First you need to write

$$\int \tan x\,\mathrm{d}x = \int \frac{\sin x}{\cos x}\,\mathrm{d}x$$

Let $u = \cos x$, then

$$\frac{\mathrm{d}u}{\mathrm{d}x} = -\sin x$$

and
$$\sin x = -\frac{\mathrm{d}u}{\mathrm{d}x}$$

Then:
$$\int \frac{\sin x}{\cos x}\,\mathrm{d}x = \int \frac{1}{u}\cdot\left(-\frac{\mathrm{d}u}{\mathrm{d}x}\right)\mathrm{d}x = -\int \frac{1}{u}\,\mathrm{d}u$$

Integrating gives:

$$\int \frac{1}{u}\,\mathrm{d}u = \ln|u| + C$$

But $\int \tan x\,\mathrm{d}x = -\int \frac{1}{u}\,\mathrm{d}u$ and $u = \cos x$.

So:
$$\int \tan x\,\mathrm{d}x = -\ln|\cos x| + C$$

But $\sec x = \dfrac{1}{\cos x}$ and

$$\ln\sec x = \ln\frac{1}{\cos x} = \ln 1 - \ln\cos x$$
$$= -\ln\cos x$$

So: $$\int \tan x \, dx = \ln|\sec x| + C$$

You can use a similar method to show that

$$\int \cot x \, dx = \ln|\sin x| + C$$

Example 8

Use the substitution $u^2 = x + 1$ to find

$$\int \frac{x}{(x+1)^{\frac{1}{2}}} \, dx$$

Differentiating $x + 1 = u^2$ gives $\frac{dx}{du} = 2u$;

also: $(x+1)^{\frac{1}{2}} = u$ and $x = u^2 - 1$

So: $$\int \frac{x}{(x+1)^{\frac{1}{2}}} \, dx = \int \frac{u^2 - 1}{u} \, dx$$

You cannot integrate a function of u with respect to x and so you must replace the 'dx' by writing

$$dx = \frac{dx}{du} \cdot du$$

This technique is analogous to the chain rule in differentiation. It changes the variable of integration from x to u.

That is $$\int \frac{x}{(x+1)^{\frac{1}{2}}} \, dx = \int \left(\frac{u^2 - 1}{u}\right)\left(\frac{dx}{du}\right) du$$

$$= \int \frac{u^2 - 1}{u} (2u) \, du$$

$$= \int (2u^2 - 2) \, du$$

Integrating with respect to u gives

$$\int \frac{x}{(x+1)^{\frac{1}{2}}} \, dx = \tfrac{2}{3}u^3 - 2u + C$$

Reverting to the variable x gives

$$\int \frac{x}{(x+1)^{\frac{1}{2}}} \, dx = \tfrac{2}{3}(x+1)^{\frac{3}{2}} - 2(x+1)^{\frac{1}{2}} + C$$

It is important to appreciate that you do not have to revert to the original variable when evaluating a *definite* integral. Suppose that you were required to evaluate

$$\int_0^3 \frac{x}{(x+1)^{\frac{1}{2}}}\,\mathrm{d}x$$

As $u^2 = x + 1$, you can see that at $x = 0$, $u = 1$ and at $x = 3$, $u = 2$.

So : $$\int_0^3 \frac{x}{(x+1)^{\frac{1}{2}}}\,\mathrm{d}x = \int_1^2 (2u^2 - 2)\,\mathrm{d}u = \left[\tfrac{2}{3}u^3 - 2u\right]_1^2$$

$$= \tfrac{16}{3} - 4 - (\tfrac{2}{3} - 2)$$

$$= \tfrac{8}{3} = 2\tfrac{2}{3}$$

Example 9

Use the substitution $x = \operatorname{cosec} t$ to evaluate

$$\int_{\frac{2}{\sqrt{3}}}^2 \frac{1}{x^2\sqrt{(x^2-1)}}\,\mathrm{d}x$$

If $x = \operatorname{cosec} t$, then $\dfrac{\mathrm{d}x}{\mathrm{d}t} = -\operatorname{cosec} t \cot t$ (see chapter 8, section 8.11)

and $\sqrt{(x^2 - 1)} = \sqrt{(\operatorname{cosec}^2 t - 1)} = \sqrt{(\cot^2 t)} = \cot t$

(from using the identity $\operatorname{cosec}^2 t \equiv 1 + \cot^2 t$).

At $x = 2$: $\operatorname{cosec} t = 2 \Rightarrow t = \dfrac{\pi}{6}$

At $x = \dfrac{2}{\sqrt{3}}$: $\operatorname{cosec} t = \dfrac{2}{\sqrt{3}} \Rightarrow t = \dfrac{\pi}{3}$

So: $$\int \frac{1}{x^2\sqrt{(x^2-1)}}\,\mathrm{d}x = \int \frac{1}{x^2\sqrt{(x^2-1)}}\,\frac{\mathrm{d}x}{\mathrm{d}t}\,\mathrm{d}t$$

and $$\int_{\frac{2}{\sqrt{3}}}^2 \frac{1}{x^2\sqrt{(x^2-1)}}\,\mathrm{d}x = \int_{\frac{\pi}{3}}^{\frac{\pi}{6}} \frac{1}{\operatorname{cosec}^2 t \cot t}\,(-\operatorname{cosec} t \cot t)\,\mathrm{d}t$$

$$= -\int_{\frac{\pi}{3}}^{\frac{\pi}{6}} \frac{1}{\operatorname{cosec} t}\,\mathrm{d}t = -\int_{\frac{\pi}{3}}^{\frac{\pi}{6}} \sin t\,\mathrm{d}t$$

Integrating gives

$$\int_{\frac{2}{\sqrt{3}}}^2 \frac{1}{x^2\sqrt{(x^2-1)}}\,\mathrm{d}x = \Big[\cos t\Big]_{\frac{\pi}{3}}^{\frac{\pi}{6}} = \cos\tfrac{\pi}{6} - \cos\tfrac{\pi}{3}$$

$$= \frac{\sqrt{3}}{2} - \frac{1}{2} = \frac{\sqrt{3}-1}{2}$$

Exercise 9C

In each of the following find the integral by using the substitution given. Your final result should be given in terms of x.

1 $\int \sin^3 x \cos x \, dx; \quad u = \sin x$

2 $\int \tan^2 x \sec^2 x \, dx; \quad u = \tan x$

3 $\int x(x^2+1)^3 \, dx; \quad u = x^2 + 1$

4 $\int x^3 \sqrt{(x^4 - 1)} \, dx; \quad u = x^4$

5 $\int \frac{x}{\sqrt{(x^2-1)}} \, dx; \quad u = x^2$

6 $\int \sec^3 x \tan x \, dx; \quad u = \sec x$

7 $\int x e^{x^2} \, dx; \quad u = x^2$

8 $\int \frac{(\ln x)^2}{x} \, dx; \quad u = \ln x$

9 $\int \left(\frac{x}{x+1}\right)^2 dx; \quad u = x + 1$

10 $\int \frac{x}{\sqrt{(x+1)}} \, dx; \quad u^2 = x + 1$

11 Use the substitution $u = x - 1$ to evaluate $\int_2^5 \frac{x}{\sqrt{(x-1)}} \, dx$.

12 Show that $\int_0^1 \frac{x}{\sqrt{(1+x)}} \, dx = \frac{2}{3}(2 - \sqrt{2})$.

13 Use the substitution $u = \sin x$ to evaluate $\int_{\frac{\pi}{6}}^{\frac{\pi}{2}} \cos 2x \cos x \, dx$.

14 By using the substitution $u = \sin x$, evaluate

(a) $\int_0^{\frac{\pi}{2}} e^{\sin x} \cos x \, dx$ (b) $\int_0^{\frac{\pi}{3}} \sin 2x \sin x \, dx$

(c) $\int_0^{\frac{\pi}{2}} \frac{\cos x}{4 + \sin x} \, dx$

15 Evaluate (a) $\int_0^{\frac{\pi}{4}} \tan x \, \mathrm{d}x$ (b) $\int_{\frac{\pi}{4}}^{\frac{\pi}{2}} \cot x \, \mathrm{d}x$ and interpret your answers geometrically.

9.4 Integration by parts

Remember the product formula for differentiation:

$$\frac{\mathrm{d}}{\mathrm{d}x}(uv) = v\frac{\mathrm{d}u}{\mathrm{d}x} + u\frac{\mathrm{d}v}{\mathrm{d}x}$$

You can rewrite it as:

$$v\frac{\mathrm{d}u}{\mathrm{d}x} = \frac{\mathrm{d}}{\mathrm{d}x}(uv) - u\frac{\mathrm{d}v}{\mathrm{d}x}$$

By integrating this equation, you obtain

■ $$\int v\frac{\mathrm{d}u}{\mathrm{d}x}\,\mathrm{d}x = uv - \int u\frac{\mathrm{d}v}{\mathrm{d}x}\,\mathrm{d}x$$

This is the formula to use when you need to integrate the product of two functions of x, v and $\frac{\mathrm{d}u}{\mathrm{d}x}$. The formula can be applied as long as you are able to differentiate v to get $\frac{\mathrm{d}v}{\mathrm{d}x}$ and you are able to integrate $\frac{\mathrm{d}u}{\mathrm{d}x}$ to get u. Even when you *can* do this, the result may not be helpful. It is regarded as helpful when $\int u\frac{\mathrm{d}v}{\mathrm{d}x}\,\mathrm{d}x$ is a less complicated expression than $\int v\frac{\mathrm{d}u}{\mathrm{d}x}\,\mathrm{d}x$. The following examples illustrate this process, which is called **integration by parts**.

Example 10

Find $\int x\mathrm{e}^x \, \mathrm{d}x$

Take $v = x$, then $\frac{\mathrm{d}v}{\mathrm{d}x} = 1$

Take $\frac{\mathrm{d}u}{\mathrm{d}x} = \mathrm{e}^x$, then $u = \mathrm{e}^x$

Using the formula

$$\int v\frac{\mathrm{d}u}{\mathrm{d}x}\,\mathrm{d}x = uv - \int u\frac{\mathrm{d}v}{\mathrm{d}x}\,\mathrm{d}x$$

you have:

$$\int x\mathrm{e}^x\,\mathrm{d}x = x\mathrm{e}^x - \int \mathrm{e}^x \cdot 1\,\mathrm{d}x$$

$$= x\mathrm{e}^x - \mathrm{e}^x + C$$

Example 11

Find $\int x \cos x \, dx$

Here you take $v = x$, then $\frac{dv}{dx} = 1$ and $\frac{du}{dx} = \cos x$, then $u = \sin x$

Applying $$\int v \frac{du}{dx} \, dx = uv - \int u \frac{dv}{dx} \, dx$$

you get: $$\int x \cos x \, dx = x \sin x - \int \sin x \cdot 1 \, dx$$
$$= x \sin x + \cos x + C$$

Example 12

Find $\int \ln x \, dx$, $x > 0$

In this case we write $v = \ln x$, then $\frac{dv}{dx} = \frac{1}{x}$

and $\frac{du}{dx} = 1$, then $u = x$

Using $$\int v \frac{du}{dx} \, dx = uv - \int u \frac{dv}{dx} \, dx$$

you get: $$\int \ln x \, dx = x \ln x - \int x \frac{1}{x} \, dx = x \ln x - x + C$$

In some cases, the process of integration by parts needs to be repeated.

Example 13

Find $\int x^2 e^{2x} \, dx$

Take $v = x^2$, then $\frac{dv}{dx} = 2x$ and $\frac{du}{dx} = e^{2x}$, then $u = \frac{1}{2}e^{2x}$

Applying $$\int v \frac{du}{dx} \, dx = uv - \int u \frac{dv}{dx} \, dx$$

you get: $$\int x^2 e^{2x} \, dx = \tfrac{1}{2}x^2 e^{2x} - \int (\tfrac{1}{2}e^{2x})(2x) \, dx$$
$$= \tfrac{1}{2}x^2 e^{2x} - \int x e^{2x} \, dx \qquad (1)$$

Now you need to integrate by parts again to find $\int xe^{2x}\,dx$.

Take $v = x$, then $\dfrac{dv}{dx} = 1$

and $\dfrac{du}{dx} = e^{2x}$, then $u = \frac{1}{2}e^{2x}$

Applying the 'parts' formula gives

$$\int xe^{2x}\,dx = \tfrac{1}{2}xe^{2x} - \int \tfrac{1}{2}e^{2x}\cdot 1\,dx$$
$$= \tfrac{1}{2}xe^{2x} - \tfrac{1}{4}e^{2x} \qquad (2)$$

By combining the results (1) and (2) you get

$$\int x^2e^{2x}\,dx = \tfrac{1}{2}x^2e^{2x} - \tfrac{1}{2}xe^{2x} + \tfrac{1}{4}e^{2x} + C$$
$$= \tfrac{1}{4}e^{2x}(2x^2 - 2x + 1) + C$$

Notice that we leave out the constant of integration until the final line.

It is easy to start applying integration by parts, and then find that the process is getting *more* complicated instead of simpler. Suppose that for $\int x^2e^{2x}\,dx$, you took

$v = e^{2x}$ giving $\dfrac{dv}{dx} = 2e^{2x}$

and $\dfrac{du}{dx} = x^2$ giving $u = \frac{1}{3}x^3$

Then, by applying 'parts' you get

$$\int x^2\,e^{2x}\,dx = \tfrac{1}{3}x^3\,e^{2x} - \int \tfrac{2}{3}x^3\,e^{2x}\,dx$$

which is *more* complicated than $\int x^2\,e^{2x}\,dx$. In this case, **you have started to work in the wrong direction**. You would have to start again with $v = x^2$ and $\dfrac{du}{dx} = e^{2x}$, as shown in Example 13.

Exercise 9D

Use integration by parts to find:

1 $\int xe^{-x}\,dx$ **2** $\int xe^{3x}\,dx$ **3** $\int x\sin x\,dx$

4 $\int x\ln x\,dx$ **5** $\int \ln(x-1)\,dx$ **6** $\int x\cos 3x\,dx$

7 $\int x(x-1)^4\,dx$ **8** $\int x\sqrt{(x-1)}\,dx$ **9** $\int x^2e^x\,dx$

10 $\int x^2\cos x\,dx$ **11** $\int x^2e^{-x}\,dx$ **12** $\int x^3\ln x\,dx$

Evaluate each of the following definite integrals:

13 $\int_0^{\pi} x\sin x\,dx$ **14** $\int_0^{\frac{\pi}{2}} x\cos\frac{1}{2}x\,dx$ **15** $\int_1^{e} x^2\ln x\,dx$

16 $\int_0^1 x(x-1)^3\,dx$ **17** $\int_0^2 (x-1)(x+1)^3\,dx$

18 $\int_1^{e} \frac{\ln x}{x^4}\,dx$ **19** $\int_1^{e} (\ln x)^2\,dx$ **20** $\int_0^{\frac{\pi}{2}} e^x\sin x\,dx$

9.5 A systematic approach to integration

At this stage, you should review the various methods of integration that you have met in this chapter and in chapter 9 of Book P1.

- **The standard forms given at the end of this chapter in the summary should all be memorised.**
- **You should be able to recognise expressions that can be integrated at once.**

Example 14

(a) $$\int (4x-1)^5\,dx = \frac{(4x-1)^6}{4\times 6} + C$$

$$= \tfrac{1}{24}(4x-1)^6 + C$$

(b) $$\int e^{-4x}\,dx = -\tfrac{1}{4}e^{-4x} + C$$

(c) $\int \sin \frac{1}{3}x \, dx = -3\cos \frac{1}{3}x + C$

- **You should be able to convert expressions to equivalent forms that *can* be integrated at once, using trigonometric identities and partial fractions.**

Example 15

(a) $\int \tan^2 3x \, dx = \int (\sec^2 3x - 1) \, dx$

$$= \tfrac{1}{3}\tan 3x - x + C$$

(b) $\int \frac{1}{x(x+1)} \, dx = \int \left(\frac{1}{x} - \frac{1}{x+1} \right) dx$

$$= \ln|x| - \ln|x+1| + C$$

- **You should recognise an expression that is given as a function and its derivative.**

Example 16

(a) $\int x e^{x^2} \, dx = \frac{1}{2} e^{x^2} + C$

$\begin{pmatrix} \text{function of } x^2 \\ \text{derivative } 2x \end{pmatrix}$ OR the substitution $u = x^2$ could be used

(b) $\int \frac{x}{(x^2+1)^{\frac{1}{2}}} \, dx = (x^2+1)^{\frac{1}{2}} + C$

$\begin{pmatrix} \text{function of } x^2 \\ \text{derivative } 2x \end{pmatrix}$ OR the substitution $u = x^2 + 1$ could be used

(c) $\int \tan^3 2x \sec^2 2x \, dx = \frac{1}{8} \tan^4 2x + C$

$\begin{pmatrix} \text{function of } \tan 2x \\ \text{derivative } 2\sec^2 2x \end{pmatrix}$ OR the substitution $u = \tan 2x$ could be used

- **Recognise when you need to use integration by parts.**

Look back at the examples given in section 9.4.

In your work you will need to identify which approach is required and choose your method accordingly. In some cases you may be given a hint about the method or the substitution to employ. Practice is the key to success and the following exercises are typical of what is expected by ULEAC examiners.

Exercise 9E

Integrate with respect to x:

1 $(4x+5)^{\frac{1}{2}}$

2 $\dfrac{1}{4x+5}$

3 $\left(1-\dfrac{1}{x}\right)^2$

4 $\cos x \sin x$

5 $\tan 3x$

6 $x \sin 3x$

7 $\dfrac{1+x}{x^{\frac{1}{2}}}$

8 $\dfrac{x}{1+x}$

9 $\sin x \cos^4 x$

10 $3 \ln x$

11 $\dfrac{x+2}{x(x-1)}$

12 $\dfrac{\sec^2 x}{(1+\tan x)^3}$

13 $\sin^2 2x$

14 $\dfrac{x^2}{x-2}$

15 $(\sin x + 2\cos x)^2$

16 $x^2 e^{\frac{1}{2}x}$

17 $\dfrac{1}{x^2-4}$

18 $\dfrac{x}{9x^2+1}$

19 $(1-x^{-2})^2$

20 $(2-3x)^{-2}$

21 $(4-5x)^{-1}$

22 $\cot 3x$

23 $\operatorname{cosec} 2x \cot 2x$

24 $\cot^2 3x$

25 $x \cos 5x$

26 $\dfrac{x}{(x-1)^{\frac{1}{2}}}$

27 $x^2 e^{-x}$

28 $\cos 2x \sin x$

29 $\sin 2x \cos x$

30 $\tan 2x \sec 2x$

31 $\dfrac{(x+1)^2}{x^2+1}$

32 $\dfrac{2}{(x-2)(x-4)}$

33 $\dfrac{1}{x^2(x-1)}$

34 $\operatorname{cosec}^2 2x + 1$

35 $\dfrac{x+4}{x-4}$

36 $\dfrac{1}{x(x^2-1)}$

37 $\dfrac{x^2}{x^3+1}$

38 $(e^x + x)^2$

39 $x^3 \ln x$

40 $x^3 e^{x^2}$

41 Use the identity $\cos^2 x + \sin^2 x \equiv 1$ and the substitution $\cos x = u$ to find $\int \sin^3 x \, dx$.

42 Find $\int \cos^3 x \, dx$ and $\int \sin^5 x \, dx$.

43 Use the identity $\sec^2 x \equiv \tan^2 x + 1$ and the substitution $\tan x = u$ to find $\int \tan^4 x \, dx$.

44 Find (a) $\int \sec^4 x \, dx$ (b) $\int \cot^4 x \, dx$.

45 Use the identity $\sin(A+B)+\sin(A-B) \equiv 2\sin A\cos B$ to find

(a) $\int 2\sin 6x \cos 4x \, dx$

(b) $\int \sin x \cos \frac{1}{2}x \, dx$

46 Evaluate $\int_0^1 \frac{x+9}{(x+2)(3-2x)} \, dx$.

47 Use the substitution $x = 3\sin t$ to show that

$$\int_0^3 x^2(9-x^2)^{\frac{1}{2}} \, dx = \tfrac{81}{16}\pi$$

48 Evaluate $\int_{\frac{\pi}{6}}^{\frac{\pi}{3}} \sec^3 x \tan x \, dx$.

49 Evaluate $\int_3^4 \frac{2x+4}{(x-2)(x^2+4)} \, dx$

50 Evaluate $\int_1^2 \frac{x}{(1+x^2)} \, dx$.

9.6 The area under a curve

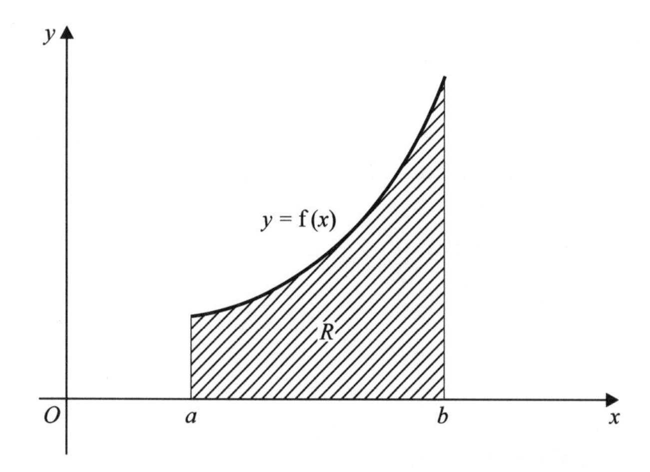

The diagram shows part of the curve $y = f(x)$ between the values $x = a$ and $x = b$. The finite region R is bounded by the curve, the lines $x = a$ and $x = b$ and the x-axis.

You can find an approximate value for the area of R by dividing R into small strips parallel to the y-axis as shown below.

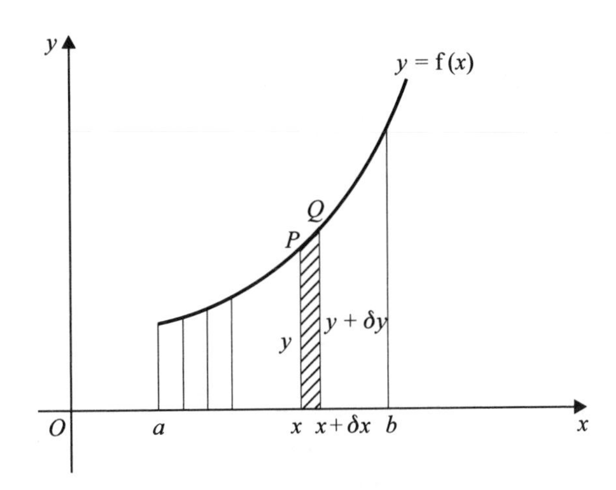

If you divide R into a *very* large number of thin strips (much more than can be shown in the diagram), the curved top of each strip will be almost a straight line. So each strip could be regarded as a trapezium and the area of R is approximately the sum of the areas of all the trapezia.

Take a typical strip by considering two neighbouring points $(x, 0)$ and $(x + \delta x, 0)$ on the x-axis. The corresponding points on the curve $y = \mathrm{f}(x)$ are P and Q with coordinates (x, y) and $(x + \delta x, y + \delta y)$ respectively. δx and δy are very small quantities (increments). If you think of the strip as a trapezium, its area is:

$$\delta x\left(\frac{y + \delta y + y}{2}\right)$$
$$= y\delta x + \tfrac{1}{2}\delta y \cdot \delta x$$
$$\approx y\delta x$$

(because $\frac{1}{2}\delta y\delta x$ is so small that it can be left out).

Then:

$$\text{area of } R \approx \sum_{x=a}^{x=b} y\,\delta x$$

that is, the sum of the areas of all the strips from $x = a$ to $x = b$.

The more strips you take, the more accurate your approximation to the area of R will be. By using more advanced mathematics than you will find in an A-level course, it can be shown that, as you increase the number of strips indefinitely, that is, making $\delta x \to 0$,

$$\lim_{\delta x \to 0} \sum_{x=a}^{x=b} y\,\delta x = \int_a^b y\,dx = \int_a^b f(x)\,dx$$

In Book P1, chapter 9, you were given this formula in order to evaluate areas of regions. Now you have been given an explanation about where the formula comes from. That is, the summation tends to the value of the definite integral $\int_a^b f(x)\,dx$, which is the limiting value **and** the actual area of R exactly.

- **Area of $R = \int_a^b f(x)\,dx$**

This result is often called the **fundamental theorem of integral calculus** and its importance is paramount because all integration theory is based on it. In your work, you are expected to assume the validity of this result, and to use it for the growing list of functions you are able to integrate.

In a similar way it can be shown that the area of the region bounded by the curve $x = g(y)$, the y-axis and the lines $y = p$ and $y = q$ is

$$\int_p^q x\,dy = \int_p^q g(y)\,dy$$

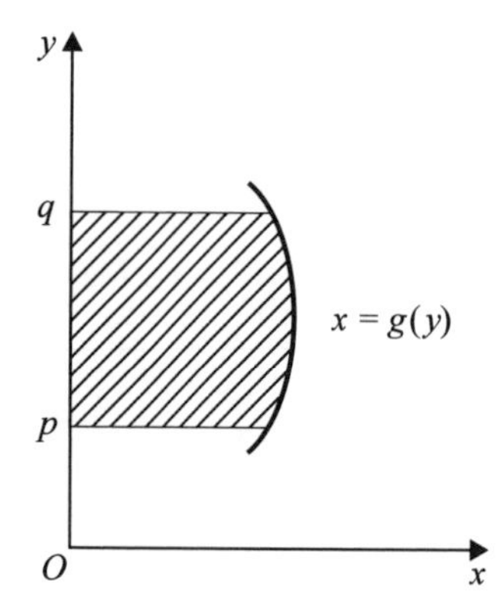

Example 17

The region R is bounded by the curve $y = \cos^2 2x$, the x-axis and the lines $x = 0$ and $x = \frac{\pi}{6}$. Find the area of R.

First draw a sketch, showing R:

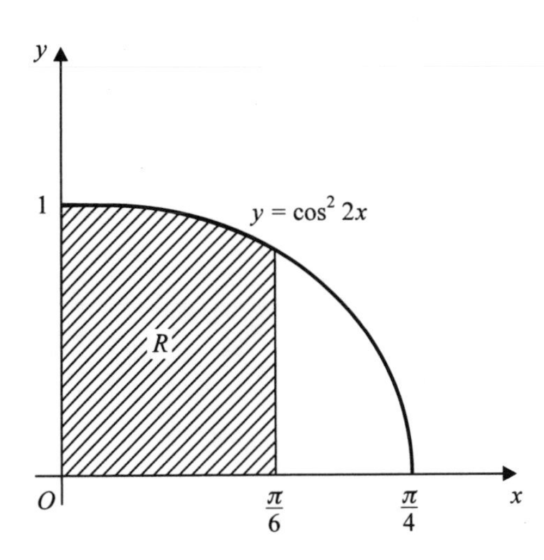

$$\text{Area of } R = \int_0^{\frac{\pi}{6}} \cos^2 2x \, \mathrm{d}x$$

Use the identity $\cos 2A \equiv 2\cos^2 A - 1$, putting $A = 2x$:

$$\cos 4x \equiv 2\cos^2 2x - 1$$

$$\Rightarrow \qquad \cos^2 2x \equiv \frac{1 + \cos 4x}{2}$$

So:

$$\text{area of } R = \int_0^{\frac{\pi}{6}} \left(\tfrac{1}{2} + \tfrac{1}{2}\cos 4x\right) \mathrm{d}x$$

$$\text{Area of } R = \left[\tfrac{1}{2}x + \tfrac{1}{8}\sin 4x\right]_0^{\frac{\pi}{6}}$$

$$= \frac{\pi}{12} + \tfrac{1}{8}\sin\frac{2\pi}{3} - (0 + 0)$$

$$= \frac{\pi}{12} + \frac{\sqrt{3}}{16} \qquad \left(\text{since } \sin\frac{2\pi}{3} = \frac{\sqrt{3}}{2}\right)$$

9.7 Volumes of revolution

You have seen that the area of the region R bounded by the curve $y = \mathrm{f}(x)$, the x-axis and the lines $x = a$ and $x = b$ can be found by dividing the region into elementary strips of width δx and summing the area of all the strips as $\delta x \to 0$. A typical strip is shown in the left-hand diagram below.

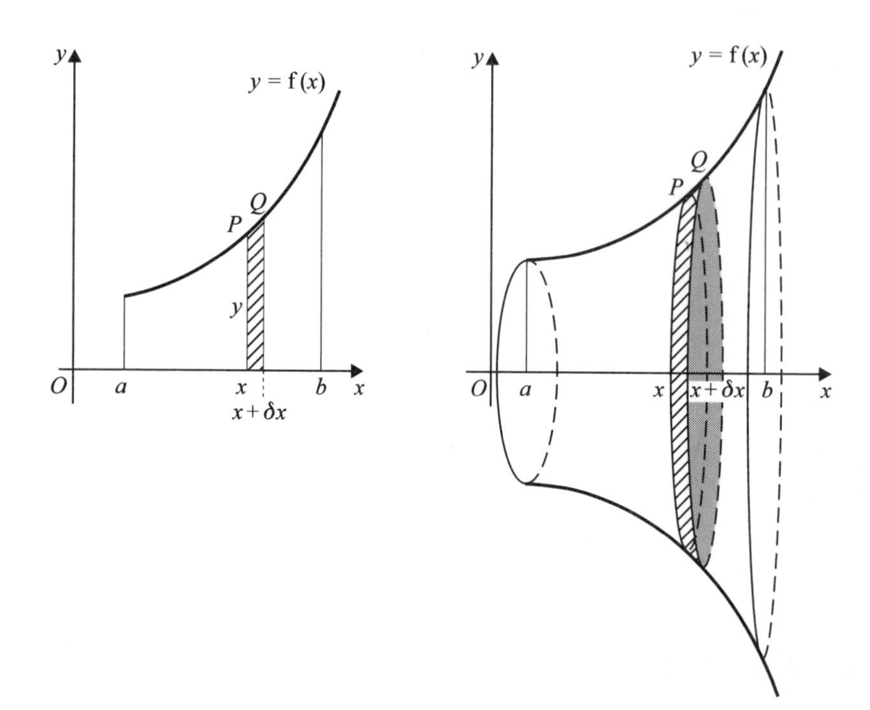

Suppose now that the region R is rotated completely (that is, through 2π radians) about the x-axis. A solid S will be formed as shown in the right-hand diagram. Each end of the solid is a circle, their radii being a and b.

Now think about what happens to the strip of length y and thickness δx when it is rotated completely about the x-axis. The rotating strip forms a circular disc of radius y and thickness δx. The volume of this disc is $\pi y^2 \delta x$.

The volume of S is the sum of all such discs from $x = a$ to $x = b$. As δx is made progressively smaller, the successive approximations to the volume of S approach the actual value of the volume and we can assume:

$$\text{Volume of } S = \lim_{\delta x \to 0} \sum_{x=a}^{x=b} \pi y^2 \, \delta x = \int_a^b \pi y^2 \, dx$$

$$= \int_a^b \pi [f(x)]^2 \, dx$$

We have then

■ **Volume of $S = \int_a^b \pi [f(x)]^2 \, dx$**

In the same way, you can think about a region bounded by the curve $x = g(y)$, the y-axis and the lines $y = p$ and $y = q$. If this region is rotated completely about the y-axis, the solid formed has volume

$$\int_p^q \pi[g(y)]^2 \, dy$$

This volume is often referred to as the **volume of revolution** and is said to be *generated* by the rotation of the curve.

Example 18

The finite region bounded by the curve $y = x - x^2$ and the x-axis is rotated through 2π radians about the x-axis. Find the volume of revolution so formed.

First, sketch the curve $y = x - x^2$:

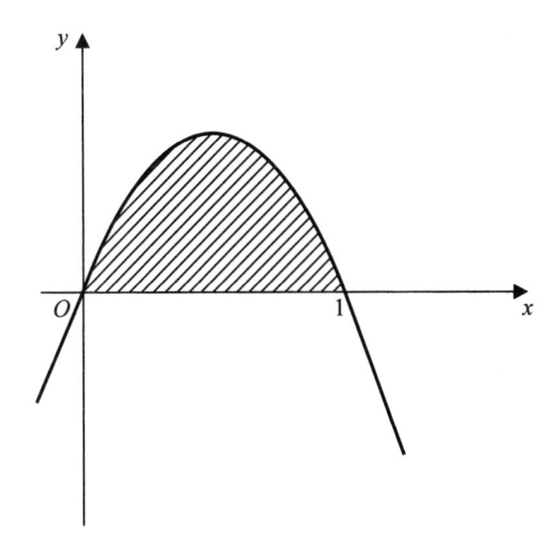

The curve meets the x-axis at points where $x - x^2 = 0$;

that is: $\quad x(1 - x) = 0$

So: $\quad x = 0$ and $x = 1$

The finite region is shown shaded.

$$\text{Volume of revolution} = \int_0^1 \pi y^2 \,dx$$

$$= \pi \int_0^1 (x - x^2)^2 \,dx$$

$$= \pi \int_0^1 (x^2 - 2x^3 + x^4) \,dx$$

$$= \pi \left[\frac{x^3}{3} - \frac{2x^4}{4} + \frac{x^5}{5} \right]_0^1$$

$$= \pi\left(\tfrac{1}{3} - \tfrac{1}{2} + \tfrac{1}{5}\right) = \frac{\pi}{30}$$

Example 19

The region R shown is bounded by the curve $y = \ln x$, the y-axis and the lines $y = 2$ and $y = 5$.

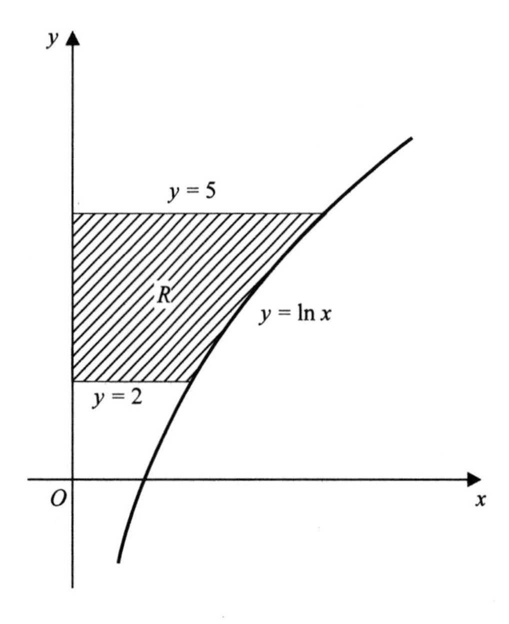

Find

(a) the area of R (b) the volume generated when R is rotated through 2π about the y-axis, giving your answers to 3 significant figures.

(a) Rewrite the equation $y = \ln x$ in the form $x = e^y$ because to find the area of R you need to evaluate

$$\int_2^5 x \,dy = \int_2^5 e^y \,dy$$

$$\text{Area of } R = \left[e^y\right]_2^5 = e^5 - e^2 \approx 141 \text{ units}^2$$

(b) The volume generated when R is rotated through 2π about the y-axis is

$$\int_2^5 \pi x^2 \, dy = \pi \int_2^5 e^{2y} \, dy$$
$$= \pi \left[\tfrac{1}{2} e^{2y}\right]_2^5$$
$$= \frac{\pi}{2}(e^{10} - e^4) \approx 34\,500 \text{ units}^3$$

Example 20

The normal to the curve $y = e^x$ at the point $B(1, e)$ meets the x-axis at the point C. The finite region bounded by the curve $y = e^x$, the line BC, the y-axis and the x-axis is rotated through 2π radians about the x-axis. Find the volume of revolution so generated.

For $y = e^x$,

$$\frac{dy}{dx} = e^x = e \text{ at } x = 1$$

So the gradient of the normal to $y = e^x$ at $B(1, e)$ is $-\dfrac{1}{e}$.

The equation of the normal to $y = e^x$ at B is

$$y - e = -\frac{1}{e}(x - 1)$$

The normal meets the x-axis at $y = 0$.

So: $$0 - e = -\frac{1}{e}(x - 1)$$

$$x - 1 = e^2$$

and: $$x = e^2 + 1$$

The point C is $(e^2 + 1, 0)$.

The region to be rotated looks like this:

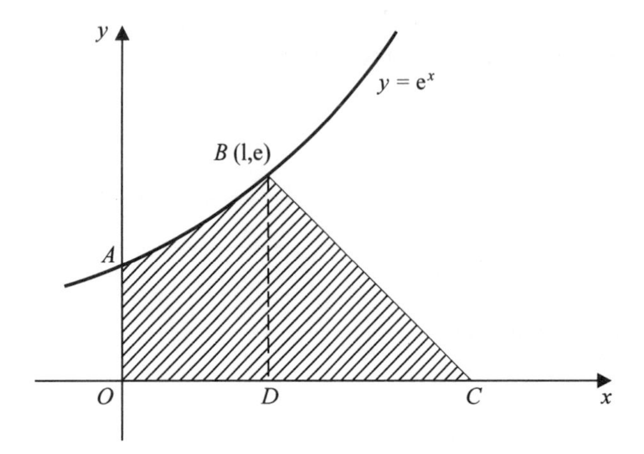

Part of the region lies under the curve from A to B and the rest lies under the line BC. The point D on the x-axis has the same x-coordinate as B; that is, it is at $(1, 0)$. When $\triangle BDC$ is rotated about the x-axis it forms a cone of base radius $BD = \mathrm{e}$ and height $DC = \mathrm{e}^2$ because

$$DC = OC - OD = (1 + \mathrm{e}^2) - 1 = \mathrm{e}^2$$

The volume of a cone is $\frac{1}{3}$ (base area) $\times$ height and the volume of the cone formed by the complete rotation of $\triangle DBC$ about the x-axis is

$$\tfrac{1}{3}\pi \times BD^2 \times DC = \tfrac{1}{3}\pi \times \mathrm{e}^2 \times \mathrm{e}^2 = \tfrac{1}{3}\pi\mathrm{e}^4$$

The volume generated when the region $ABDO$ is rotated completely about the x-axis is:

$$\begin{aligned}\int_0^1 \pi(\mathrm{e}^x)^2\,\mathrm{d}x &= \int_0^1 \pi\mathrm{e}^{2x}\,\mathrm{d}x \\ &= \tfrac{1}{2}\pi\left[\mathrm{e}^{2x}\right]_0^1 \\ &= \tfrac{1}{2}\pi(\mathrm{e}^2 - 1)\end{aligned}$$

$$\begin{aligned}\text{Total volume generated} &= \tfrac{1}{3}\pi\mathrm{e}^4 + \tfrac{1}{2}\pi\mathrm{e}^2 - \tfrac{1}{2}\pi \\ &= \tfrac{1}{6}\pi(2\mathrm{e}^4 + 3\mathrm{e}^2 - 3)\end{aligned}$$

Example 21

The curve given parametrically by

$$x = \tan t,\ y = \sin t,\ 0 \leqslant t < \tfrac{\pi}{2}$$

is sketched in the diagram.

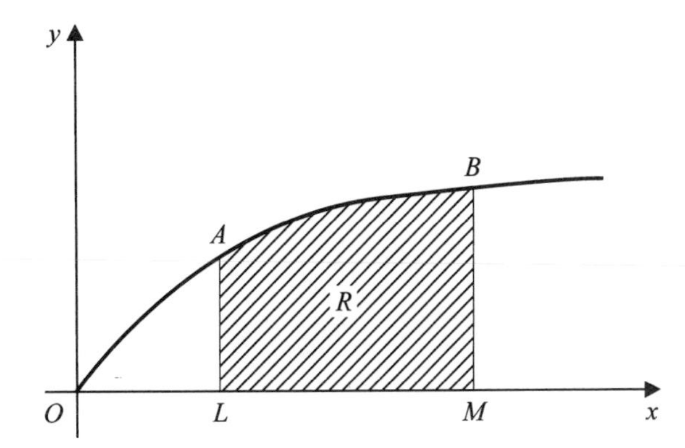

At the points A and B on the curve $t = \frac{\pi}{6}$ and $\frac{\pi}{3}$ respectively. The lines AL and BM are drawn parallel to the y-axis, as shown. Determine the area of the finite region R, shown shaded, and find the volume generated when R is rotated completely about the x-axis.

Area of $R = \int_{t=\frac{\pi}{6}}^{t=\frac{\pi}{3}} y\,\mathrm{d}x$ and we need to express both y and $\mathrm{d}x$ in terms of t.

We know that $y = \sin t$ and that, by differentiation, $\frac{\mathrm{d}x}{\mathrm{d}t} = \sec^2 t$.

Also, the integral $\int y\,\mathrm{d}x$ can be written as

$$\int y \cdot \frac{\mathrm{d}x}{\mathrm{d}t} \cdot \mathrm{d}t$$

by using the chain rule, as you saw in examples 8 and 9 (integration by substitution).

So:

$$\text{Area of } R = \int_{\frac{\pi}{6}}^{\frac{\pi}{3}} \sin t \sec^2 t\,\mathrm{d}t$$

$$= \int_{\frac{\pi}{6}}^{\frac{\pi}{3}} \frac{\sin t}{\cos^2 t}\,\mathrm{d}t$$

This integral is a function of $\cos t$ and its derivative $-\sin t$.

Putting $u = \cos t$, $\frac{\mathrm{d}u}{\mathrm{d}t} = -\sin t$

and when $t = \frac{\pi}{3}$, $u = \frac{1}{2}$; $t = \frac{\pi}{6}$, $u = \frac{\sqrt{3}}{2}$

So:

$$\text{Area of } R = \int_{\frac{\sqrt{3}}{2}}^{\frac{1}{2}} \frac{1}{u^2}(-\mathrm{d}u)$$

$$= -\int_{\frac{\sqrt{3}}{2}}^{\frac{1}{2}} u^{-2}\,\mathrm{d}u$$

$$= \left[u^{-1}\right]_{\frac{\sqrt{3}}{2}}^{\frac{1}{2}} = 2 - \frac{2}{\sqrt{3}}$$

$$\text{Volume generated} = \int_{t=\frac{\pi}{6}}^{t=\frac{\pi}{3}} \pi y^2\,\mathrm{d}x$$

$$= \pi \int_{t=\frac{\pi}{6}}^{t=\frac{\pi}{3}} y^2 \frac{\mathrm{d}x}{\mathrm{d}t} \cdot \mathrm{d}t$$

$$= \pi \int_{\frac{\pi}{6}}^{\frac{\pi}{3}} \sin^2 t \sec^2 t\,\mathrm{d}t$$

$$= \pi \int_{\frac{\pi}{6}}^{\frac{\pi}{3}} \frac{\sin^2 t}{\cos^2 t}\,\mathrm{d}t = \pi \int_{\frac{\pi}{6}}^{\frac{\pi}{3}} \tan^2 t\,\mathrm{d}t$$

You can use the identity $\sec^2 t \equiv 1 + \tan^2 t$

so: $\tan^2 t \equiv \sec^2 t - 1$

Thus:

$$\text{Volume generated} = \pi \int_{\frac{\pi}{6}}^{\frac{\pi}{3}} (\sec^2 t - 1)\,\mathrm{d}t$$

$$= \pi \left[\tan t - t\right]_{\frac{\pi}{6}}^{\frac{\pi}{3}}$$

$$= \pi \left[\tan \frac{\pi}{3} - \frac{\pi}{3} - \left(\tan \frac{\pi}{6} - \frac{\pi}{6}\right)\right]$$

$$= \pi \left[\sqrt{3} - \frac{\pi}{3} - \frac{1}{\sqrt{3}} + \frac{\pi}{6}\right] = \pi \left[\frac{2}{\sqrt{3}} - \frac{\pi}{6}\right]$$

Exercise 9F

In questions 1–10, find the area of the region bounded by the curve $y = \mathrm{f}(x)$, the x-axis and the lines $x = a$ and $x = b$.

1 $\mathrm{f}(x) = \cos x$, $a = \frac{\pi}{6}$, $b = \frac{\pi}{3}$

2 $f(x) = \sec^2 x$, $a = 0$, $b = \dfrac{\pi}{4}$

3 $f(x) = xe^x$, $a = 1$, $b = 3$

4 $f(x) = \ln x$, $a = 2$, $b = 5$

5 $f(x) = \sin^2 x$, $a = \dfrac{\pi}{12}$, $b = \dfrac{5\pi}{12}$

6 $f(x) = \tan^2 x$, $a = \dfrac{\pi}{8}$, $b = \dfrac{3\pi}{8}$

7 $f(x) = x\cos 2x$, $a = \dfrac{\pi}{6}$, $b = \dfrac{\pi}{5}$

8 $f(x) = xe^{x^2}$, $a = -2$, $b = -1$

9 $f(x) = \dfrac{\sin x}{2 + \cos x}$, $a = \dfrac{\pi}{2}$, $b = \dfrac{2\pi}{3}$

10 $f(x) = \tan^2 x \sec^2 x$, $a = \dfrac{\pi}{6}$, $b = \dfrac{\pi}{3}$

In questions 11–20, the finite region R is bounded by the curve $y = f(x)$, the x-axis and the lines $x = a$ and $x = b$. Find the volume generated when R is rotated completely about the x-axis.

11 $f(x) = x^{\frac{1}{2}}$, $a = 0$, $b = 4$

12 $f(x) = 2x^{\frac{1}{4}}$, $a = 1$, $b = 16$

13 $f(x) = \sin x$, $a = 0$, $b = \pi$

14 $f(x) = x^{-\frac{1}{2}}$, $a = 2$, $b = 5$

15 $f(x) = x^{\frac{1}{2}}e^x$, $a = 1$, $b = 2$

16 $f(x) = x^2 - 4$, $a = 3$, $b = 5$

17 $f(x) = \cot x$, $a = \frac{\pi}{4}$, $b = \frac{\pi}{2}$

18 $f(x) = \ln x$, $a = 1$, $b = 3$

19 $f(x) = \dfrac{x + 1}{x}$, $a = 1$, $b = 4$

20 $f(x) = x\sqrt{(4 - x^2)}$, $a = 0$, $b = 2$

21 The finite region bounded by the curve $y = \tan\frac{1}{2}x$, the line $x = \frac{\pi}{2}$ and the coordinate axes is rotated through 2π radians about the x-axis. Show that the volume generated is $\frac{\pi}{2}(4 - \pi)$.

22 By rotating the semi-circle for which $y \geqslant 0$ from the circle with equation $x^2 + y^2 = a^2$ completely about the x-axis, show that the volume of a sphere, of radius a, is $\frac{4}{3}\pi a^3$.

23 A triangular region is bounded by the line $y = \frac{r}{h}x$, where r and h are positive constants, the line $x = r$ and the x-axis. By considering the complete rotation of this triangle about the x-axis, show that the volume of a cone of height h and base radius r is $\frac{1}{3}\pi r^2 h$.

24 Find the area of the region bounded by the curve $y^3 = x$, the lines $y = 2$, $y = 4$ and $x = 0$. This region is rotated completely about the y-axis. Find the volume generated.

25 The region R shown in the figure is bounded by the curve $y = \cos x - \sin x$ and the coordinate axes.

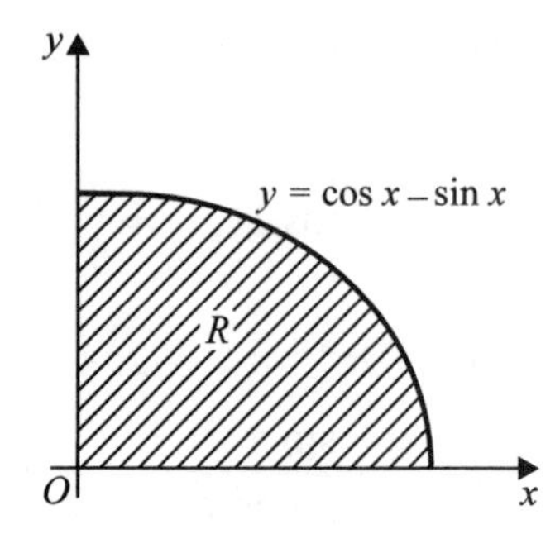

(a) Find the area of R.

The region R is rotated completely about the x-axis.

(b) Find the volume generated.

26 The diagram shows a sketch of the curve

$$x = t^2 + 1, \; y = t$$

where t is a parameter for $t > 0$.

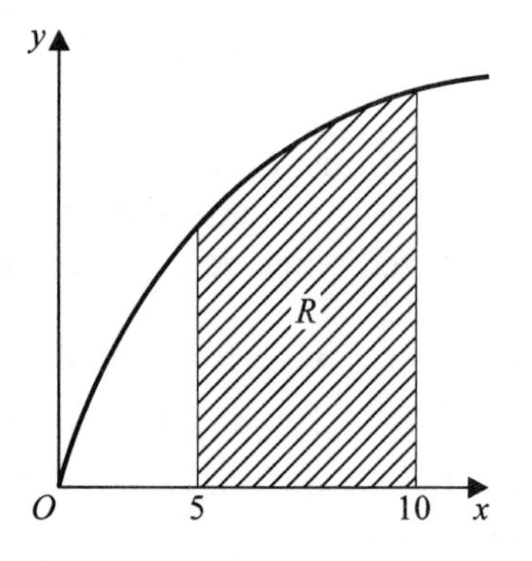

The shaded region R is bounded by the curve, the x-axis and the lines $x = 5$ and $x = 10$.

(a) Find the area of R.

(b) Find the volume generated when R is rotated through 2π about the x-axis.

27 Find the area of the finite region R bounded by the curve with parametric equations $x = 1 + t$, $y = 4 - t^2$, the lines $x = 3$ and $x = 6$ and the x-axis.

Find also the volume generated when R is rotated completely about the x-axis.

28 Find the area of the finite region R bounded by the curve with

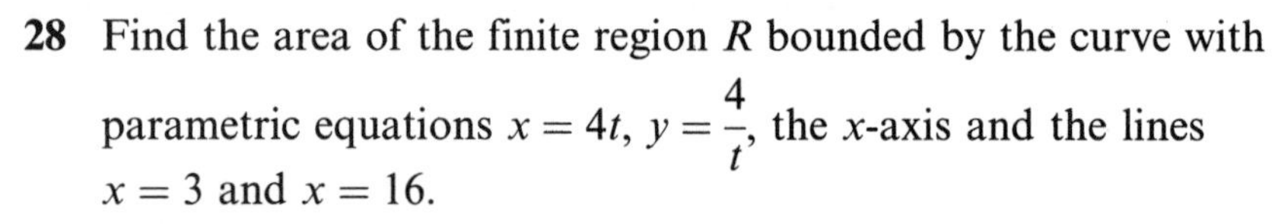

parametric equations $x = 4t$, $y = \frac{4}{t}$, the x-axis and the lines $x = 3$ and $x = 16$.

The region R is rotated through 2π about the x-axis. Find the volume of the solid generated.

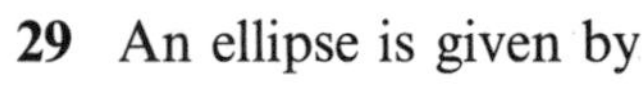

29 An ellipse is given by

$$x = 3\cos t, \; y = 2\sin t, \; 0 \leqslant t < 2\pi$$

(a) Find the area of the finite region bounded by the ellipse and the positive x- and y-axes.

(b) This region is rotated completely about the y-axis to form a solid of revolution. Find the volume of this solid.

9.8 Forming and solving simple differential equations

Any equation involving derivatives of one variable with respect to another variable is called a **differential equation**. Simple examples are

$$\frac{dy}{dx} = x^2 - 5 \quad \text{and} \quad \frac{dy}{dx} = \sin y \cos x, \text{ etc.}$$

These equations are called **first order differential equations** because the highest derivative each contains is the first derivative of y with respect to x, that is $\frac{dy}{dx}$.

A differential equation of the **second order is like this:**

$$\frac{d^2y}{dx^2} + x^2 \frac{dy}{dx} = y$$

In Book P2, you will only be concerned with first order differential equations of a special simple type. In Books P3 and P4, you will meet other differential equations.

In many practical situations, the rate at which one variable is changing with respect to another is expressed by a **physical law**. A differential equation can be set up from the experimental data and possible solutions of this equation are found using integration. One of the most important and widely occurring relationships between variables is called the **law of natural growth or decay** (see also chapter 3). Here are a few examples.

(a) **Radioactive decay**

In a mass of radioactive material, where the atoms are disintegrating spontaneously, the average rate of disintegration is proportional to the number of atoms present. At time t, there are N atoms present and this situation is described by the differential equation

$$\frac{dN}{dt} = -kN$$

where k is a constant. The minus sign indicates that the rate of change is *decreasing*.

(b) **Newton's law of cooling**

The rate of change of the temperature of a cooling body is proportional to the excess temperature over the surroundings. If the excess temperature is θ at time t, then this situation is described by the differential equation

$$\frac{d\theta}{dt} = -k\theta$$

where k is a positive constant.

(c) **Chemical reactions**

Some chemical reactions follow a law which states that the rate of change of the reacting substance is proportional to its concentration. If the concentration is C at time t, then this situation is described by the differential equation

$$\frac{dC}{dt} = -kC$$

where k is a positive constant.

As you can see one common differential equation can be used to model the rates of change of several quite different variables with respect to time. These variables come from a whole range of different physical situations.

In your exam, you may be given the description of a law in words and you will be required to form a differential equation from the description. You will not require any special knowledge to do this, except to recognise that the rate of change of a variable y with respect to x is $\frac{dy}{dx}$ and that if A is proportional to B, then $A = kB$ where k is a constant.

Example 22

The length y cm of a leaf during the period of its growth is proportional to the amount of water it contains. During this period the leaf retains a similar shape; that is, the ratio of its length to its width remains constant. The leaf absorbs water from its 'parent' plant at a rate proportional to y and it loses water by evaporation at a rate proportional to the area of the leaf at the time when its length is y cm. Form a differential equation to describe the growth of the leaf.

Assume that the leaf has length y cm at time t days after it was first observed.

The rate at which the leaf is receiving water is $k_1 y$ where k_1 is a positive constant.

The area of the leaf at time t days is proportional to y^2, since it maintains its shape.

So the leaf is losing water at a rate of k_2y^2, where k_2 is another positive constant.

The rate of growth of the leaf is given by $\frac{dy}{dt}$, the rate of change of its length.

$$\frac{dy}{dt} = k_1y - k_2y^2$$

is a differential equation describing the growth of the leaf.

In this chapter you will learn to solve differential equations of the first order in which the variables are separable. This type of differential equation is of the form

$$\frac{dy}{dx} = f(x)g(y)$$

First, let's look at two simpler cases.

(i) Suppose that $g(y) = 1$, then:

$$\frac{dy}{dx} = f(x)$$

By direct integration:

$$y = \int f(x)\,dx + C$$

where C is a constant.

$y = \int f(x)\,dx + C$ is called **the general solution of the differential equation** $\frac{dy}{dx} = f(x)$, once the integration is completed.

Example 23

Solve the differential equation $\frac{dy}{dx} = \ln x$, $x > 0$, given that $y = 2$ at $x = 1$.

We have $y = \int \ln x + C$, where C is a constant. From example 12 of this chapter,

$$\int \ln x \, dx = x \ln x - x$$

and therefore $y = x \ln x - x + C$ is the general solution of the differential equation.

At $y = 2$, $x = 1 \Rightarrow 2 = 1 \cdot \ln 1 - 1 + C$

So: $$C = 3$$

The solution of the differential equation is:

$$y = x \ln x - x + 3$$

(ii) Suppose now that $f(x) = 1$ so that the differential equation is

$$\frac{dy}{dx} = g(y)$$

$\frac{dy}{dx} = \frac{1}{\frac{dx}{dy}}$, so:

$$\frac{dx}{dy} = \frac{1}{g(y)}$$

and

$$x = \int \frac{1}{g(y)} \, dy + C$$

where C is a constant.

That is, provided you can integrate $\frac{1}{g(y)}$ with respect to y, you have found the general solution.

Example 24

Given that $y > -\frac{1}{2}$, solve the differential equation $\frac{dy}{dx} = 2y + 1$. Express the general solution in the form $y = f(x)$.

$$\frac{dy}{dx} = 2y + 1 \Rightarrow \frac{dx}{dy} = \frac{1}{2y + 1}$$

Integrating: $$x = \int \frac{1}{2y + 1} \, dy = \tfrac{1}{2} \ln |2y + 1| + C$$

The general solution is then

$$x = \tfrac{1}{2}\ln|2y + 1| + C$$

You now want y in terms of x.

Rearranging: $\ln|2y + 1| = 2(x - C)$

$\Rightarrow$ $2y + 1 = e^{2(x-C)}$

and $y = \tfrac{1}{2}(e^{2x-2C} - 1)$ is the form required.

The differential equation

$$\frac{dy}{dx} = f(x)g(y)$$

can be written as

$$\frac{1}{g(y)}\frac{dy}{dx} = f(x)$$

Integrating with respect to x gives:

$$\int \frac{1}{g(y)}\frac{dy}{dx} \cdot dx = \int f(x)\,dx + C$$

- **That is,** $$\int \frac{1}{g(y)}\,dy = \int f(x)\,dx + C$$
  is the general solution, provided that $\frac{1}{g(y)}$ can be integrated with respect to y and $f(x)$ can be integrated with respect to x.

Differential equations of the type $\frac{dy}{dx} = f(x)g(y)$ are known as **first order separable** because, as you have seen, they can be solved by *separating* the variables and integrating.

Example 25

Express y in terms of x, given that $\frac{dy}{dx} = (y + 2)(2x + 1)$ and that $y = 2$ at $x = 0$.

Rewrite the differential equation as:

$$\frac{1}{y + 2}\frac{dy}{dx} = 2x + 1$$

Integrating gives $\ln|y + 2| = x^2 + x + C$

as the general solution.

At $y = 2$, $x = 0 \Rightarrow \ln 4 = 0 + 0 + C$

So: $$C = \ln 4$$

and: $$\ln|y + 2| - \ln 4 = x^2 + x$$

$$\ln\left|\frac{y+2}{4}\right| = x^2 + x$$

$$\frac{y+2}{4} = e^{x^2+x}$$

And so $$y = 4e^{x^2+x} - 2$$

is the required solution.

Exercise 9G

Find the general solutions of the differential equations in questions 1–10.

1 $\dfrac{dy}{dx} = e^{2x-1}$

2 $\dfrac{dy}{dx} = e^{2y-1}$

3 $\dfrac{dy}{dx} = \cos^2 x$

4 $\dfrac{dy}{dx} = \cos^2 y$

5 $\dfrac{dy}{dx} = xy$

6 $\dfrac{dy}{dx} = e^{x+y}$

7 $\dfrac{dy}{dx} = \sec y \ln x$

8 $(x + 1)\dfrac{dy}{dx} = y + 2$

9 $x\dfrac{dy}{dx} = y + x^2 y$

10 $y\dfrac{dy}{dx} + \cot x \operatorname{cosec} x = 0$

Obtain the solution that satisfies the given conditions of the differential equations in questions 11–20.

11 $\dfrac{dy}{dx} = x^2 + x, \quad y = 0$ at $x = 0$

12 $\dfrac{dy}{dx} = \sin^2 x \cos x, \quad y = 0$ at $x = \dfrac{\pi}{2}$

13 $\dfrac{dy}{dx} = 3y + 1, \quad y = 0$ at $x = 1$

14 $\dfrac{dy}{dx} = xe^y, \quad y = 0$ at $x = 1$

15 $\dfrac{dy}{dx} = \tan x \tan y, \quad y = \dfrac{\pi}{4}$ at $x = \dfrac{\pi}{4}$

16 $\frac{dy}{dx} = \sin^2 x \cos^2 y, \quad y = 0 \text{ at } x = 0$

17 $x^2 \frac{dy}{dx} = \operatorname{cosec} y \sec y, \quad y = \frac{\pi}{3} \text{ at } x = 1$

18 $y\frac{dy}{dx} = \sec^2 x(2\tan x + 1), \; y = 3 \text{ at } x = \frac{\pi}{4}$

19 $y \sin y \frac{dy}{dx} = x \cos x, \; y = 0 \text{ at } x = \frac{\pi}{2}$

20 $\frac{dy}{dx} = (y^2 - 1)\cot x, \; y = 2 \text{ at } x = \frac{\pi}{4}$

21 Newton's law of cooling states that the rate of change of the temperature of a cooling liquid is proportional to the excess temperature over the room temperature. The law is given by the differential equation

$$\frac{d\theta}{dt} = -k\theta$$

where θ is the excess temperature at time t. At $t = 0$, $\theta = \theta_0$. Show that

$$\theta = \theta_0 e^{-kt}$$

22 The temperature of a liquid in a room, where the temperature is constant at 20 °C, was observed to be 80 °C and 7 minutes later it was 60 °C. Calculate, to the nearest minute, using Newton's law of cooling,

(a) the time taken for the temperature to fall from 80 °C to 40 °C

(b) the temperature of the liquid 10 minutes after it was 80 °C.

23 A lump of radioactive substance is disintegrating. At time t days after it was first observed to have mass 10 grams, its mass is m grams and

$$\frac{dm}{dt} = -km$$

where k is a positive constant.

Find the time, in days, for the substance to reduce to 1 gram in mass, given that its half-life is 8 days.

(The half-life is the time in which half of any mass of the substance will decay.)

24 Given that $x\dfrac{dy}{dx} = (1 - 2x^2)y$, where $x > 0$, and that $y = 1$ at $x = 1$, find y in terms of x.

25 The gradient at any point (x, y), where $x > 0$, on a curve is $\ln x$ and $y = e$ at $x = 1$. Find the equation of the curve.

SUMMARY OF KEY POINTS

1 Standard results (to be memorised)

$$\int x^n \, dx = \frac{1}{n+1} x^{n+1} + C$$

$$\int e^{ax+b} \, dx = \frac{1}{a} e^{ax+b} + C$$

$$\int (ax+b)^n \, dx = \frac{1}{a(n+1)} (ax+b)^{n+1} + C \quad (n \neq -1)$$

$$\int (ax+b)^{-1} \, dx = \frac{1}{a} \ln |ax+b| + C$$

$$\int \sin(ax+b) \, dx = -\frac{1}{a} \cos(ax+b) + C$$

$$\int \cos(ax+b) \, dx = \frac{1}{a} \sin(ax+b) + C$$

$$\int \tan(ax+b) \, dx = \frac{1}{a} \ln |\sec(ax+b)| + C$$

$$\int \cot(ax+b) \, dx = \frac{1}{a} \ln |\sin(ax+b)| + C$$

$$\int \ln x \, dx = x(\ln x - 1) + C$$

2 Integration of sums and differences:

$$\int (u \pm v) \, dx = \int u \, dx \pm \int v \, dx$$

3 Integration by parts:

$$\int v \frac{du}{dx} \, dx = uv - \int u \frac{dv}{dx} \, dx$$

4 Integration by substitution:

$$\int f(x)\,dx = \int f[g(t)]\,\frac{dx}{dt}\,dt, \quad x = g(t)$$
$$= \int fg(t)g'(t)\,dt$$

5

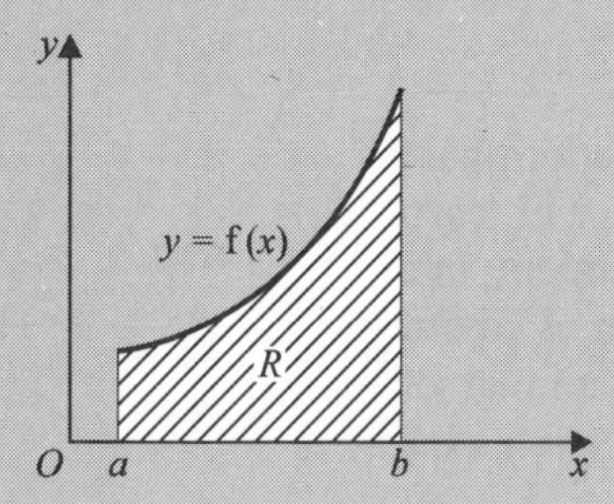

Area of $R = \int_a^b f(x)\,dx$

Volume generated when R is rotated completely about the x-axis is:

$$\pi \int_a^b [f(x)]^2\,dx$$

6 The general solution of the differential equation

$$\frac{dy}{dx} = f(x)g(y)$$

is:

$$\int \frac{1}{g(y)}\,dy = \int f(x)\,dx + C$$

provided that the integrals can be found.

Coordinate geometry 10

You have already been introduced to the basics of coordinate geometry in both two and three dimensions. Now you will learn how to find the coordinates of the mid-point of a straight line joining two given points. The remainder of the chapter is about sketching curves. In section 10.2 you will learn the techniques of sketching curves given by cartesian equations, while section 10.3 looks at sketches of curves given by parametric equations.

10.1 The coordinates of the mid-point of the line segment joining two given points

Consider the points $A(x_1, y_1)$ and $B(x_2, y_2)$ and let $M(X, Y)$ be the mid-point of the line AB.

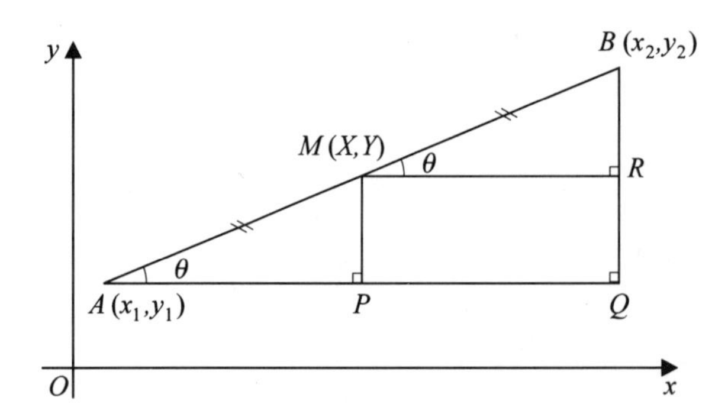

Draw MP and BQ parallel to the y-axis and draw APQ and MR parallel to the x-axis.

In triangles AMP and MBR,

$$\angle MPA = \angle BRM = 90^\circ$$

$$\angle MAP = \angle BMR \text{ (because } MR \text{ is parallel to } AQ)$$

So: $$\angle AMP = \angle MBR \text{ (angle sum of a triangle)}$$

But $AM = MB$ (because M is the mid-point of AB). So triangles AMP and MBR are congruent. That is, the two triangles are exactly the same size and shape.

Now $\quad AP = X - x_1$

and $\quad MR = x_2 - X$

But $AP = MR$ (congruent triangles)

So: $\quad X - x_1 = x_2 - X$

i.e. $\quad X + X = x_1 + x_2$

$$2X = x_1 + x_2$$

and $\quad X = \dfrac{x_1 + x_2}{2}$

Similarly, $\quad MP = Y - y_1$

and $\quad BR = y_2 - Y$

But $MP = BR$

So: $\quad Y - y_1 = y_2 - Y$

and $\quad Y = \dfrac{y_1 + y_2}{2}$

The coordinates of M are $\left(\dfrac{x_1 + x_2}{2}, \dfrac{y_1 + y_2}{2}\right)$.

- **So the coordinates of the mid-point of the line segment joining (x_1, y_1) to (x_2, y_2) are**

$$\left(\frac{x_1 + x_2}{2}, \frac{y_1 + y_2}{2}\right)$$

Example 1

Find the coordinates of the mid-points of the line segment joining (a) $(3, 7)$ to $(2, 5)$ (b) $(3, 4)$ to $(-2, 6)$ (c) $(-1, -4)$ to $(-2, 3)$

(a) The mid-point is

$$\left(\frac{3+2}{2}, \frac{7+5}{2}\right) = (2.5, 6)$$

(b) The mid-point is

$$\left(\frac{3-2}{2}, \frac{4+6}{2}\right) = (0.5, 5)$$

(c) The mid-point is

$$\left(\frac{-1-2}{2}, \frac{-4+3}{2}\right) = (-1.5, -0.5)$$

Example 2

The straight line joining A and B has mid-point M. Calculate the coordinates of B for:

(a) $A(1, 2)$, $M(3, 4)$ (b) $A(-1, 5)$, $M(4, -7)$
(c) $A(-1, -3)$, $M(3, -1)$.

(a) Let B have coordinates (x, y).

Then
$$\frac{1+x}{2} = 3$$
$$1 + x = 6$$
$$x = 5$$

and
$$\frac{2+y}{2} = 4$$
$$2 + y = 8$$
$$y = 6$$

B is the point $(5, 6)$.

(b) Let B have coordinates (x, y).

Then
$$\frac{x+(-1)}{2} = 4$$
$$x - 1 = 8$$
$$x = 9$$

and
$$\frac{y+5}{2} = -7$$
$$y + 5 = -14$$
$$y = -19$$

B is the point $(9, -19)$.

(c) Let B have coordinates (x, y).

Then
$$\frac{x+(-1)}{2} = 3$$
$$x - 1 = 6$$
$$x = 7$$

and
$$\frac{y+(-3)}{2} = -1$$
$$y - 3 = -2$$
$$y = 1$$

B is the point $(7, 1)$.

Example 3

Find an equation of the perpendicular bisector of the line joining $A(2, -3)$ to $B(-5, 1)$.

The perpendicular bisector of the line joining A and B is perpendicular to AB and passes through the mid-point of AB. You therefore need the gradient of a line perpendicular to AB and the coordinates of the mid-point of AB.

The gradient of $AB = \dfrac{1+3}{-5-2} = \dfrac{4}{-7} = -\frac{4}{7}$

So the gradient of a line perpendicular to AB is

$$\frac{-1}{-\frac{4}{7}} = \frac{7}{4}$$

The mid-point of AB is

$$\left(\frac{2-5}{2}, \frac{-3+1}{2}\right) = (-1.5, -1)$$

So an equation of the perpendicular bisector is

$$y + 1 = \tfrac{7}{4}(x + 1.5)$$
$$4y + 4 = 7x + 10.5$$
$$4y = 7x + 6.5$$

or

$$8y = 14x + 13$$

Exercise 10A

1 Find the coordinates of the mid-points of the straight lines joining each of the following pairs of points:

(a) $(1, 2)$ and $(3, 6)$ (b) $(7, 5)$ and $(2, 4)$
(c) $(1, -2)$ and $(3, 7)$ (d) $(4, -2)$ and $(-2, 5)$
(e) $(-1, -3)$ and $(-7, -6)$ (f) $(-2, -9)$ and $(3, -1)$
(g) $(-3, 7)$ and $(6, -4)$ (h) $(-5, -2)$ and $(-9, -4)$
(i) $(-3, -6)$ and $(4, -7)$ (j) $(5, -\frac{1}{2})$ and $(-\frac{1}{4}, 2)$

2 The straight line joining A and B has mid-point M. Calculate the coordinates of B for:

(a) $A(2, 3)$, $M(5, 1)$ (b) $A(6, 2)$, $M(3, 4)$
(c) $A(3, 7)$, $M(4, 3)$ (d) $A(-5, 2)$, $M(3, 1)$
(e) $A(2, -4)$, $M(5, 3)$ (f) $A(2, -7)$, $M(-6, 3)$
(g) $A(5, 1)$, $M(-2, -4)$ (h) $A(-3, -1)$, $M(-4, -3)$
(i) $A(-3, -2)$, $M(2, 5)$ (j) $A(1, 4)$, $M(-2, -1)$

3 Find an equation of the perpendicular bisector of the line joining A to B:

(a) $A(1, 2)$, $B(3, 1)$ (b) $A(2, 3)$ $B(5, 2)$

(c) $A(1, -7)$, $B(2, 4)$ (d) $A(-2, 3)$, $B(2, -1)$

(e) $A(-1, -2)$, $B(6, -5)$ (f) $A(-3, -2)$, $B(-1, -5)$

(g) $A(3, 6)$, $B(-2, -4)$ (h) $A(1, -6)$, $B(2, 7)$

(i) $A(3, -4)$, $B(7, -3)$ (j) $A(1, 6)$, $B(-6, 3)$

4 The points A and B have coordinates $(6, 4)$ and $(1, -8)$ respectively. Find the coordinates of the centre and the radius of the circle drawn on AB as diameter.

5

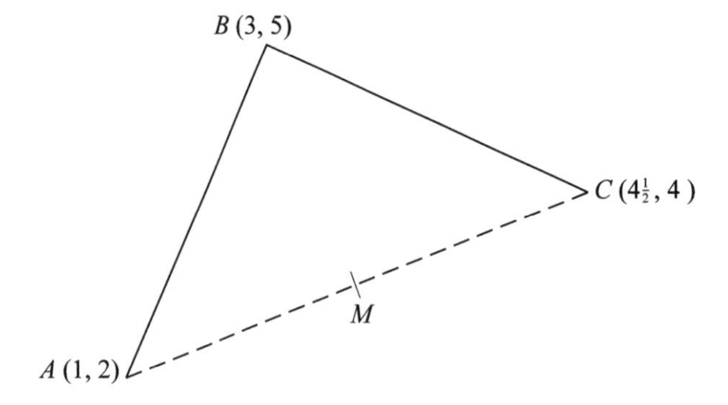

For the triangle ABC, show that

(a) AB is perpendicular to BC

(b) $BM = CM$, where M is the mid-point of AC.

6 The centre of a square is at $(3, 4)$ and one of its vertices is at $(7, 1)$. Find the coordinates of the other three vertices.

10.2 Sketching curves given by cartesian equations

In your studies for GCSE you will probably have plotted, and accurately drawn, the graphs of many functions. However, at AS and A level you will meet questions which ask you not to plot and draw accurately the graph of a function but merely to draw a *sketch* of the function. What is then required is a freehand sketch of the curve that shows the main features of that curve. If you are asked to sketch a curve in an examination you will, in general, waste a great deal of time and effort (and marks!) if you try to plot the graph accurately.

The technique of curve sketching is used not only when demanded by an examination question. You should always use it in questions where you are asked to calculate the area under a curve. Unless you can see a picture of the curve first, and so know what area it is you are trying to calculate, it is very difficult to work out the correct limits for the integration and so to obtain a correct answer.

The form of some graphs should be familiar to you. The graph of $y = ax + b$ is a straight line. So a sketch of $y = 3x + 6$ looks like this:

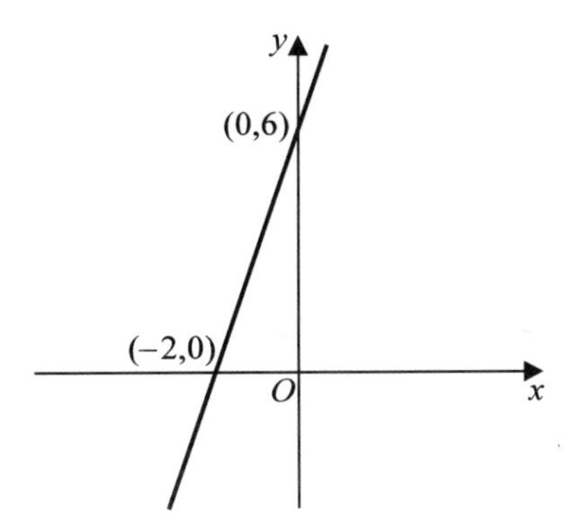

The line has positive gradient and meets the coordinate axes at the points $(-2, 0)$ and $(0, 6)$ as shown.

A sketch of $y = -4x - 8$ looks like this:

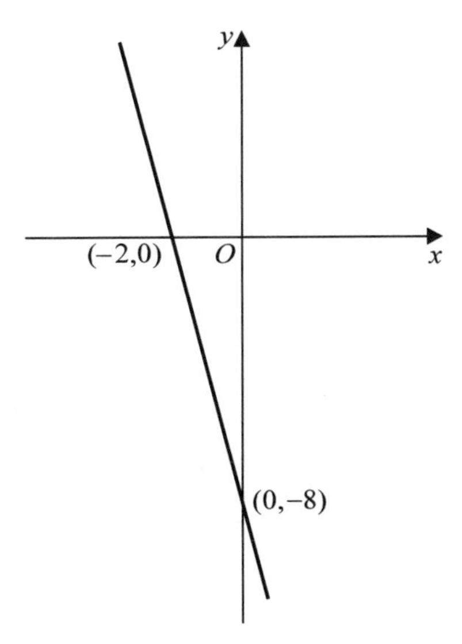

The line has negative gradient and meets the coordinate axes at $(-2, 0)$ and $(0, -8)$ as shown.

The graph of $y = ax^2 + bx + c$ is always a parabola.

When $a > 0$ the curve has a 'valley' shape: it is of the $\cup$ form.

When $a < 0$ the curve has a 'hill' shape: It is of the $\cap$ form.

Any curve of the form $y = \mathrm{f}(x)$ cuts the x-axis where $y = 0$, so the graph of $y = ax^2 + bx + c$ cuts the x-axis where $ax^2 + bx + c = 0$. That is, where

$$x = \frac{-b \pm \sqrt{(b^2 - 4ac)}}{2a}$$

as shown in section 2.3 of Book P1.

If $b^2 - 4ac < 0$, then the equation $ax^2 + bx + c = 0$ has no real roots because a negative number has no real square root. That is, the graph of $y = ax^2 + bx + c$ does *not* cross the x-axis.

If $b^2 - 4ac = 0$, then $x = \frac{-b}{2a}$. That is, the equation $ax^2 + bx + c = 0$ has just *one* root. So the graph of $y = ax^2 + bx + c$ just touches the x-axis.

If $b^2 - 4ax > 0$, the then the equation $ax^2 + bx + c = 0$ has two real roots and so the graph of $y = ax^2 + bx + c$ cuts the x-axis at two distinct points.

The possibilities are summarised in the diagrams.

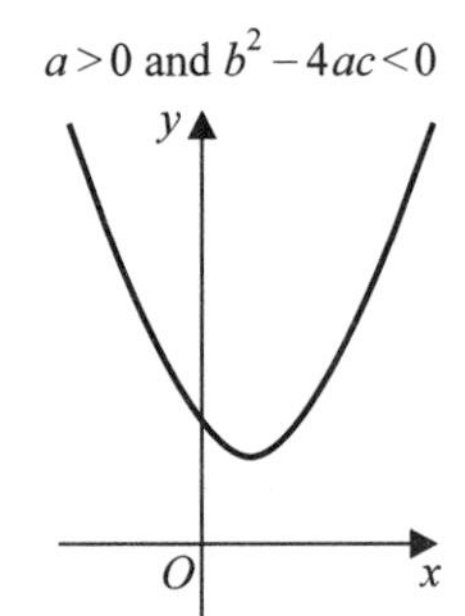

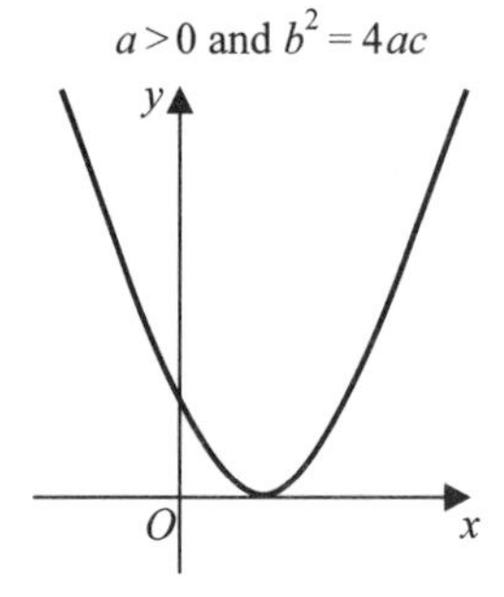

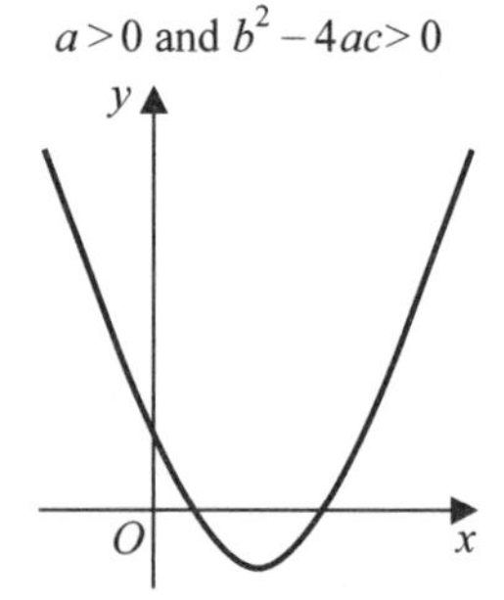

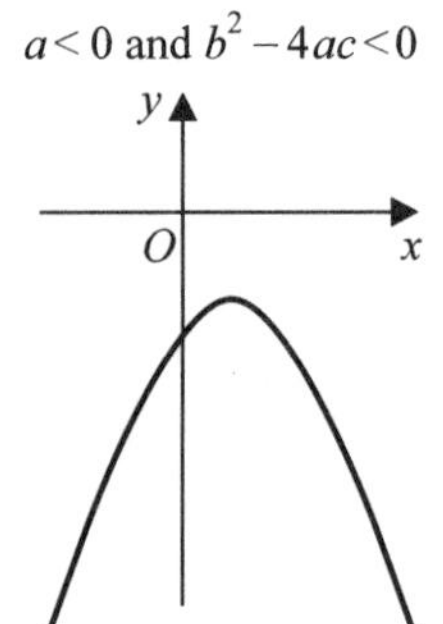

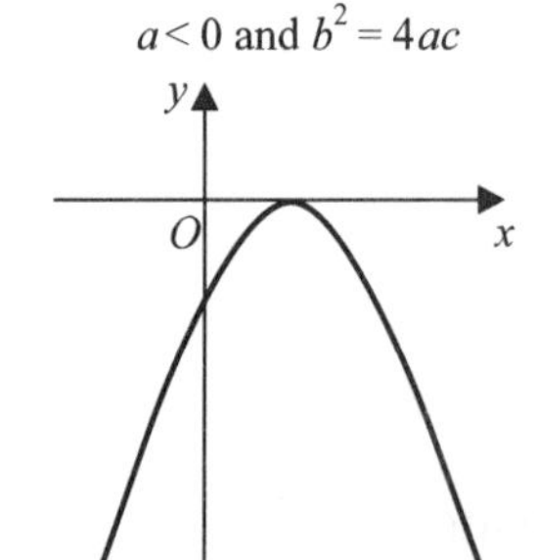

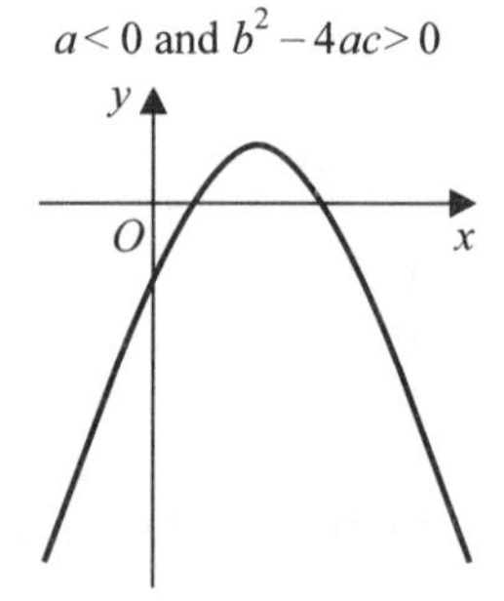

So a sketch of $y = x^2 - x - 2 = (x + 1)(x - 2)$ looks like this:

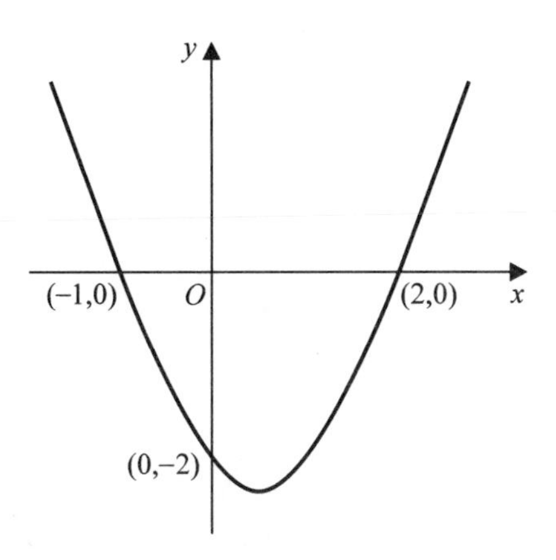

A sketch of $y = 6 + x - x^2 = (3 - x)(2 + x)$ looks like this:

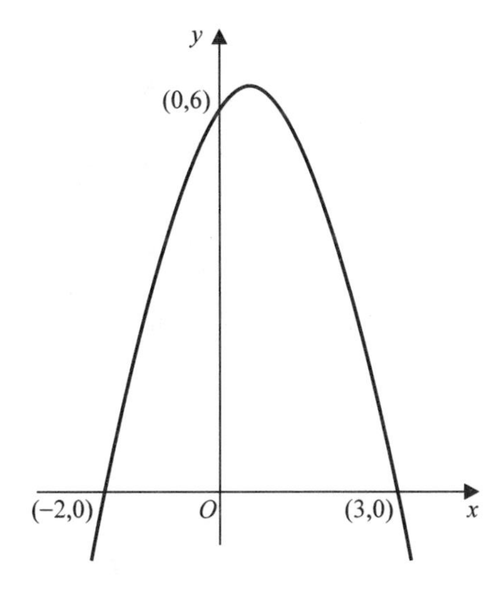

The graph of $y = ax^3 + bx^2 + cx + d$ either consists of one 'hill' and one 'valley', that is it has one maximum point and one minimum point, or sometimes these two merge to form a point of inflexion. In the equation $y = ax^3 + bx^2 + cx + d$, the ax^3 term dominates. So if $a > 0$, then:

as $x \to +\infty$, $y \to +\infty$

and as $x \to -\infty$, $y \to -\infty$

The graph therefore takes the form ∿ or ⟋.

If $a < 0$, then:
as $x \to +\infty$, $y \to -\infty$
and as $x \to -\infty$, $y \to +\infty$
The graph therefore takes the form or .

Here are some examples:

The graph $y = (x + 3)(x - 1)(x - 4)$
i.e. $y = x^3 - 2x^2 - 11x + 12$
looks like this:

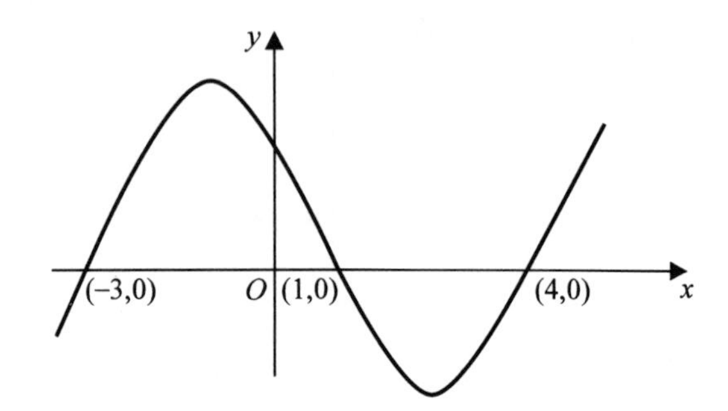

The graph of $y = (2 - x)(3 + x)(1 + x)$
i.e. $y = 6 + 5x - 2x^2 - x^3$
looks like this:

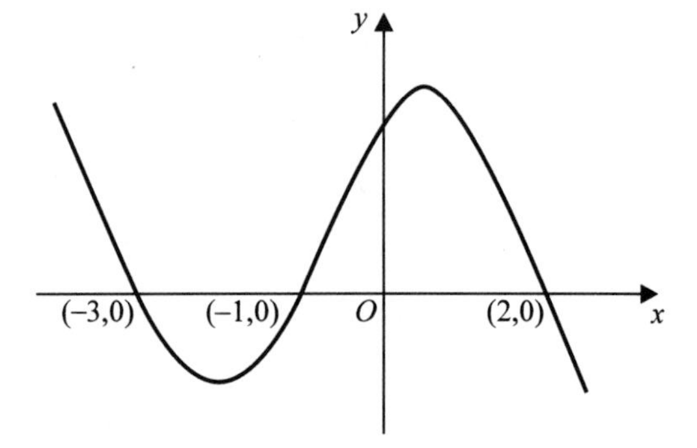

The graph of $y = (x - 1)^3$
i.e. $y = x^3 - 3x^2 + 3x - 1$
looks like this:

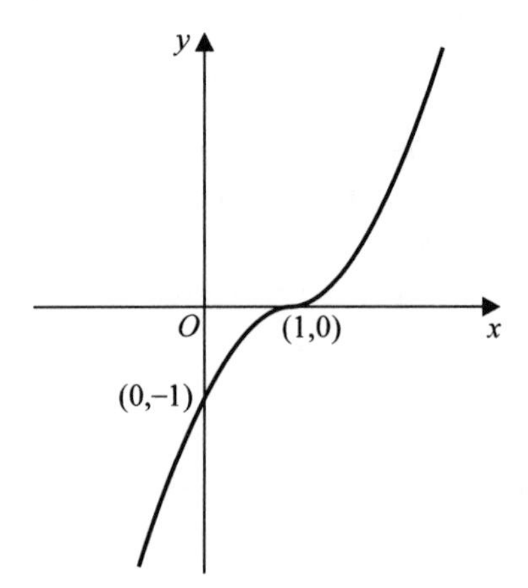

However, the shape of the graphs of most equations will be unfamiliar to you and so you need to have some general strategy for sketching a curve. There are a number of steps that you need to take.

(1) Where does the curve cut the axes? To find out where it cuts the x-axis you put $y = 0$ in the equation of the curve. To find out where it cuts the y-axis you put $x = 0$ in the equation of the curve.

(2) Are there any asymptotes and where do they occur?

(3) What happens as $x \to \pm\infty$?

(4) Has the graph any symmetry? You should know from Book P1 that $f(x)$ is an even function if $f(-x) = f(x)$. If $f(x)$ is an even function then the graph of $y = f(x)$ is symmetrical about the y-axis. Similarly, if the equation is $x = g(y)$ and $g(y)$ is an even function, then the graph of $x = g(y)$ is symmetrical about the x-axis. If $f(x)$ is an odd function i.e. $(f(-x) = -f(x))$ then the graph of $y = f(x)$ has rotational symmetry of 180° about the origin.

(5) Are there any values of x for which $f(x)$ is not defined?

(6) Where do the stationary points occur and are they maxima, minima or points of inflexion?

Example 4

Sketch the curve with equation $y = \dfrac{x+2}{x-2}$.

(1) When $x = 0$, $y = -1$

So the curve cuts the y-axis at $(0, -1)$.

When $y = 0$, $x + 2 = 0 \Rightarrow x = -2$

So the curve cuts the x-axis at $(-2, 0)$.

(2) $x = 2 \Rightarrow x - 2 = 0$

So at $x = 2$, y is undefined. Thus $x = 2$ is an asymptote parallel to the y-axis.

(3) $\dfrac{x+2}{x-2} = 1 + \dfrac{4}{x-2}$

So as $x \to \pm\infty$, $\dfrac{4}{x-2} \to 0$ and $y \to 1$

So $y = 1$ is an asymptote, parallel to the x-axis.

(4) $\dfrac{x+2}{x-2}$ is neither odd nor even.

(5) The curve exists for all x except $x = 2$.

(6) $$\frac{\mathrm{d}y}{\mathrm{d}x} = \frac{(x-2)\cdot 1 - (x+2)\cdot 1}{(x-2)^2}$$

$$= -\frac{4}{(x-2)^2} < 0$$ for all x except $x = 2$, where it is undefined.

So the curve has no stationary points.

It looks like this:

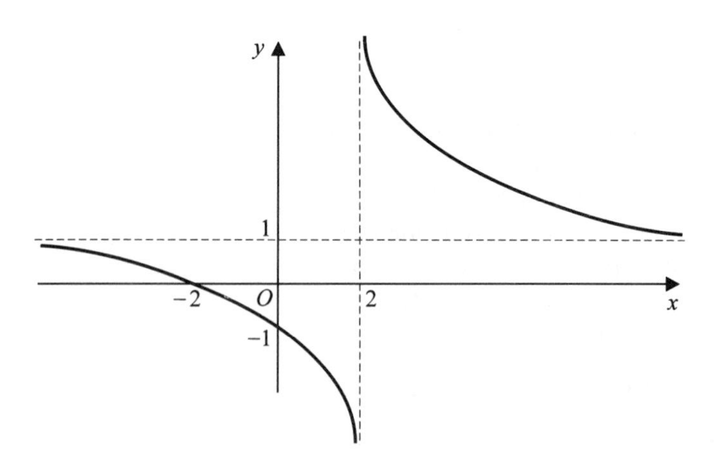

Example 5

Sketch the curve with equation $y = \frac{x}{4 - x^2}$.

(1) When $x = 0$, $y = 0$

So the curve passes through the origin.

(2) When $x = \pm 2$, $4 - x^2 = 0$ and so y is infinite. Thus $x = \pm 2$ are asymptotes.

(3) As $x \to +\infty$, $y \to 0$

As $x \to -\infty$, $y \to 0$

(4) The function is odd and so the graph has rotational symmetry about the origin.

(5) The graph exists for all x except $x = \pm 2$.

(6) $$\frac{\mathrm{d}y}{\mathrm{d}x} = \frac{(4 - x^2)\cdot 1 - x(-2x)}{(4 - x^2)^2}$$

$$= \frac{x^2 + 4}{(4 - x^2)^2}$$

$$\frac{\mathrm{d}y}{\mathrm{d}x} = 0 \Rightarrow x^2 + 4 = 0 \Rightarrow x^2 = -4$$

No real value of x exists such that $x^2 = -4$ so the curve has no stationary point. It looks like this:

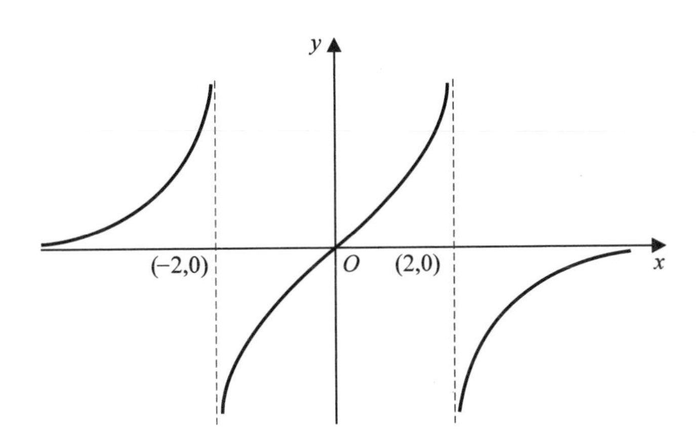

Example 6

Sketch the graph of $y = \dfrac{x^2+9}{x^2-1}$

(1) At $x = 0$, $y = \dfrac{9}{-1} = -9$

The curve cuts the y-axis at $y = -9$.

At $y = 0$, $x^2 + 9 = 0 \Rightarrow x^2 = -9$

No real values of x exist such that $x^2 = -9$. So the graph does not cut the x-axis.

(2) At $x = 1$ or $x = -1$, $x^2 - 1 = 0$

Thus y is infinite for these values of x.

Therefore the lines $x = -1$ and $x = 1$ are asymptotes.

(3) $\dfrac{x^2+9}{x^2-1} = \dfrac{x^2-1+10}{x^2-1} = 1 + \dfrac{10}{x^2-1}$

So as $x \to \pm\infty$, $\dfrac{10}{x^2-1} \to 0$ and $y \to 1$

So $y = 1$ is an asymptote.

(4) $\dfrac{x^2+9}{x^2-1}$ is an even function, so the graph is symmetrical about the y-axis.

(5) $y = \dfrac{x^2+9}{x^2-1}$ is defined for all x except $x = \pm 1$.

(6) $\dfrac{dy}{dx} = \dfrac{(x^2-1)2x - (x^2+9)2x}{(x^2-1)^2}$

$$\frac{dy}{dx} = 0 \Rightarrow -2x - 18x = 0$$

$$\Rightarrow x = 0$$

Since $\frac{dy}{dx} = \frac{-20x}{(x^2 - 1)^2}$,

when x is just negative, $\frac{dy}{dx} > 0$

When x is just positive, $\frac{dy}{dx} < 0$

So $(0, -9)$ is a maximum.

The curve $y = \frac{x^2 + 9}{x^2 - 1}$ looks like this:

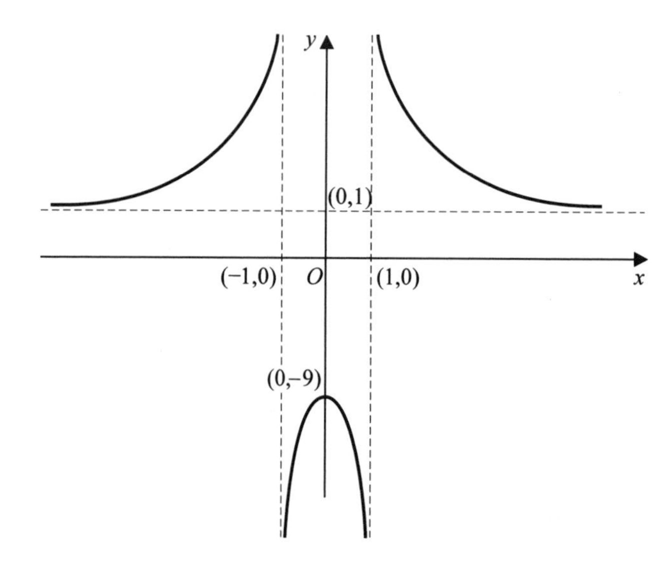

Modulus functions

You met the modulus function in section 3.4 of Book P1. You should remember that $|f(x)|$ is always positive or zero. So $|f(x)| = f(x)$ when $f(x)$ is positive and $|f(x)| = -f(x)$ when $f(x)$ is negative.

If you need to sketch the graph of $y = |f(x)|$, then where $f(x)$ is positive you sketch $f(x)$ and where $f(x)$ is negative you sketch the reflection of $f(x)$ in the x-axis. There is no part of the graph that lies in the region of the xy-plane for which $y < 0$.

Example 7
Sketch the graph of $y = x + 1$ and hence sketch the graph of $y = |x + 1|$.

The graph of $y = x + 1$ has positive gradient, cuts the x-axis at $(-1, 0)$ and cuts the y-axis at $(0, 1)$. It looks like this:

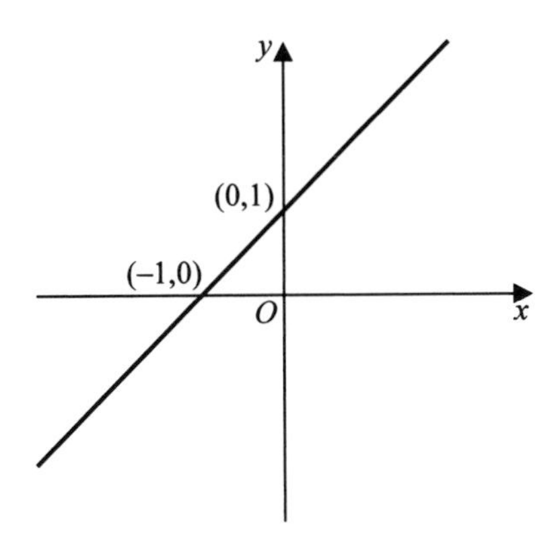

So the sketch of $y = |x + 1|$ looks like this:

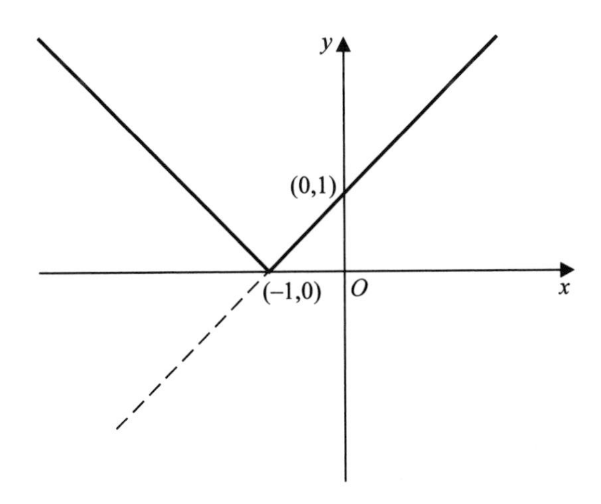

The dotted line shows the y-negative part of the graph of $y = x + 1$ which has been reflected in the x-axis.

Example 8
Sketch the graph of $y = (2 - x)(4 + x)$ and hence sketch the graph of $y = |(2 - x)(4 + x)|$.

The equation $\quad y = (2 - x)(4 + x)$

can be written $\quad y = 8 - 2x - x^2$

So its graph is a parabola of the 'hill' variety (since the coefficient of x^2 is negative), it cuts the x-axis at $(2, 0)$ and $(-4, 0)$ and it cuts the y-axis at $(0, 8)$. So its graph looks like this:

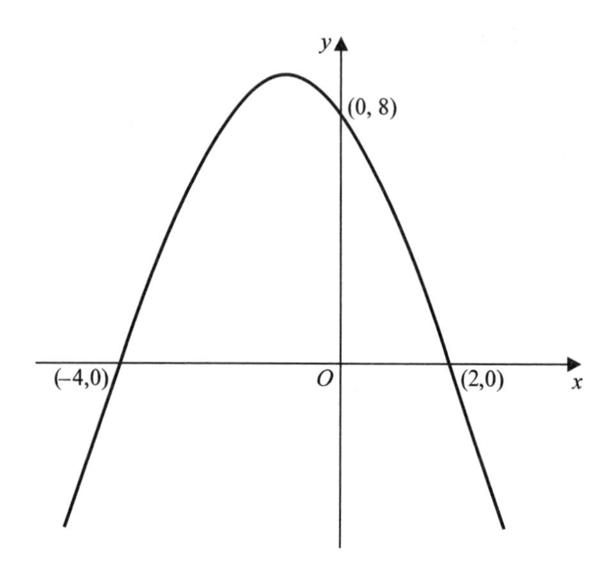

The graph of $y = |8 - 2x - x^2|$ looks like this:

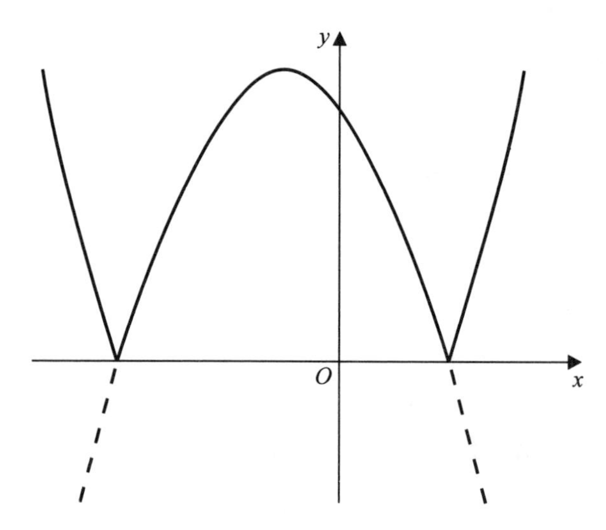

Again, the dotted lines represent the y-negative part of the graph of $y = 8 - 2x - x^2$ which has been reflected in the x-axis.

Example 9
Sketch the graph of $y = \frac{1}{x}$ and hence sketch the graph of $y = \left|\frac{1}{x}\right|$.

For the equation $y = \frac{1}{x}$, when $x = 0$, y is infinite and when $y = 0$, x is infinite: that is, the x- and y-axes are asymptotes.

$\frac{dy}{dx} = -\frac{1}{x^2}$ and since x^2 is always positive or zero, $-\frac{1}{x^2}$ is always negative. In particular it is never zero. So the graph of $y = \frac{1}{x}$ has no stationary points.

As $x \to \pm\infty$, $y \to 0$ and if $f(x) = \frac{1}{x}$ then $f(-x) = -\frac{1}{x} = -f(x)$. So f is an odd function and the graph has rotational symmetry of 180° about the origin. The graph looks like this:

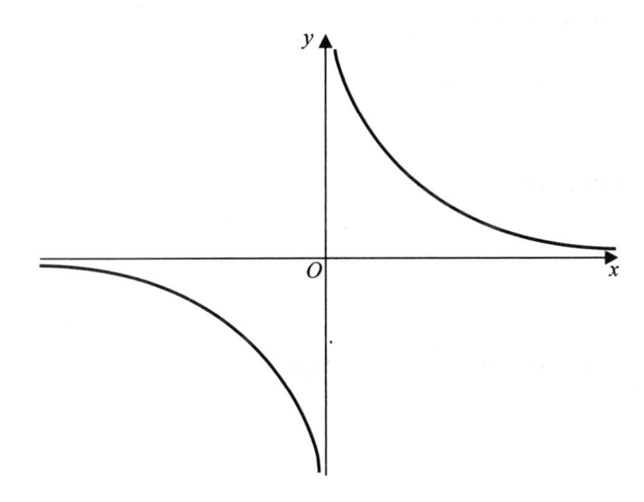

So the graph of $y = \left|\frac{1}{x}\right|$ looks like this:

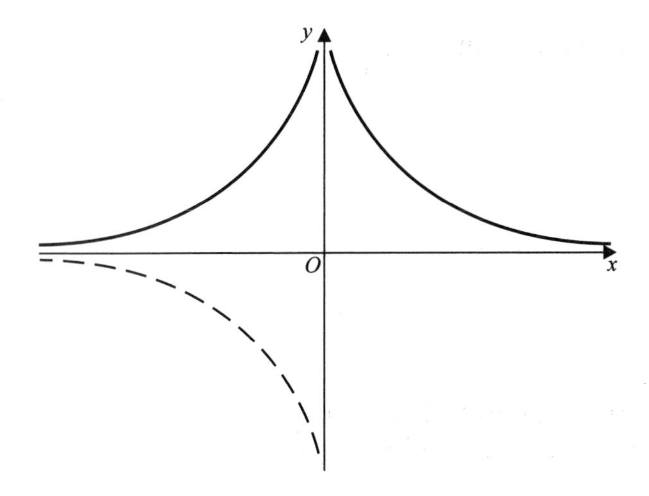

Sketching the graph of $y = \frac{1}{f(x)}$

If you are given, or are able to draw for yourself, a sketch of the graph of $y = f(x)$, then it is relatively simple to sketch the graph of $y = \frac{1}{f(x)}$ if you remember the following:

(1) If at $x = a$, the value of $f(x)$ is b, $b \neq 0$, then at $x = a$, the value of $f(x) = \frac{1}{b}$.

(2) If, at $x = a$, the value of $f(x)$ is 0, then as $x \to a$, $\frac{1}{f(x)} \to \infty$; that is, for the graph of $y = \frac{1}{f(x)}$, $x = a$ is an asymptote.

(3) If $x = a$ is an asymptote to the graph of $y = f(x)$, then, at $x = a$, $\frac{1}{f(x)} = 0$.

(4) If at $x = a$, $f(x) > 0$, then $\frac{1}{f(x)} > 0$ and if $f(x) < 0$ then $\frac{1}{f(x)} < 0$.

(5) If at $x = a$, there is a local maximum on the graph of $y = f(x)$, then at $x = a$ there is a local minimum on the graph of $y = \frac{1}{f(x)}$, and vice versa.

Example 10
On the same axes sketch the graphs of $y = (x + 1)(x - 2)$ and $y = \frac{1}{(x + 1)(x - 2)}$.

The graph of $y = (x + 1)(x - 2)$ is a parabola of the 'valley' variety which cuts the x-axis at $x = -1$ and $x = 2$ and cuts the y-axis at $y = -2$. So its graph looks like this:

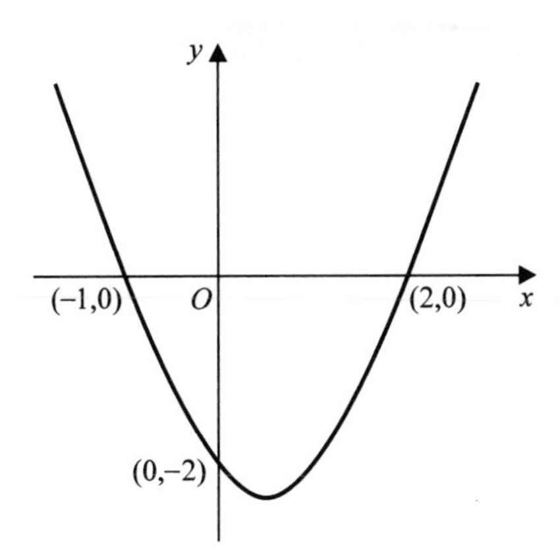

The value of y is 0 at $x = -1$ and at $x = 2$; so the graph of $y = \dfrac{1}{(x+1)(x-2)}$ has asymptotes at $x = -1$ and at $x = 2$.

The graph of $y = (x+1)(x-2)$ cuts the y-axis at $y = -2$; so the graph of $y = \dfrac{1}{(x+1)(x-2)}$ cuts the y-axis at $y = -\frac{1}{2}$.

The graph of $y = (x+1)(x-2)$ has a minimum value of $-\frac{9}{4}$ at $x = \frac{1}{2}$; so the graph of $y = \dfrac{1}{(x+1)(x-2)}$ has a maximum value of $-\frac{4}{9}$ at $x = \frac{1}{2}$.

On the graph of $y = (x+1)(x-2)$ as $x \to \pm\infty$, $y \to +\infty$; so on the graph of $y = \dfrac{1}{(x+1)(x-2)}$, as $x \to \pm\infty$, $y \to 0$.

The graph of $y = \dfrac{1}{(x+1)(x-2)}$ looks like this:

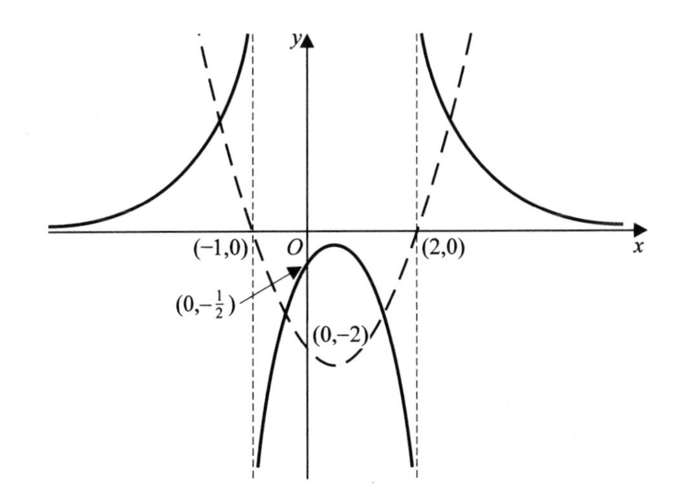

Exercise 10B

Sketch the curves with the following equations:

1 $y = x^3 - 3x^2$

2 $y = x^4 - 9x^2$

3 $y = \frac{1}{x+3}$

4 $y = \frac{x}{x-2}$

5 $y = \frac{x+3}{x+5}$

6 $y = \frac{3x-1}{x+1}$

7 $y = \frac{1}{x^2-16}$

8 $y = \frac{x^2}{x^2-16}$

9 $y = \frac{x^2}{4-x^2}$

10 $y = \frac{1}{x^2-x-2}$

11 $y = \frac{x}{(x+2)(x-3)}$

12 $y = |x|$

13 $y = |x+1|$

14 $y = |x-2|$

15 $y = 2 - |x+3|$

16 $y = |x-6| + 3$

17 $y = |3x+2|$

18 $y = |4-3x|$

19 $y = \left|\frac{x+6}{x}\right|$

20 $y = |(x-1)(x+2)|$

21 $y = \left|1 - \frac{2}{x^2}\right|$

22 $y = \frac{1}{(x+3)(x+4)}$

23 $y = \frac{1}{(x-1)(x+4)}$

24 $y = \frac{1}{(2x-3)(x+2)}$

25 $y = \frac{1}{(x+2)(3-x)}$

26 $y = (x+4)^2$

27 $y = x(x-5)$

28 $y = 4 - x^2$

10.3 Sketching curves given by parametric equations

You met the idea of a parameter in chapter 8 and you learned there how to find the equation of a tangent and of a normal to a curve given by parametric equations.

The parametric equations of a curve are generally used when they are of a simpler form than the cartesian equation of the curve. Under these circumstances the mathematics involved in working with the parametric equations becomes much easier to handle.

Parametric equations that easily transform to a cartesian equation

It is often possible to transform the parametric equations of a curve into the corresponding cartesian equation. This is done by eliminating the parameter, t, say, between the parametric equations $x = \text{f}(t)$ and $y = \text{g}(t)$.

For example, if $x = 2t^2$ and $y = 4t$ then $t^2 = \frac{x}{2}$ and $t = \frac{y}{4}$.

So: $$\frac{x}{2} = \left(\frac{y}{4}\right)^2$$

That is: $$\frac{x}{2} = \frac{y^2}{16}$$

or $$y^2 = 8x$$

The technique is to take all the t's (or whatever the parameter is) to the left-hand side of each of the parametric equations and everything else to the right-hand side. The two equations can then be linked as in the example above. When $t^2 = \frac{x}{2}$ and $t = \frac{y}{4}$, then also $t^2 = \left(\frac{y}{4}\right)^2$. So if $t^2 = \frac{x}{2}$ and $t^2 = \left(\frac{y}{4}\right)^2$ then

$$\frac{x}{2} = \left(\frac{y}{4}\right)^2 = \frac{y^2}{16} \Rightarrow y^2 = 8x$$

It is of the utmost importance when you do this that you make sure that the rearranged equation in x has no parameter on the right-hand side and, similarly, that the rearranged equation in y has no parameter on the right-hand side. Unless you do this you will find that your 'cartesian' equation will still involve a parameter. So you will have an equation with *three* variables – x, y and a parameter!

If you can transform the parametric equations into the cartesian equation you will often find that from the cartesian equation you can sketch the graph using the techniques you have already been shown.

Example 11

Sketch the curve given by the parametric equations

$$x = 3t, \quad y = 1 + t^2$$

From the first equation:

$$t = \frac{x}{3}$$

From the second equation:

$$t^2 = y - 1$$

So:
$$y - 1 = \left(\frac{x}{3}\right)^2$$

i.e.
$$y = \frac{x^2}{9} + 1$$

This represents a parabola with a 'valley'. If the equation of the curve were $y = \frac{x^2}{9}$, the curve would have its minimum at the origin. So the graph of $y = \frac{x^2}{9} + 1$ has its minimum at $(0, +1)$, using the result found in section 3.6 of Book P1.

The graph looks like this:

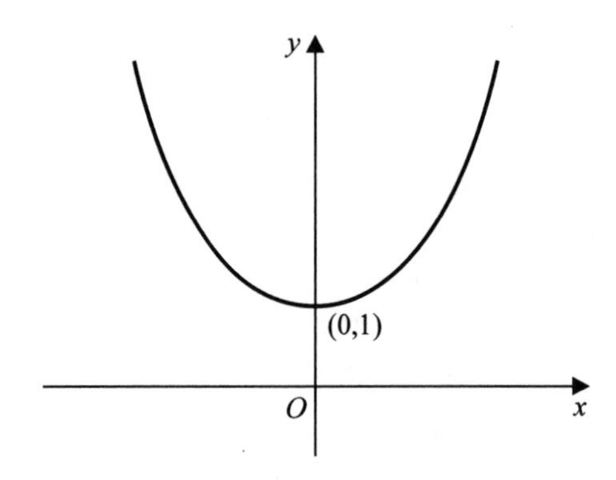

Example 12
Sketch the curve with parametric equations

$$x = t + 2, \quad y = \frac{1}{t}$$

From the first equation $t = x - 2$. From the second equation $t = \frac{1}{y}$.

So: $$x - 2 = \frac{1}{y}$$

or $$y = \frac{1}{x - 2}$$

Consider the equation $y = \frac{1}{x}$.

As $x \to 0$, $y \to \infty$

As $x \to +\infty$, $y \to 0$

As $x \to -\infty$, $y \to 0$

$$\frac{dy}{dx} = -\frac{1}{x^2}$$

This can never be zero so the function has no stationary points. The graph looks like this:

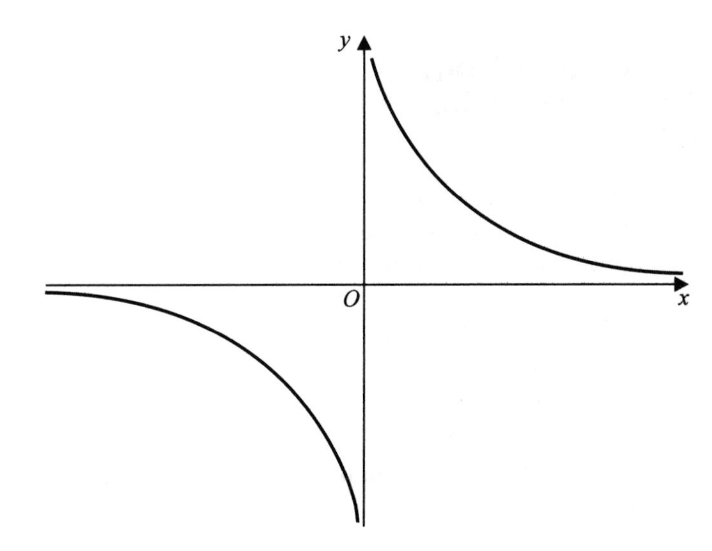

Now if $f(x) \equiv \frac{1}{x}$

then: $$f(x-2) \equiv \frac{1}{x-2}$$

So, using the results from section 3.6 of Book P1, you can see that the graph of $y = \frac{1}{x-2}$ looks like this:

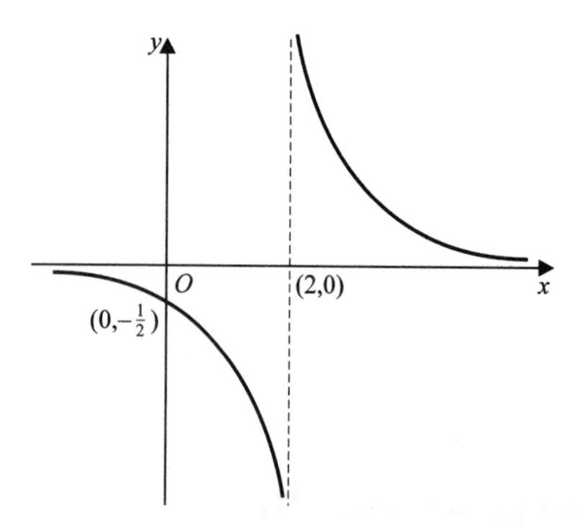

Curves that cannot easily be sketched from their cartesian equation or whose cartesian equation is difficult to obtain

It is always worth trying to rewrite the parametric equations of a curve to obtain the cartesian equation. However, frequently the cartesian equation turns out to be as complicated, if not more complicated, than the parametric equations of the curve. So, by going down this route, you often end up with a problem that is as complicated as, or more complicated than the original question. In these circumstances you need an alternative strategy. This is to try to sketch the curve directly from the parametric equations. Before you try to do this, you need to be sure that you can plot (rather than sketch) a curve given by parametric equations.

Example 13

Draw the curve given by the parametric equations

$$x = t^2, \quad y = t^3$$

t	-4	-3	-2	-1	0	1	2	3	4
$x = t^2$	16	9	4	1	0	1	4	9	16
$y = t^3$	-64	-27	-8	-1	0	1	8	27	64

The curve is called a semi-cubical parabola and looks like this:

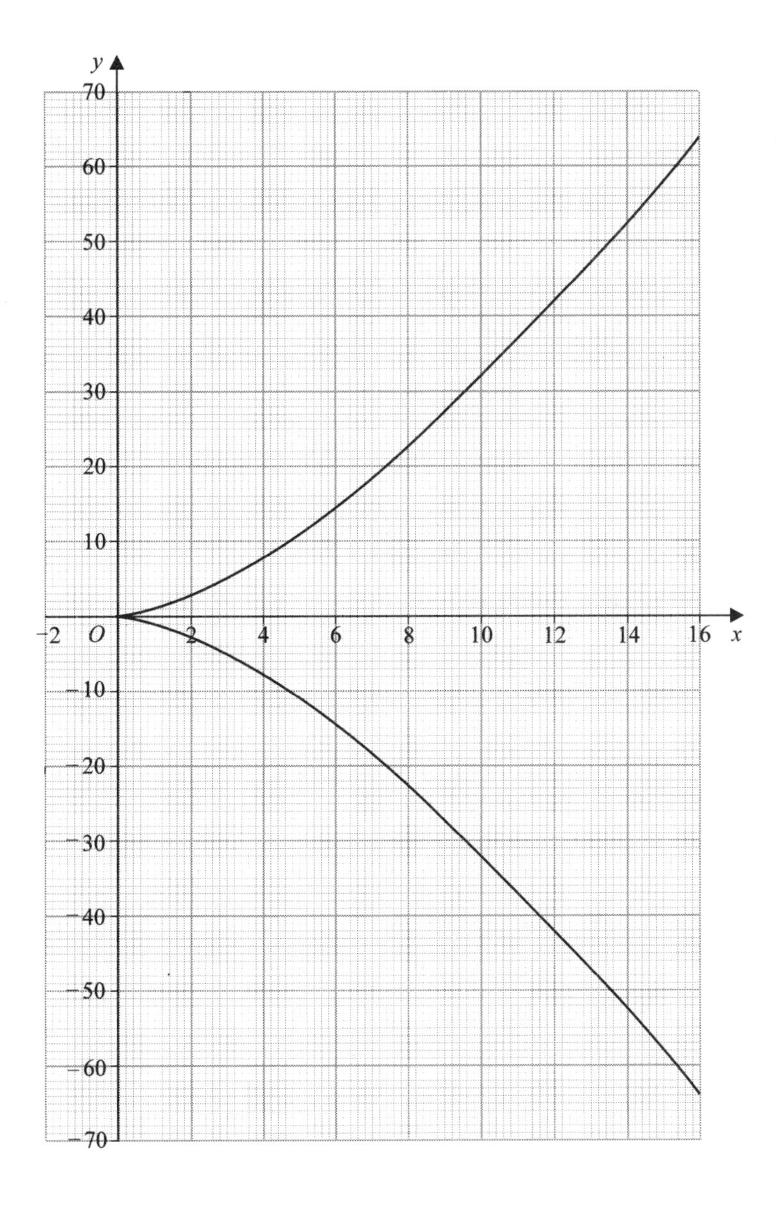

Example 14
Draw the curve given by the parametric equations

$$x = 3\sin\theta, \quad y = \cos\theta$$

for $0 \leqslant \theta < 2\pi$

θ	0	$\frac{\pi}{6}$	$\frac{\pi}{3}$	$\frac{\pi}{2}$	$\frac{2\pi}{3}$	$\frac{5\pi}{6}$	π	$\frac{7\pi}{6}$	$\frac{4\pi}{3}$	$\frac{3\pi}{2}$	$\frac{5\pi}{3}$	$\frac{11\pi}{6}$
$x = 3\sin\theta$	0	1.5	2.60	3	2.60	1.5	0	−1.5	−2.60	−3	−2.60	−1.5
$y = \cos\theta$	1	0.87	0.5	0	−0.5	−0.87	−1	−0.87	−0.5	0	0.5	0.87

The graph is called an ellipse and looks like this:

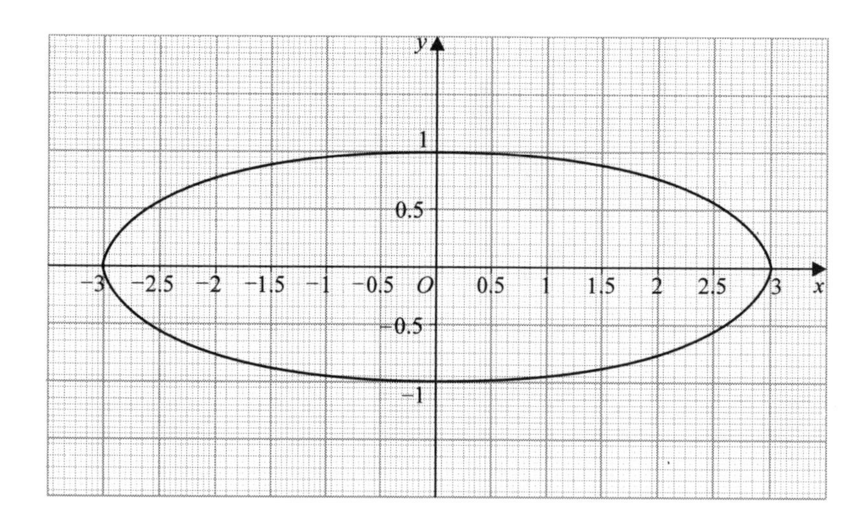

Now that you know how to draw a curve given by its parametric equations you can try to *sketch* such a curve. The things to look for are:

(1) Where does the curve cut the x- and y-axes?

(2) Has the curve any symmetry?

(3) Are there any restrictions placed on x and/or y by the parameter?

(4) If all else fails, plot a few points.

Example 15
Sketch the curve given by the parametric equations

$$x = t^2, \quad y = t^3 - t$$

(1) The curve cuts the x-axis where $y = 0$

i.e. where: $t^3 - t = 0$

$\Rightarrow$ $t(t^2 - 1) = 0$

$\Rightarrow$ $t(t - 1)(t + 1) = 0$

So: $t = 0, 1, -1$

For $t = 0$, $x = 0$. That is, the curve passes through $(0, 0)$.

For $t = \pm 1$, $x = 1$. That is, the curve passes through $(1, 0)$.

The curve cuts the y-axis where $x = 0$; that is where

$$t^2 = 0 \Rightarrow y = 0$$

So it cuts the y-axis only at $(0, 0)$.

(2) If $t = a$ (where a is positive);

then: $x = a^2$ and $y = a^3 - a$

If $t = -a$, then:

$$x = a^2 \text{ (again) but } y = -a^3 + a = -(a^3 - a)$$

So the curve passes through $(a^2, a^3 - a)$ and through $(a^2, -a^3 + a)$. That is, the curve is symmetrical about the x-axis.

(3) Since for any real value of t, t^2 is positive, and since $x = t^2$, x can only be positive or zero. So there is no curve to the left of the y-axis.

Also, when t is very large $|t^3|$ is much larger than $|t|$. So:

as $t \to +\infty$, $x \to +\infty$ and $y \to +\infty$

and as $t \to -\infty$, $x \to +\infty$ and $y \to -\infty$

The graph therefore looks like this:

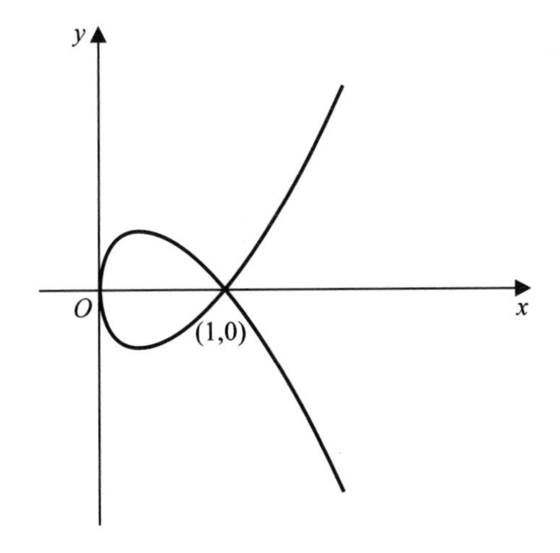

Example 10C

Sketch the curves given by the parametric equations:

1 $x = t^2 + 1,\ y = t$

2 $x = 1 + t,\ y = 4 - t^2$

3 $x = 4t,\ y = \dfrac{4}{t}$

4 $x = t - 1,\ y = t^2$

5 $x = 3\cos\theta,\ y = 2\sin\theta,\ 0 \leqslant \theta < 2\pi$

6 $x = t^2 - 1,\ y = t^3 - t$

7 $x = 2t + 3,\ y = t^2 - 1$

8 $x = 2t + 3,\ y = (t + 2)(t - 3)$

9 $x = \theta - \sin\theta,\ y = 1 - \cos\theta,\ 0 \leqslant \theta < 2\pi$

10 $x = \sin\theta,\ y = \cos 2\theta,\ 0 < \theta < \dfrac{\pi}{2}$

11 $x = 2\cos\theta,\ y = 2\sin\theta,\ 0 \leqslant \theta < 2\pi$

12 $x = 4\cos\theta,\ y = 3\sin\theta,\ 0 \leqslant \theta < 2\pi$

13 $x = \sin\theta,\ y = \sin 2\theta,\ 0 \leqslant \theta < 2\pi$

14 $x = t^2 - t,\ y = 2t$

15 $x = 2t^3,\ y = 3t^2$

16 $x = t^2 - 2,\ y = t^3 - 6t$

17 $x = t^2,\ y = t^3$

18 $x = t + 1,\ y = t^2 - 1$

19 $x = \dfrac{1}{t},\ y = t + 3$

20 $x = t^2 - 1,\ y = t^4 + 1$

SUMMARY OF KEY POINTS

1 The coordinates of the mid-point joining (x_1, y_1) and (x_2, y_2) are

$$\left(\frac{x_1 + x_2}{2}, \frac{y_1 + y_2}{2}\right)$$

2 Before attempting to sketch a curve given by a cartesian equation you should consider the following:

(a) Where does the curve cut the axes?

(b) Where are the asymptotes, if any?

(c) What happens as $x \to \pm\infty$?

(d) Has the curve any symmetry?

(e) Are there any points for which the curve is undefined?

(f) Where do the stationary points occur?

3 If you are asked to sketch a curve given by parametric equations, try to eliminate the parameter between $x = \mathrm{f}(t)$, $y = \mathrm{g}(t)$ to obtain the cartesian equation of the curve and then proceed as in (2).

4 If you are asked to sketch a curve given by parametric equations and you either cannot obtain the cartesian equation or if the cartesian equation is more complicated than the parametric equations, then sketch the curve from the parametric equations, asking the following questions:

(a) Where does the curve cut the axes?

(b) Does the curve have any symmetry?

(c) Are there any points for which the curve is undefined?

If all else fails, plot a few points.

Probability

11

This is the area of mathematics where you may be invited to try to estimate the chances of winning a raffle prize, or perhaps, the footballs pools or even the National Lottery. It is appropriate, perhaps, that this important topic of study is often introduced through games of chance because the initial studies in the subject began with questions arising from betting on dice at the French court in the middle part of the seventeenth century.

11.1 Sets

We often refer to sets of people, animals, items or objects by special names. We speak about a *herd* of cattle, a *crowd* at a football match, a *swarm* of bees or a *collection* of stamps, and so on. In mathematics, we use the one word **set** for all words such as herd, swarm, collection, etc. A set, then, is a list of objects placed together by some common quality. The objects in a set are called **members** or **elements**.

One way of naming a set is to list all its members inside a curly bracket:

{Sunday, Monday, Tuesday, Wednesday, Thursday, Friday, Saturday, Sunday}

A capital letter such as W is used to name the set of the days of the week and you could also write:

$$W = \{\text{the days of the week}\}$$

This is the second way you can identify the set: by using a general description. In words you would say 'W is the set of the days of the week.' So a set can be defined by listing all its members or by a general description.

Another thing you can say about the set W is that it has seven members. This is written as $n(W) = 7$ and in words 'the number of members in set W is 7'. Similarly the set M could be

$$M = \{\text{the months of the year}\}$$

and $n(M) = 12$, because there are 12 months in the year.

By using a general description you can define sets clearly and without ambiguity. For example, the set of positive integers between 2 and 11 inclusive can be written as:

$$\{2, 3, 4, 5, 6, 7, 8, 9, 10, 11\}$$

More often, though, we write this set as:

$$\{n : n \text{ is an integer such that } 2 \leqslant n \leqslant 11\}$$

and this is read in words as 'the set of integers n, where n lies between 2 and 11 inclusive'.

This second, more general, way of defining a set has obvious advantages. It is clearly much the better way of defining set A, where

$$A = \{n : n \text{ is an integer and } 0 < n < 1000\}$$

Writing out every member of A is both tedious and unnecessary!

You will also meet many sets where it will prove *impossible* to list all the elements, because there is an infinite number of them. For example, the set E, where

$$E = \{n : n \text{ is an integer and even}\}$$

has an infinite number of elements. Also the set P where

$$P = \left\{(x, y) : \begin{array}{l}(x, y) \text{ is a point in the triangle bounded by the} \\ x\text{-axis, the } y\text{-axis and the line } x + y = 4\end{array}\right\}$$

You should notice, however, that there is no problem about identifying whether, or not, any item is a member of E or of P. For example, 4 is a member of E and 7 is not. Similarly, $(1.5, 0.5)$ is a member of P whereas $(3, 2)$ is not.

There are several special sets of numbers that are important and they have special symbols. These are listed at the back of the book in the 'List of symbols and notation'. We briefly mention here the symbols $\mathbb{N}$, $\mathbb{Z}$, $\mathbb{Q}$ and $\mathbb{R}$, which represent the sets of positive integers and zero, the integers including zero, the rational numbers and real numbers, respectively. You met these in Book P1, chapter 3.

$\mathbb{N} = \{\text{positive integers and zero}\} = \{0, 1, 2, 3, \ldots\}$

$\mathbb{Z} = \{\text{integers}\} = \{0, \pm 1, \pm 2, \pm 3, \ldots\}$

$\mathbb{Q} = \{\text{rational numbers}\} = \left\{\dfrac{m}{n} : m, n \in \mathbb{Z}, n \neq 0\right\}$

$\mathbb{R} = \{\text{real numbers}\}$

The set of real numbers $\mathbb{R}$ includes all rational and irrational numbers. In any question, unless you are specifically told otherwise, you should assume that the set of real numbers is being used, just as you did when studying functions in Book P1, chapter 3.

You have also met the set inclusion symbol $\in$ which is used like this:

$$x \in \mathbb{R}$$

and is equivalent to saying in words 'x is a member of the set of real numbers'. The symbol $\in$ means 'is a member of'. The symbol $\notin$ is also used:

$$y \notin \mathbb{R}$$

is equivalent to saying 'y is not a member of the set of real numbers'. $\notin$ means 'is not a member of'.

Disjoint sets

Two sets that have no members in common are called disjoint. For example, the sets

$$A = \{3, 5, 6, 8\}$$
$$B = \{2, 4, 7, 9\}$$

are disjoint because they have no members in common.

The empty set

The empty set is the set that has no members. The empty set is denoted by $\varnothing$. An example of the empty set is the set of odd numbers exactly divisible by 4, because this set has no members.

Equal sets

Two sets are **equal** if, and only if, the members in each are exactly the same. For example, the sets A and B are equal when

$$A = \{1, 2\}$$
$$B = \{x : x \text{ is a solution of the equation } x^2 - 3x + 2 = 0\}$$

because the solutions of $x^2 - 3x + 2 = 0 \Rightarrow (x - 1)(x - 2) = 0$ are 1 and 2.

So: $$B = \{1, 2\} = A$$

Subsets

You will often find that a set you are considering is only part of another set that has a greater number of members. For example, the set $A = \{3, 5, 6, 8\}$ is contained in the set $C = \{1, 2, 3, 4, 5, 6, 7, 8, 9\}$

because all the members of A are also in C. A is said to be a **subset** of C and this can be written as $A \subset C$, where $\subset$ stands for 'is a subset of'.

Example 1

List all the possible subsets of the set A, where

$$A = \{a, b\}$$

One subset of A is the empty set $\varnothing$.

There are two subsets $\{a\}$ and $\{b\}$ with just one member each.

There is one subset $\{a, b\}$ with two members, and that is the set A itself.

In all then, A has four subsets.

Any subset that is *not* the entire set is called a **proper subset**. Finally then, the set A has three proper subsets and four subsets in all.

The universal set

The **universal set**, denoted by the letter $\mathscr{E}$, is the set of all the members in all the sets under discussion. All other sets considered in that situation are subsets of $\mathscr{E}$. For example, if $\mathscr{E}$ is the set of all people in your school and T is the set of teachers in your school, then T is a subset of $\mathscr{E}$ and we write $T \subset \mathscr{E}$.

11.2 Venn diagrams

Venn diagrams are used to illustrate sets. They were first used by the English mathematician John Venn (1834–1923). It is usual to represent the universal set $\mathscr{E}$ by a rectangle. An enclosed space, usually a circle, within this rectangle will represent a subset of $\mathscr{E}$. For example, if

$$\mathscr{E} = \{n : n \in \mathbb{Z},\ 1 \leqslant n \leqslant 9\}$$

and $$A = \{3, 5, 6, 8\}$$

then the corresponding Venn diagram could look like this:

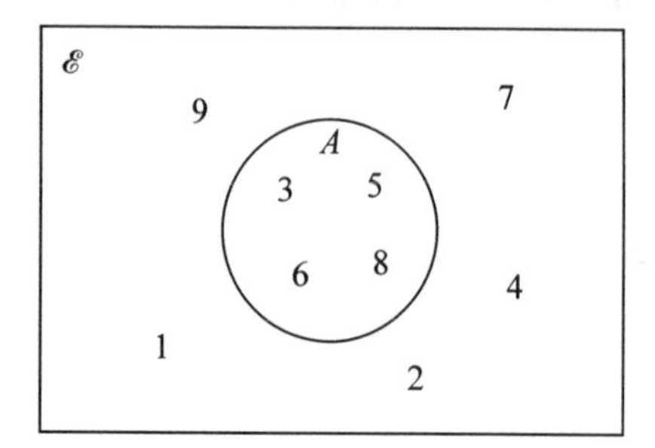

11.3 Operations on sets

Complement

For the universal set $\mathscr{E}$ and the subset A, the complement A' of A is a set containing all those members of $\mathscr{E}$ which are not members of A.

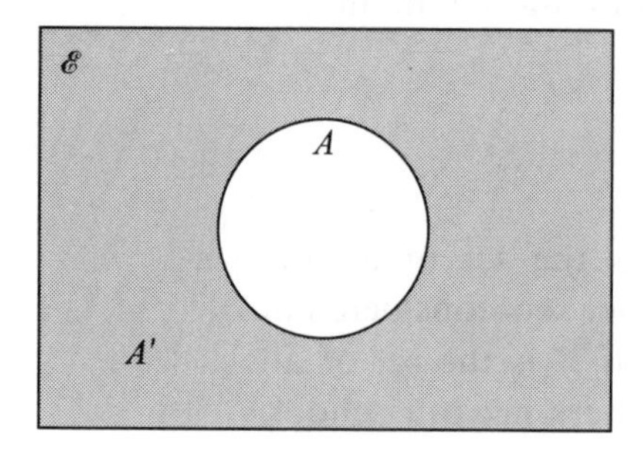

In the diagram, the set A' is shaded. Clearly

$$n(A) + n(A') = n(\mathscr{E})$$

Union

The union of two sets A and B, written $A \cup B$, or $B \cup A$, is the set of members that belong to either of the two sets, or to both.

Intersection

The intersection of two sets A and B, written $A \cap B$, or $B \cap A$, is the set of members that belong to both set A and set B.

There are several possible cases:

(1) Suppose A and B have some common members, then:

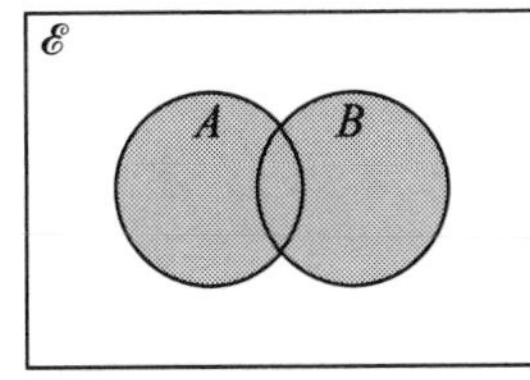

Set $A \cup B$ shaded

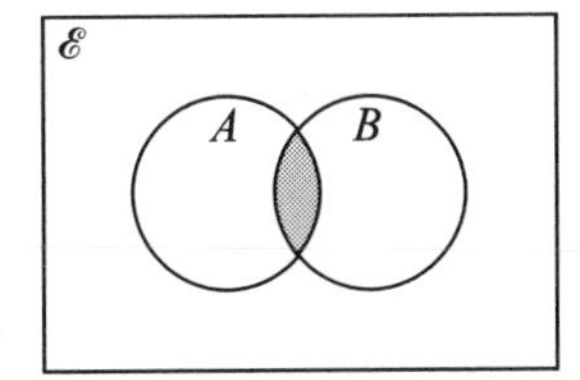

Set $A \cap B$ shaded

Also: $$\mathrm{n}(A) + \mathrm{n}(B) - \mathrm{n}(A \cap B) = \mathrm{n}(A \cup B)$$

(2) Suppose that $B \subset A$, then:

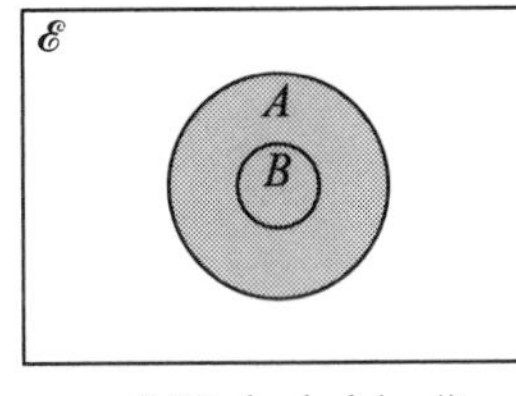

$A \cup B$ shaded $(=A)$

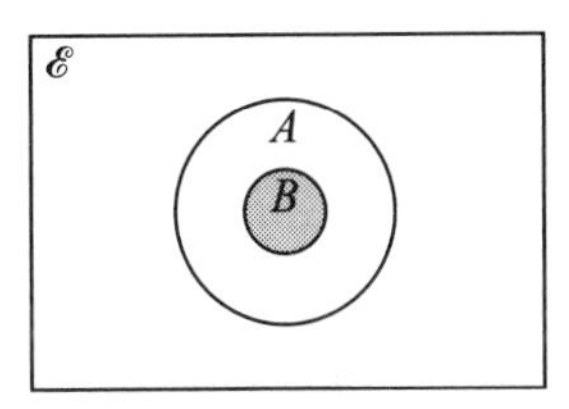

$A \cap B$ shaded $(=B)$

(3) Suppose that A and B are disjoint sets. Then:

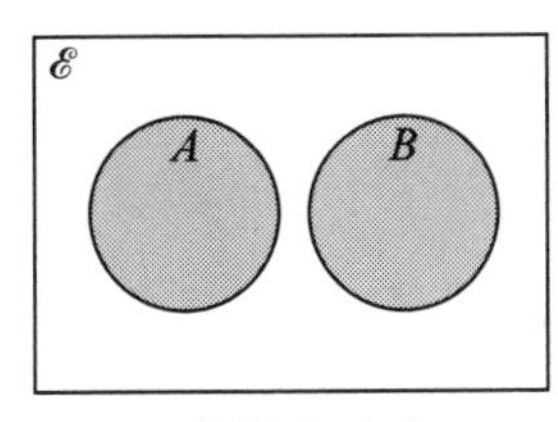

$A \cup B$ shaded

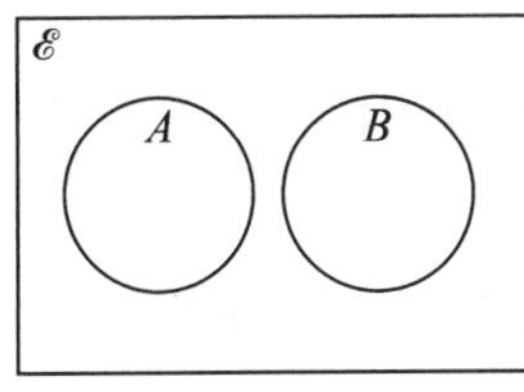

$A \cap B = \varnothing$

Also:

$$\mathrm{n}(A \cup B) = \mathrm{n}(A) + \mathrm{n}(B) \quad \text{and} \quad \mathrm{n}(A \cap B) = 0$$

Example 2

There are 25 students in a class and they study either Mathematics or English or both Mathematics and English. M is the set of students studying Mathematics and E is the set of those who study English. Given that $\mathrm{n}(M) = 12$ and $\mathrm{n}(E) = 17$, find $\mathrm{n}(M \cap E)$.

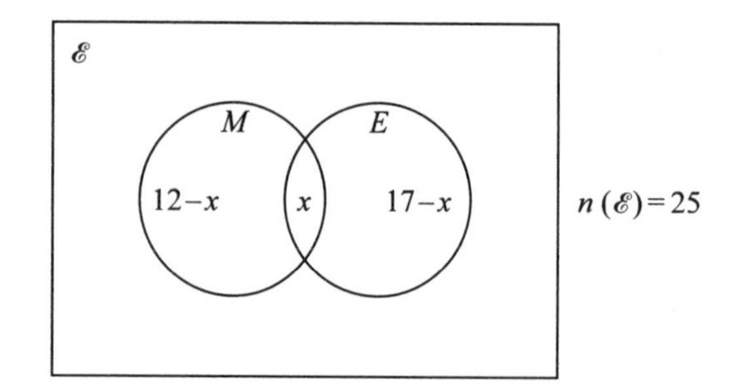

Let $n(M \cap E) = x$, then you can complete the Venn diagram showing $12 - x$ students studying Maths and not English and $17 - x$ students studying English and not Maths. Since all the 25 students study Maths or English or both,

$$(12 - x) + (17 - x) + x = 25$$
$$29 - 25 = x$$
$$x = 4$$

The number of students in the set $M \cap E$ is 4.

Example 3

The universal set contains all the positive integers from 1 to 10 inclusive. $A = \{1, 2, 5, 10\}$ and $B = \{1, 4, 6, 8, 10\}$.

List the members of (a) $A \cup B$ (b) $A \cap B$ (c) A'

(a) The set $A \cup B$ has members that are in A or in B or in both A and B; that is

$$A \cup B = \{1, 2, 4, 5, 6, 8, 10\}$$

(b) The set $A \cap B$ has members that are in both A and B; that is

$$A \cap B = \{1, 10\}$$

(c) The set A' contains all those elements that are in the universal set but not in A; that is

$$A' = \{3, 4, 6, 7, 8, 9\}$$

Exercise 11A

1 The universal set $\mathscr{E} = \{4, 5, 6, 7, 8, 9, 10, 11, 12\}$,
$A = \{5, 6, 9, 11\}$, $B = \{4, 6, 8, 10, 12\}$.
Display this information on a Venn diagram.
Find the sets A', B', $A \cup B$, $A' \cap B'$.

2 The sets A and B are such that $n(A) = 17$, $n(B) = 5$.
(a) If B is a subset of A, find $n(A \cup B)$ and $n(A \cap B')$.

(b) If $n(A \cup B) = 20$, find $n(A \cap B)$.

3 The set A contains all integers between 1 and 100 inclusive which are even and the set B contains all those integers between 1 and 100 inclusive which are multiples of either 3 or 5. Find (a) $n(A)$ (b) $n(B)$ (c) $n(A \cap B)$ (d) $n(A \cup B)$.

4 The universal set $\mathscr{E} = \{1, 2, 3, 4, 5, 6, 7, 8, 9, 10, 11, 12\}$.
$P = \{2, 4, 5, 6, 8, 9, 11\}$, $Q = \{1, 2, 3, 5, 6, 7, 10, 11, 12\}$.
Show these sets in a Venn diagram.
List the members of the sets P', $P \cap Q$, $P' \cap Q$.

5 The number x is a member of the set of real numbers.
$A = \{x : -2 \leqslant x \leqslant 5\}$, $B = \{x : -4 \leqslant x \leqslant 3\}$
Writing the sets in a similar notation, find $A \cap B$ and $A \cup B$.

6 The universal set is the set of real numbers.
$P = \{x : -4 \leqslant x \leqslant 6\}$, $Q = \{y : 0 \leqslant y \leqslant 8\}$
Find the sets $P \cap Q$ and $P \cup Q$.

7 All the students in a class of 30 musicians play either the piano or the violin. In this class, 22 play the piano and 17 play the violin. Display this information in a Venn diagram and hence find how many students play either the piano or the violin but not both.

11.4 Events and sample spaces

You will need to know the meaning of the following terms: **sample space** or **possibility space**, **simple event**, **compound event**, **outcome** and **trial**. The meanings are best understood by looking at some examples.

Suppose a die is thrown on a table and the number showing on the top face is recorded. The die could then be thrown again, if required.

Each time the die is thrown, we say that a **trial** takes place. Each trial has six possible **outcomes** which are: a score of 1 or 2 or 3 or 4 or 5 or 6 turns up. Each possible outcome is called a **simple event**. For one throw of the die then, one of six possible simple events will occur. These simple events are the only members of a **set** called the **sample space** or the **possibility space** of the trial.

Another way of describing the outcome of a trial would be to record the number turning up as odd or even. That is, when 1 or 3 or 5 turns up, the outcome is odd. In a case like this, the event 'the outcome is odd' is clearly a **compound event** which contains three simple events as members. You will soon appreciate how necessary it is to distinguish between simple events and a compound event that is the union of several simple events.

Now suppose that you draw a card at random from a normal pack of 52 playing cards which contains four suits: hearts, spades, clubs and diamonds. Each suit has an ace, a king, a queen, a jack and cards numbered 2,3, . . . 10. So there are 13 cards in each suit.

As each card can be distinguished from any other, one of 52 possible simple events can occur when a card is drawn at random. The compound event 'a heart is drawn' is the union of 13 simple events. Hearts and diamonds are red while spades and clubs are black. The compound event 'a red card is drawn' implies that a diamond or a heart is drawn; that is, this compound event is the union of 26 simple events.

11.5 Probability

If an event can take place in n different ways which are all equally likely, and if r of these ways are considered to be a success, then the probability of a success is defined to be $\frac{r}{n}$. This definition has some limitations but it will cover most of the cases you meet.

The probability that an event occurs is measured on a scale from 0 to 1. When the event *always* happens, its probability of occurring is 1 and similarly when an event *never* occurs, its probability is 0. Most events in a possibility space happen sometimes, and their probability is then a number x, where $0 < x < 1$.

Games of chance are often used to explain probability theory because it is possible to give fixed values to the probabilities of the outcomes. For example, the probability of scoring 3 in a single throw of an unbiased die is $\frac{1}{6}$. The probability of drawing the ace of hearts at random from a pack of cards is $\frac{1}{52}$; the probability of drawing a heart is $\frac{13}{52} = \frac{1}{4}$, and so on. You need, then, to be able to count up the number of simple events r, which are successes, and the total number of simple events n in the sample space; assuming that all simple events are equally likely, you can summarise this as:

$$\text{probability of success} = \frac{r}{n}$$

11.6 Relative frequency and probability

If you have a very large number of events, it will not be possible to check every one to see whether it is a 'success' or not. But you *can* adapt this approach so that you can find an *approximate* value for the probability that any event selected is a success. To do this you must select a smaller number (a **sample**) of events from the whole sample space (or **population**). But you must be careful to make sure that your selection is random so that the sample is representative of the whole population.

Suppose that you make n trials and that r of those trials are successes. Then, based on this evidence, an estimate for the probability of success is $\frac{r}{n}$. In general, estimates of the probability of success get progressively more accurate as the number of trials n increases. So when n is large, we assume that $\frac{r}{n}$ gives a good estimate of the probability of success in any single random trial. This approach to estimating probabilities is sometimes referred to as the **relative frequency approach**.

Example 4

A bag contains a large number of beads, of which 25% are blue and the rest are white. A bead is to be selected at random from the bag. Find the probability that a white bead will be selected.

You know that 75% of the beads are white.

$$\text{P(white selected)} = \frac{75}{100} = 0.75$$

Exercise 11B

1 Explain why there are 36 simple events in the sample space when two dice are thrown together. Find the probability that when two dice are thrown together

(a) a total score of 7 is obtained

(b) a double is obtained

(c) a total score of either 3 or 5 or 9 or 11 is obtained.

2 A card is drawn at random from a pack of 52 playing cards. Find the probability that the card will be
(a) an ace (b) a king or a queen or a jack
(c) a card whose face value lies between 3 and 9 inclusive.

3 In a class, there are 14 girls and 12 boys. A child is selected at random. Find the probability that
(a) a girl is selected (b) the oldest boy is selected.

4 The universal set is $\mathscr{E} = \{x : x \text{ is an integer}, 1 \leqslant x \leqslant 22\}$
$A = \{m : m \text{ is an even integer}, 2 \leqslant m \leqslant 18\}$
$B = \{n : n \text{ is an integer}, 3 \leqslant n \leqslant 22\}$
An integer is selected at random from $\mathscr{E}$. Find the probability that this integer is a member of the set
(a) A (b) B (c) $A \cap B$ (d) $A \cup B$

5 Two unbiased coins are spun. List the four possible outcomes. Find the probability that one, and only one, tail is obtained. Another coin is biased so that it is twice as likely to show a head as a tail when spun. Find, for this coin, when it is spun twice, the probability that either two heads or two tails are obtained.

6 A rectangle measures 5 cm by 6 cm. A circle of radius 2 cm is drawn to lie completely inside the rectangle. A point P is chosen at random to lie inside the rectangle. Find the probability that P lies inside the circle. Find also the probability that P does not lie inside the circle. (Give your answers to 2 decimal places.)

11.7 Combining events

Events may be combined in a very similar way to sets. The possibility space, or the sample space, is equivalent to the universal set. The union of all the simple events under consideration is the possibility space and it has probability 1 because one of its events must occur at each trial.

Events A, B, C, etc. are subsets of the possibility space and their probabilities of occurring are written $\mathrm{P}(A)$, $\mathrm{P}(B)$, $\mathrm{P}(C)$, etc. Events such as A', the complement of A, $A \cup B$, the union of A and B and $A \cap B$, the intersection of the events A and B, are synonymous with the sets A', $A \cup B$ and $A \cap B$ in set theory. The probabilities $\mathrm{P}(A')$, $\mathrm{P}(A \cup B)$ and $\mathrm{P}(A \cap B)$ are synonymous with the number of members in the sets. Probability theory is an important application of set theory and your understanding of one will help you to understand the other.

For any two events A and B of a possibility space S, you should learn and memorise this important formula (see section 11.3):

- $\mathbf{P}(\boldsymbol{A}) + \mathbf{P}(\boldsymbol{B}) - \mathbf{P}(\boldsymbol{A} \cap \boldsymbol{B}) = \mathbf{P}(\boldsymbol{A} \cup \boldsymbol{B})$

Example 5

The events A and B are members of a possibility space S. $\mathrm{P}(A) = 0.4$, $\mathrm{P}(B) = 0.35$ and $\mathrm{P}(A \cap B) = 0.15$. Show this information in a Venn diagram and use it to find the values of (a) $\mathrm{P}(A \cup B)$ (b) $\mathrm{P}(A' \cap B)$.

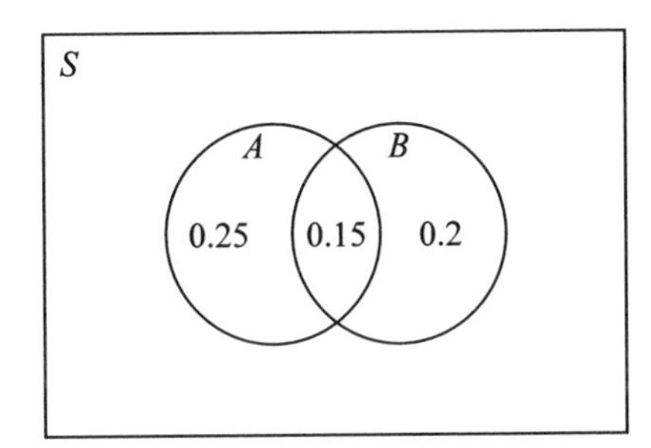

Draw a rectangle with two intersecting circles inside for the Venn diagram and write in the probabilities for each region, starting with $\mathrm{P}(A \cap B) = 0.15$. Using the information given, you can fill in the other two probabilities as shown.

(a) Using the formula,

$$\begin{aligned}\mathrm{P}(A \cup B) &= \mathrm{P}(A) + \mathrm{P}(B) - \mathrm{P}(A \cap B) \\ &= 0.4 + 0.35 - 0.15 = 0.6\end{aligned}$$

(b) The region in the diagram which is shaded represents the event $A' \cap B$.

$$\mathrm{P}(A' \cap B) = 0.35 - 0.15 = 0.2$$

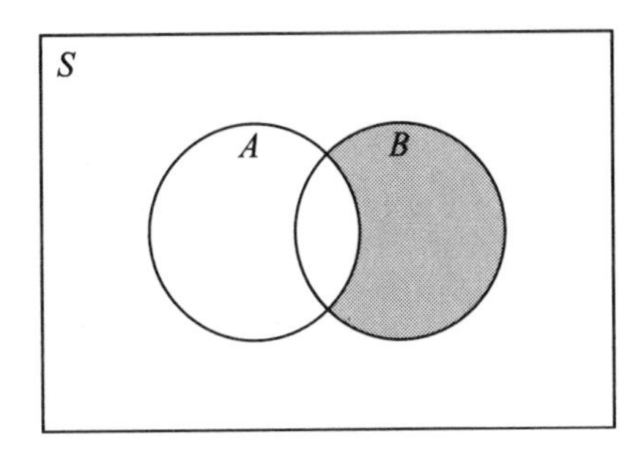

Mutually exclusive events

A possibility space S is given and two of its events, A and B, have no simple events in common. This is equivalent to saying that $P(A \cap B) = 0$ and therefore you have the formula

$$P(A) + P(B) = P(A \cup B)$$

In this case, the events A and B are called **mutually exclusive**. That is, if A occurs, then B cannot occur and vice versa. This formula is sometimes called the **addition rule of probability** but, remember, the events A and B must be mutually exclusive.

This Venn diagram illustrates the situation:

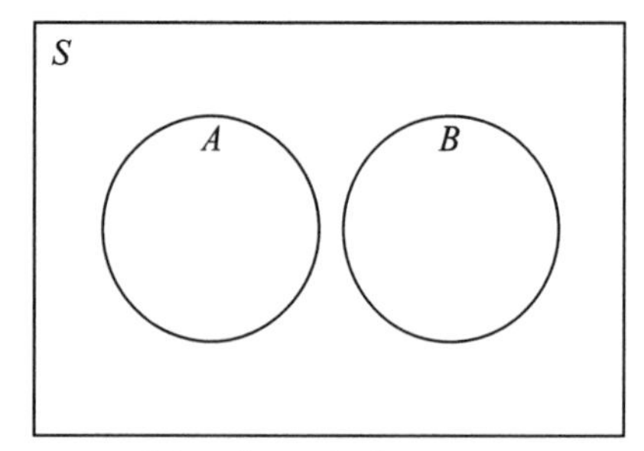

Mutually exclusive events
A and B

Complementary events

A possibility space S is given, together with an event A. The event A' contains all those simple events in S that are *not* members of A and A' is called the **complement** of A. The events A and A' are mutually exclusive and

$$P(A) + P(A') = 1$$

These events A and A' are sometimes called **exhaustive** because their union is the whole possibility space.

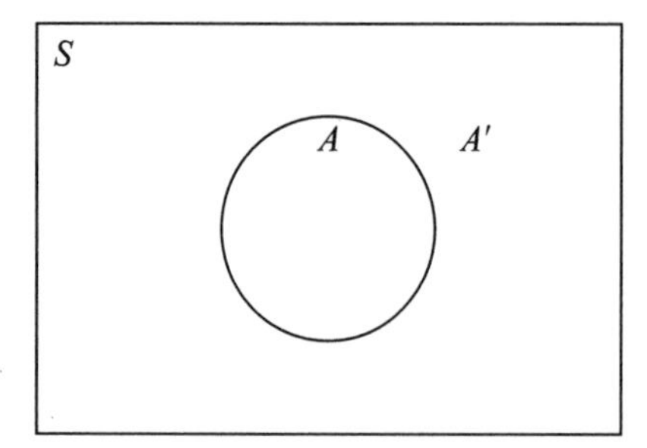

The events A and A'

Example 6

Each of ten counters has a different member of the set $\{1, 2, 3, 4, 5, 6, 7, 8, 9, 10\}$ written on it. A counter is selected at random. The event A is 'the number on the selected counter is a multiple of 3'. The event B is 'the number on the selected counter is a multiple of 5'. Find $\mathrm{P}(A)$, $\mathrm{P}(A')$, $\mathrm{P}(B)$, $\mathrm{P}(A \cup B)$.

The event A has three simple events as members, namely 3,6,9, so you can write

$$\mathrm{P}(A) = \tfrac{3}{10}$$

Also $\mathrm{P}(A') = 1 - \mathrm{P}(A) = 1 - \tfrac{3}{10} = \tfrac{7}{10}$

The event B has two simple events, namely 5, 10, and

$$\mathrm{P}(B) = \tfrac{2}{10}$$

The simple events in $A \cup B$, are 3,5,6,9,10 and $\mathrm{P}(A \cup B) = \tfrac{5}{10}$.

As A and B have no simple events in common, the events A and B are mutually exclusive and you know that $\mathrm{P}(A) + \mathrm{P}(B) = \mathrm{P}(A \cup B)$ in this case. The following diagrams would be useful to you when building up your own solution:

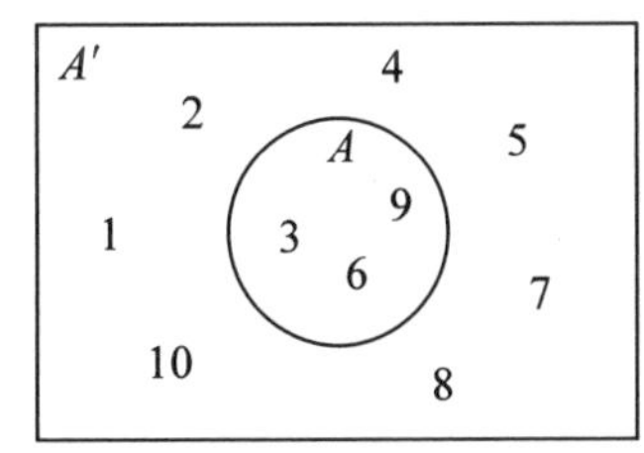

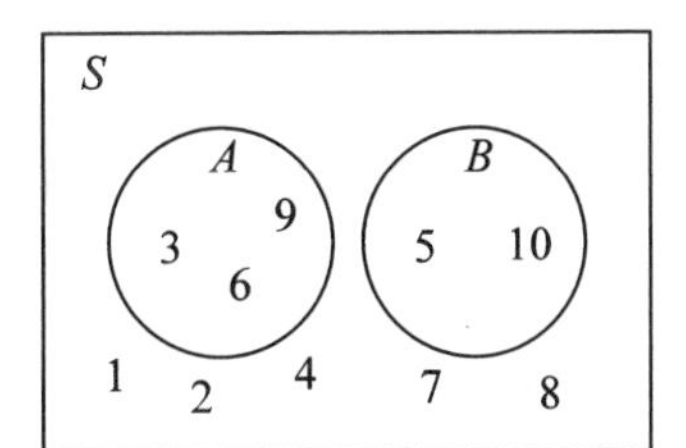

Independent events

- **Provided that neither $\mathrm{P}(A)$ nor $\mathrm{P}(B)$ is zero, the events A and B of a sample space S are *independent* if and only if**
$$\mathrm{P}(A \cap B) = \mathrm{P}(A)\,\mathrm{P}(B)$$

This is sometimes called the **product rule of probability** but, remember, it is only true when A and B are independent events.

Example 7

A card is selected at random from a pack of 52 playing cards. The card is replaced and a second card is selected at random. The event H is the first card is a heart. The event D is the second card is a diamond. Find $\mathrm{P}(H)$, $\mathrm{P}(D)$ and $\mathrm{P}(H \cap D)$. Comment on your answers.

There are 52 cards in the pack, of which 13 are hearts; so

$$\mathrm{P(H)} = \tfrac{13}{52} = \tfrac{1}{4}$$

Similarly, $\mathrm{P}(D) = \tfrac{1}{4}$

There are 52×52 possible pairings of the first and second cards to be selected and, of these, 13×13 are (heart, diamond) pairs.

So: $$\mathrm{P}(H \cap D) = \frac{(13 \times 13)}{(52 \times 52)} = \tfrac{1}{16}$$

Notice that you have $\mathrm{P}(H)\,\mathrm{P}(D) = \mathrm{P}(H \cap D)$, confirming, as you would expect, that H and D are independent events.

Example 8

A red die and a black die are thrown. There are 36 simple events in the possibility space and each of these can be uniquely identified by the number pair (x, y), where x and y are integers between 1 and 6 inclusive and where x is the score on the red die and y is the score on the black die. For example, $(5, 6)$ is the event 'the red die scores 5 and the black die scores 6'. The event A is '$x = 3$ or 4' and the event B is '$y = 5$'. Discuss whether or not A and B are independent events.

The event A has 12 simple events as members $\{(3, 1), (3, 2) \ldots (3, 6), (4, 1), (4, 2) \ldots (4, 6)\}$

So: $$\mathrm{P}(A) = \tfrac{12}{36} = \tfrac{1}{3}$$

The event B has 6 simple events as members $\{(1, 5), (2, 5) \ldots (6, 5)\}$ and

$$\mathrm{P}(B) = \tfrac{6}{36} = \tfrac{1}{6}$$

The event $A \cap B$ has two members, namely $(3, 5)$ and $(4, 5)$ and so

$$\mathrm{P}(A \cap B) = \tfrac{2}{36} = \tfrac{1}{18}$$

You can now work out the product

$$\mathrm{P}(A)\,\mathrm{P}(B) = \tfrac{1}{3} \times \tfrac{1}{6} = \tfrac{1}{18}$$

As this is the same as $\mathrm{P}(A \cap B)$, you can conclude that A and B are independent events.

Example 9

The events A and B are independent.

$$\mathrm{P}(A) = 2x, \quad \mathrm{P}(\overline{B}) = 3x, \quad \mathrm{P}(A \cup B) = \tfrac{2}{3}$$

Find the value of x.

Using the formula $\mathrm{P}(A) + \mathrm{P}(B) = \mathrm{P}(A \cup B) + \mathrm{P}(A \cap B)$,

$$\mathrm{P}(A \cap B) = 2x + 3x - \tfrac{2}{3} = 5x - \tfrac{2}{3}$$

As the events A and B are independent,

$$\mathrm{P}(A)\,\mathrm{P}(B) = \mathrm{P}(A \cap B)$$

That is:

$$(2x)(3x) = 5x - \tfrac{2}{3}$$
$$6x^2 = 5x - \tfrac{2}{3}$$
$$18x^2 - 15x + 2 = 0$$
$$(3x - 2)(6x - 1) = 0$$

giving $x = \frac{1}{6}$ or $\frac{2}{3}$

But as $\mathrm{P}(B) = 3x < 1$, the solution $x = \frac{2}{3}$ is not possible.

Therefore $x = \frac{1}{6}$ is the solution.

11.8 Conditional probability

In a college there are 100 students in all. Of these, 49 study Maths (set M) and 19 study Geography (set G). There are 12 students who study both Maths and Geography (set $M \cap G$). We can display this information on a Venn diagram.

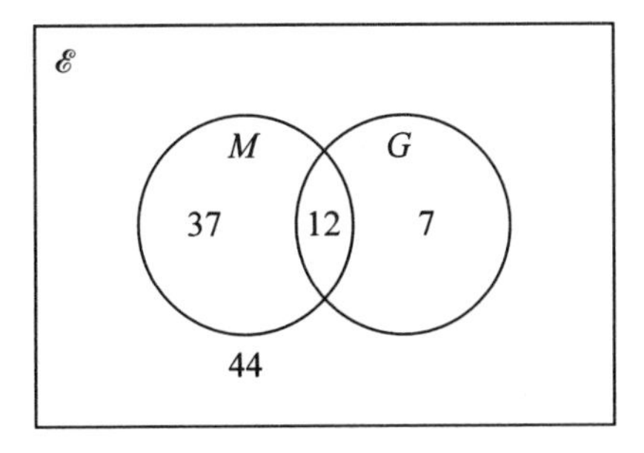

The probability of selecting at random a student studying Geography from all 100 students is $\frac{19}{100}$. That is:

$$\mathrm{P}(G) = \frac{19}{100}$$

Also we have

$$\mathrm{P}(M \cap G) = \frac{12}{100}$$

from the data given.

However, if you visit the Geography class of just 19 students and select a student at random, the probability of selecting a student who also studies Maths is $\frac{12}{19}$. This probability is written as $\mathrm{P}(M|G)$ and read as 'the probability of the student studying Maths given that the student studies Geography'.

This is an example of **conditional probability**, the probability of M (the student studying Maths) conditional upon the event G (the student studying Geography) having already happened.

In this example, $P(M|G) = \frac{12}{19}$. Notice that the 12 in the numerator represents the students in the set $M \cap G$ and the 19 in the denominator represents the students in the set G. So this could be written as

$$P(M|G) = \frac{P(M \cap G)}{P(G)} = \frac{\frac{12}{100}}{\frac{19}{100}} = \frac{12}{19}$$

and this can also be written as

$$P(M \cap G) = P(M|G)\,P(G)$$

- **So for two events A and B we have the rule**
  $$P(A \cap B) = P(A|B)\,P(B)$$
  which relates the conditional probability of A, given B, $P(A|B)$, with $P(A \cap B)$ and $P(B)$.

Also, by interchanging A and B you get:

$$P(B \cap A) = P(B|A)\,P(A)$$

As $P(B \cap A) = P(A \cap B)$ this rule is also

- **$P(A \cap B) = P(B|A)\ P(A)$**

Finally then, we have $P(A \cap B) = P(B|A)\,P(A) = P(A|B)\,P(B)$

Example 10

The events A and B of a sample space are such that $P(A) = \frac{2}{3}$, $P(B|A) = \frac{1}{5}$ and $P(A \cup B) = \frac{9}{10}$. Find $P(B)$ and $P(A|B)$.

Using the formula $P(A \cap B) = P(B|A)\,P(A)$

you get: $\quad P(A \cap B) = \frac{1}{5} \times \frac{2}{3} = \frac{2}{15}$

Using the formula $P(A) + P(B) - P(A \cap B) = P(A \cup B)$

you get: $\quad \frac{2}{3} + P(B) - \frac{2}{15} = \frac{9}{10}$

So: $\quad P(B) = \frac{9}{10} + \frac{2}{15} - \frac{2}{3} = \frac{11}{30}$

Using the formula $P(A \cap B) = P(A|B)\,P(B)$

you get: $\quad \frac{2}{15} = P(A|B)\,\frac{11}{30}$

So: $\mathrm{P}(A|B) = \frac{2}{15} \div \frac{11}{30}$

$= \frac{2}{15} \times \frac{30}{11} = \frac{4}{11}$

Tree diagrams

Probability problems can get quite complicated and diagrams can help to sort out the various outcomes and their probabilities. One such type of diagram is called a **tree diagram** or a **probability tree**.

Consider the possibilities when you toss an unbiased coin twice. The first toss results in either a head (H) or a tail (T), each having probability $\frac{1}{2}$.

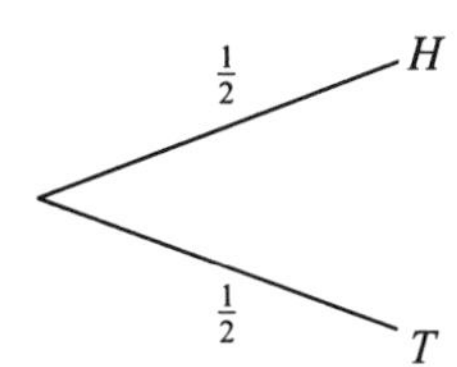

Two lines are drawn as shown, the outcomes are written at the end and the probabilities of each outcome are written beside the lines. The process is repeated for the second toss, except that it is done twice, starting at each end of the outcomes shown by the first toss.

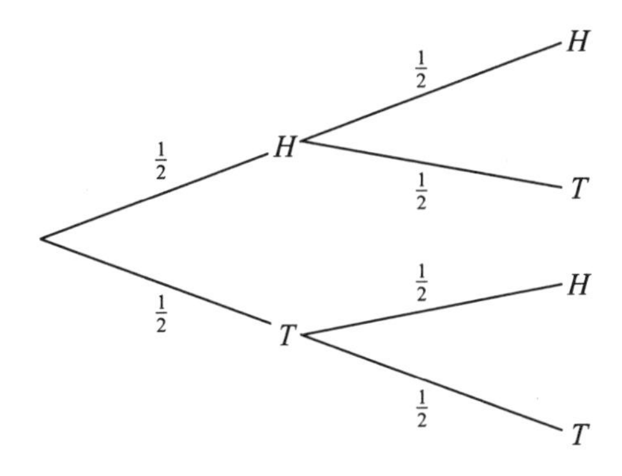

To find the probability of a specific outcome, say H followed by T, the probabilities of the two steps are multiplied together:

$$\mathrm{P}(HT) = \tfrac{1}{2} \times \tfrac{1}{2} = \tfrac{1}{4}$$

A useful rule when using tree diagrams is '*multiply along* the branches but *add between* the branches'.

The probability of getting just one head when two unbiased coins are tossed is equivalent to getting either HT or TH in the tree diagram, that is

P(one head) $= (\frac{1}{2} \times \frac{1}{2}) + (\frac{1}{2} \times \frac{1}{2})$ (multiply along but add between)
$= \frac{1}{4} + \frac{1}{4}$
$= \frac{1}{2}$

Example 11
A bag contains four yellow balls and five green balls. Two balls are drawn at random, one after another, without replacement. Draw a tree diagram to show the possible outcomes and write in the probabilities on the branches. Hence find the probability that

(a) two yellow balls are obtained

(b) the second ball is yellow given that the first is green

(c) a yellow ball and a green ball are obtained

(d) both balls are of the same colour.

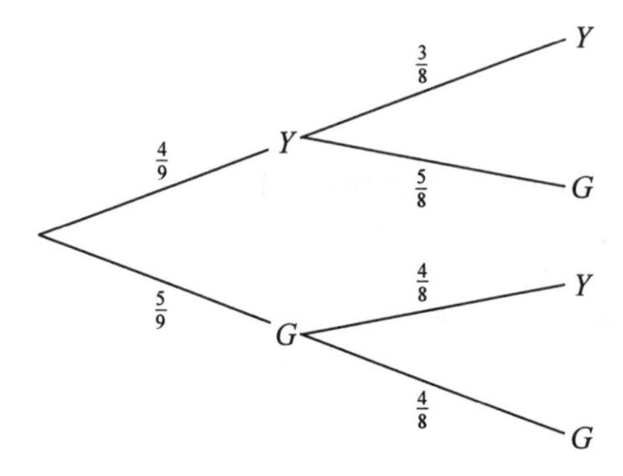

First draw the tree diagram, as shown, where Y is the event 'a yellow ball is drawn' and G is the event 'a green ball is drawn'.

(a) $\mathrm{P}(YY) = \frac{4}{9} \times \frac{3}{8} = \frac{1}{6}$

(b) $\mathrm{P}(Y|G) = \frac{4}{8} = \frac{1}{2}$

(c) $\mathrm{P}(YG \text{ or } GY) = (\frac{4}{9} \times \frac{5}{8}) + (\frac{5}{9} \times \frac{4}{8}) = \frac{40}{72} = \frac{5}{9}$

(d) $\mathrm{P}(GG \text{ or } YY) = (\frac{5}{9} \times \frac{4}{8}) + (\frac{4}{9} \times \frac{3}{8}) = \frac{32}{72} = \frac{4}{9}$

Exercise 11C

1 The events A and B are mutually exclusive with $\mathrm{P}(A) = 0.3$ and $\mathrm{P}(B) = 0.44$. Find $\mathrm{P}(A \cup B)$ and sketch a Venn diagram. Hence find $\mathrm{P}(A' \cap B')$.

2 The independent events L and M have $\mathrm{P}(L) = 0.6$ and $\mathrm{P}(L \cap M) = 0.3$. Find $\mathrm{P}(L \cup M)$.

3 The events A and B are such that $P(A) = 0.72$, $P(B) = 0.56$ and $P(A \cap B) = 0.3$. Find $P(A \cup B)$ and $P(B|A)$.

4 The events E and F are such that $P(E) = 0.6$, $P(F) = 0.3$ and $P(E|F) = 0.5$. Find $P(E \cap F)$ and $P(F|E)$.

5 The events A and B are such that $P(A) = 0.6$, $P(B) = 0.3$ and $P(A \cup B) = 0.8$. Find the probability that

(a) both A and B will occur

(b) just one of the events A and B will occur

(c) B occurs, given that A has occurred.

6 The independent events C and D are such that $P(C) = \frac{1}{2}$ and $P(C \cap D) = \frac{1}{6}$.

(a) Find $P(D)$.

The events C and E are mutually exclusive and $P(C \cup E) = \frac{3}{4}$.

(b) Find $P(E)$.

Given that $P(D \cap E) = \frac{1}{12}$, show all the events in a Venn diagram.

(c) Find $P(E \cap C' \cap D')$.

7 The independent probabilities that Ann and Carol will score a point from a free throw in netball are $\frac{3}{5}$ and $\frac{2}{3}$ respectively. They each have one throw. Find the probability that one, and only one, girl scores a point.

8 The events A and B are independent, with $P(A) = P(B)$. Given further that $P(A \cap B) = \frac{4}{9}$, find $P(A \cup B)$.

9 The events C and D of a sample space are such that $P(C) = \frac{3}{7}$, $P(D) = \frac{5}{14}$ and $P(C \cap D) = \frac{1}{7}$.

Find $P(C \cup D)$, $P(C|D)$ and $P(D|C)$.

10 Five cards are marked 1, 2, 3, 4 and 5. These cards are placed in a bag and a card is drawn at random from the bag.

The event A is 'the number on the card chosen is less than 4'.

The event B is 'the number on the card chosen is odd'.

Find $P(A)$, $P(B)$, $P(A \cup B)$, $P(A|B)$, $P(B|A)$.

11 Using the same five cards as in question 10, two cards are drawn from the bag one after the other with the first card being replaced before the second is drawn.

(a) Find the probability that the sum of the digits on the cards selected is even.

(b) Find the probability that the digit on the second card selected is greater than the digit on the first card selected.

12 There are six equally eligible students available for the four vacancies on a College Council. Given that the four are to be selected at random, find the probability that

(a) both the oldest student and the youngest student are selected

(b) the youngest student is selected given that the oldest is selected.

13 In a group of 40 children 19 are wearing black shoes, 24 are wearing anoraks and 14 are wearing neither black shoes nor an anorak. A child is selected at random.
The event A is 'the child selected is wearing an anorak'.
The event B is 'the child selected is wearing black shoes'.
Find $\mathrm{P}(A)$, $\mathrm{P}(B)$, $\mathrm{P}(A \cup B)$, $\mathrm{P}(A|B)$.
Two children are selected at random and without replacement. Find the probability that one is wearing black shoes and no anorak and the other is wearing an anorak and not black shoes.

14 A bag contains 12 balls in all, of which 5 are white and the rest are red. Two balls are selected at random and without replacement. Draw a tree diagram to show the possible outcomes and write in on the branches the corresponding probabilities. Hence find the probability that the two balls selected will be

(a) both red

(b) both of the same colour.

15 Three girls Claire, Debbie and Emily have independent probabilities of $\frac{1}{2}$, $\frac{2}{3}$ and $\frac{3}{4}$ respectively of solving a problem. Find the probability that

(a) just one of the three solves the problem

(b) they all fail to solve the problem.

16 John travels to school by train each day. On any day the probability that his train is delayed is 0.4.
If his train is delayed, the probability that he is late for school is 0.8. If his train is not delayed, the probability that he is late for school is 0.3. Draw a tree diagram to show the possible outcomes. Find the probability that John is not late for school tomorrow.

17 A bag contains three red balls and two white balls. A box contains five red balls and four white balls. A coin is spun, and if a head is obtained a ball is chosen at random from the bag. If a tail is obtained a ball is chosen at random from the box.

(a) By drawing a tree diagram find the probability that the ball drawn is white.

(b) Given that the ball drawn is red, find the probability that it was originally in the bag.

18 At a race meeting the probabilities of the favourite winning the 2:30, the 3 o'clock and the 3:30 races are $\frac{3}{4}$, $\frac{3}{5}$ and $\frac{2}{3}$ respectively. Find the probability that one favourite wins and the other two favourites lose.

19 Show that there are 15 different selections of a pair of cards from six different cards coloured red, white, blue, yellow, orange and green.

Two cards are chosen, one at a time and without replacement, from the six cards. Find the probability that one of the chosen cards is yellow.

20 Two cards are selected at random and without replacement from a pack of 52 playing cards. Find the probability that

(a) both cards are jacks

(b) the cards are of different suits

(c) one, and only one, of the cards is a king.

SUMMARY OF KEY POINTS

1 A set is a collection of objects, called members or elements, that have some common feature.

2 The number of members in the set A is denoted by $n(A)$.

3 Two sets with no common members are called disjoint.

4 The empty set has no members and is denoted by $\varnothing$.

5 The universal set includes all members in all sets under discussion and is denoted by $\mathscr{E}$.

6 If two sets have precisely the same elements, they are equal.

7 Venn diagrams are used to illustrate sets and possibility spaces.

8 A' is the complement of set A and contains all elements in $\mathscr{E}$ which are not in A.

9 $A \cup B$ is the set called the union of sets A and B and it contains all elements that are in A, in B or in both A and B.

10 $A \cap B$ is the set called the intersection of sets A and B and it contains only those elements which are members of both A and B.

11 For any two sets A and B:

$$\mathrm{n}(A) + \mathrm{n}(B) - \mathrm{n}(A \cap B) = \mathrm{n}(A \cup B).$$

12 The set whose elements are all the simple events (outcomes) of a trial is called the possibility space or the sample space.

13 If there are n simple events in the possibility space, all equally likely, and r of these events are members of a compound event A, then the probability of A occurring is written $\mathrm{P}(A)$ and

$$\mathrm{P}(A) = \frac{r}{n}$$

14 For any two events A and B of a possibility space,

$$\mathrm{P}(A) + \mathrm{P}(B) - \mathrm{P}(A \cap B) = \mathrm{P}(A \cup B)$$

15 The events A and B are mutually exclusive if $\mathrm{P}(A \cap B) = 0$ and

$$\mathrm{P}(A) + \mathrm{P}(B) = \mathrm{P}(A \cup B)$$

16 The events A and B are independent if, and only if, $\mathrm{P}(A)\,\mathrm{P}(B) = \mathrm{P}(A \cap B)$, where $\mathrm{P}(A)$ and $\mathrm{P}(B)$ are not zero.

17 $\mathrm{P}(A|B)$ is the conditional probability of event A occurring, given that event B has occurred. In particular

$$\mathrm{P}(A \cap B) = \mathrm{P}(A|B)\,\mathrm{P}(B) = \mathrm{P}(B|A)\,\mathrm{P}(A)$$

18 Venn diagrams and tree diagrams can be helpful in building a solution and investigating a possibility space.

Numerical methods 12

From your GCSE studies and the work done in Book P1 you should know how to solve linear equations and quadratic equations, either by factorising, by formula or by completing the square. However, although it is possible to solve cubic equations (polynomial equations of degree 3), quartic equations (polynomial equations of degree 4) and so on, if they factorise, it is difficult to solve them precisely otherwise. Equally it is difficult to solve equations such as $\sin\theta - \theta = 1$ precisely. Indeed, it is a fact that the majority of equations cannot be solved in any precise manner and so we have to solve them by using **iterative** procedures. An iterative procedure is a repetitive process that produces a sequence of approximations. The sequence may converge to a limiting value such as the root of an equation or the sum of a series. Sometimes, however, a sequence diverges: the differences between successive approximations get bigger and bigger instead of smaller. (Divergent sequences were mentioned in chapter 6 of Book P1.)

12.1 Approximate solutions of equations

Consider the equation $x^2 - 5x + 2 = 0$. The graph of $y = x^2 - 5x + 2$ is shown on page 366.

As you can see, one root of the equation $x^2 - 5x + 2 = 0$ lies between 0 and 1 and another between 4 and 5. You could also have found this by the methods shown in section 10.4 of Book P1. That is, if $f(x) \equiv x^2 - 5x + 2$, then

$f(0) = 2 > 0$ and $f(1) = -2 < 0$, so a root lies between 0 and 1.

$f(4) = -2 < 0$, $f(5) = +2 > 0$, so a root lies between 4 and 5.

Rearrange the equation $x^2 - 5x + 2 = 0$ so that:

$$x^2 = 5x - 2$$

and hence

$$x = \pm\sqrt{(5x - 2)}$$

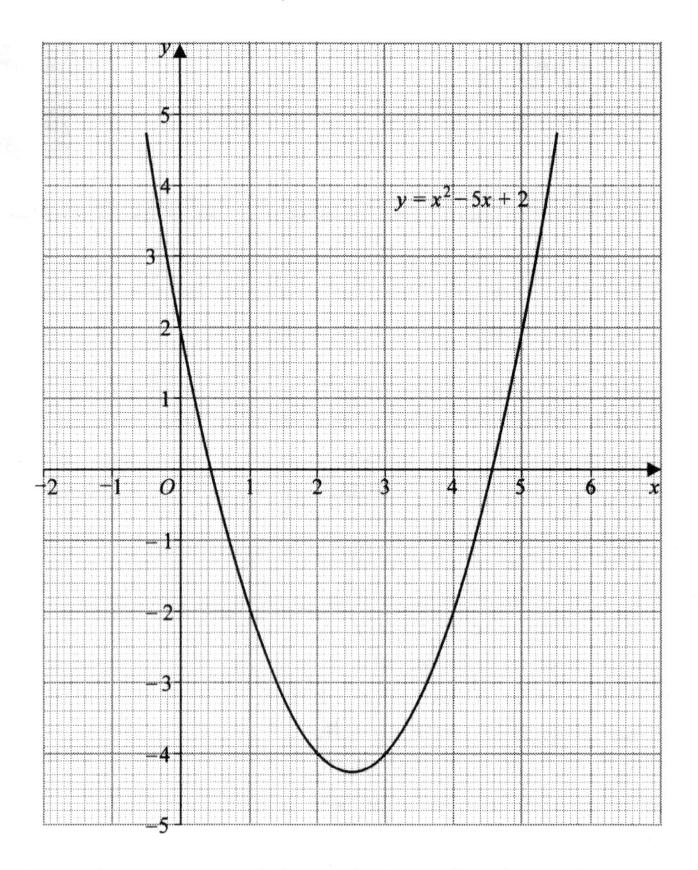

Take the positive root and form the iteration formula

$$x_{n+1} = \sqrt{(5x_n - 2)}$$

Using the formula, you can now find a sequence of approximations $x_0, x_1, x_2, \ldots$ that gets progressively closer to one of the roots of the equation $x^2 - 5x + 2 = 0$. As one root lies between 4 and 5, you could take the starting value as $x_0 = 4$. Using the iteration formula:

$$x_1 = \sqrt{[(5 \times 4) - 2]} = \sqrt{18} = 4.242\,640\,687$$
$$x_2 = \sqrt{[(5 \times 4.242\,640\,687) - 2]} = 4.383\,286\,83$$
$$x_3 = 4.462\,783\,229$$
$$x_4 = 4.507\,096\,199$$
$$x_5 = 4.531\,609\,095$$
$$x_6 = 4.545\,112\,262$$
$$x_7 = 4.552\,533\,505$$
$$x_8 = 4.556\,607\,019$$
$$x_9 = 4.558\,841\,42$$

$$x_{10} = 4.560\,066\,568$$
$$x_{11} = 4.560\,738\,19$$

where the calculations are given to 9-figure accuracy.

So one root of the equation is 4.56 (2 decimal places), which is the same as can be obtained by using the quadratic formula.

The question now arises as to why this method works. The graphs of $y = \sqrt{(5x - 2)}$ and $y = x$ look like this:

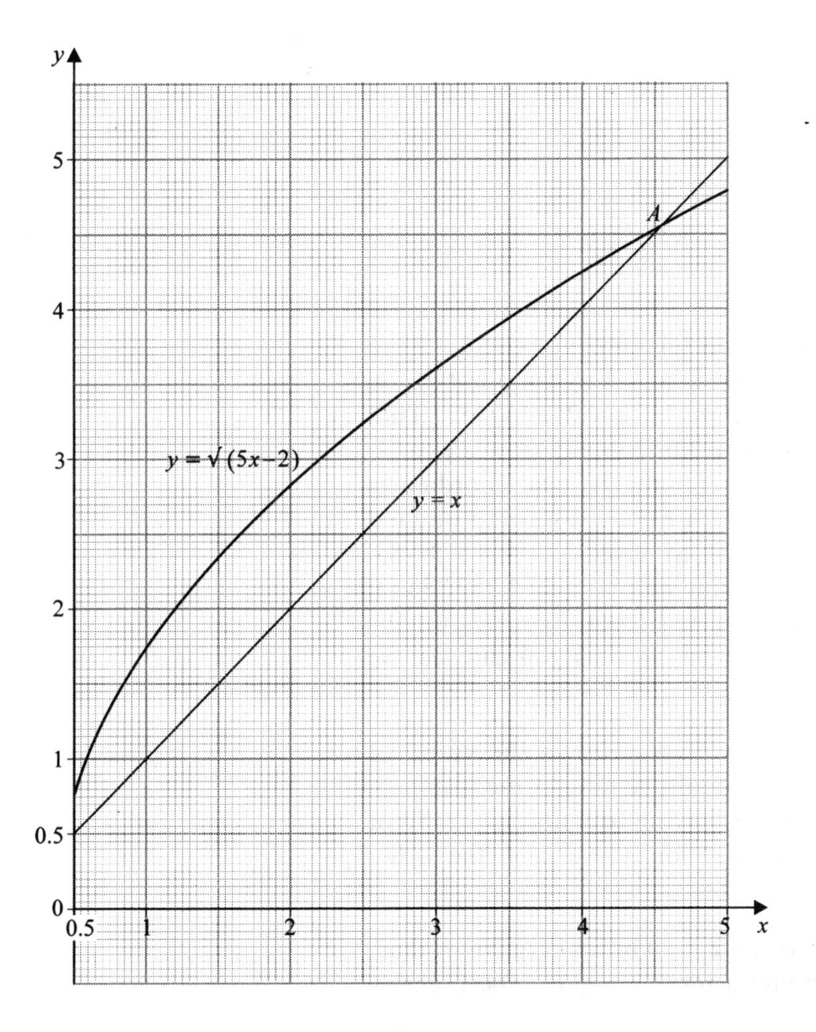

Given that the equation $x = \sqrt{(5x - 2)}$ is another form of the equation $x^2 - 5x + 2 = 0$, a solution of $x^2 - 5x + 2 = 0$ occurs where the graph of $y = x$ crosses the graph of $y = \sqrt{(5x - 2)}$, because when these two equations are satisfied simultaneously then $x = \sqrt{(5x - 2)}$. (Refer back to Book P1 section 2.2 if you do not remember this.) So you need to find a sequence of approximations that gets closer and closer to the x-coordinate of the point A.

Consider those parts of the graphs that lie between $x = 4$ and the point A:

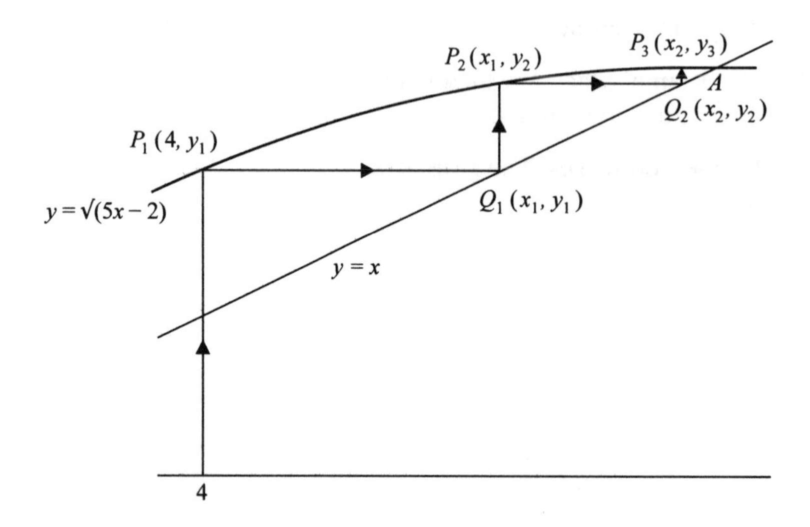

We started with $x_0 = 4$ and substituted this into $y = \sqrt{(5x-2)}$ to give $y_1 = 4.242\,640\,687$. So we went from $x = 4$ to the point P_1 on the graph of $y = \sqrt{(5x-2)}$. Then we put $x_1 = 4.242\,640\,687$. That is we put the value $y_1 = x_1$. So we went horizontally from P_1 to Q_1 on the graph of $y = x$. Thus the point Q_1 has coordinates $(4.242\,640\,687, 4.242\,640\,687)$. We then substituted $x_1 = 4.242\,640\,687$ into $y = \sqrt{(5x-2)}$ to give $y_2 = 4.383\,286\,83$. So we went from Q_1 to P_2. At this point we put $x_2 = 4.383\,286\,83$. That is, we put $y_2 = x_2$ and so we moved from P_2 to Q_2 on the graph. Then we moved to P_3, and so on, and, as you can see, we got closer and closer to the point A.

This method clearly involves going vertically up to the graph of $y = \sqrt{(5x-2)}$ and then going horizontally across to the graph of $y = x$. It depends on one of the graphs having equation $y = x$. So of the two graphs, one must have equation $y = x$ and the other $y = \mathrm{f}(x)$.

- **When trying to solve an equation $\mathbf{f}(x) = 0$ by an iterative method, you first rearrange $\mathbf{f}(x) = 0$ into a form $x = \mathbf{g}(x)$. The iteration formula is then**

$$\boldsymbol{x_{n+1} = \mathrm{g}(x_n)}$$

This is all very well, but there is no unique form $x = \mathrm{g}(x)$ into which the equation $\mathrm{f}(x) = 0$ can be rearranged. We rearranged the equation $x^2 - 5x + 2 = 0$ above into the form $x = \sqrt{(5x-2)}$. However, we could have rearranged it into the form

$$5x = x^2 + 2 \text{ and hence } x = \frac{x^2 + 2}{5}$$

OR $$x^2 = 5x - 2 \text{ and hence } x = 5 - \frac{2}{x}$$

OR $$x(x - 5) = -2 \text{ and hence } x = \frac{-2}{x - 5}$$

This raises the question of whether it matters how we rearrange the equation.

We know that the formula $x_{n+1} = \sqrt{(5x_n - 2)}$, starting with $x_0 = 4$, converges to the larger root, which is 4.56 (2 d.p.).

Let's try $x_{n+1} = \frac{x_n^2 + 2}{5}$, starting with $x_0 = 4$. Then:

$$\begin{aligned}
x_1 &= 3.6 \\
x_2 &= 2.992 \\
x_3 &= 2.190\,412\,8 \\
x_4 &= 1.359\,581\,647 \\
x_5 &= 0.769\,692\,45 \\
x_6 &= 0.518\,485\,293 \\
x_7 &= 0.453\,765\,4 \\
x_8 &= 0.441\,180\,607 \\
x_9 &= 0.438\,928\,065 \\
x_{10} &= 0.438\,531\,569
\end{aligned}$$

This root is 0.44 (2 d.p.).

So this iteration formula, starting at $x_0 = 4$, leads to the *smaller* root of the equation $x^2 - 5x + 2 = 0$.

Now try $x_{n+1} = 5 - \frac{2}{x_n}$ starting at $x_0 = 4$:

$$\begin{aligned}
x_1 &= 4.5 \\
x_2 &= 4.555\,555\,56 \\
x_3 &= 4.560\,975\,61 \\
x_4 &= 4.561\,497\,326 \\
x_5 &= 4.561\,547\,479
\end{aligned}$$

That is 4.56 (2 d.p.), and so we are back to the first root again.

Finally, consider $x_{n+1} = \frac{-2}{x_n - 5}$ starting at $x_0 = 4$:

$$x_1 = 2$$
$$x_2 = 0.666\,666\,666$$
$$x_3 = 0.461\,538\,461$$
$$x_4 = 0.440\,677\,966$$
$$x_5 = 0.438\,661\,71$$
$$x_6 = 0.438\,467\,807$$

That is 0.44 (2 d.p.) and you are back to the other root.

So some of the iteration formulae lead to one root and some to the second root, even though you have the same starting point. Of course, using each iteration formula with a given starting point will only ever lead to at most *one* root of the equation. If an equation has three roots, you will need to find at least three iterative sequences to obtain all of them. If an equation has four roots, you will need to find at least four iterative sequences to obtain all of them, and so on.

Example 1

Show that the equation $x^3 - 3x - 5 = 0$ can be written in the form $x = \sqrt[3]{(3x+5)}$.

Using the formula $x_{n+1} = \sqrt[3]{(3x_n + 5)}$, find, to 3 decimal places, a root of the equation $x^3 - 3x - 5 = 0$, starting with $x_0 = 2$.

$x^3 - 3x - 5 = 0 \Rightarrow x^3 = 3x + 5$

So: $$x = \sqrt[3]{(3x+5)}$$

Using $x_{n+1} = \sqrt[3]{(3x+5)}$ and $x_0 = 2$:

$$x_1 = 2.223\,980\,091$$
$$x_2 = 2.268\,372\,388$$
$$x_3 = 2.276\,967\,161$$
$$x_4 = 2.278\,623\,713$$
$$x_5 = 2.278\,942\,719$$

So a root of the equation is 2.279 (3 d.p.)

12.2 Starting points for iteration

At the beginning of this chapter we drew the graph of $x^2 - 5x + 2 = 0$ and so we were able to see that one root of the equation lay between 0 and 1 and the other between 4 and 5. This then allowed us to say that since 4 was close to a root we could use $x_0 = 4$ as a starting point for our iteration.

When you select a starting point for the iteration it is sensible to choose a point that lies reasonably close to the root. The further the starting point is from the root, the more calculating you are likely to have to do. It is therefore going to save you time and energy if you choose a point reasonably close to the root.

As a general rule, try to find an interval in which a root lies and then choose as a starting point, x_0, either

(i) one of the end-points of the interval, or

(ii) the mean value of the two end-points of the interval.

However, it is not sensible, in general, to draw the graph of $y = \mathrm{f}(x)$ in order to find an interval within which a root lies: this is time-consuming and, if the graph is drawn accurately it will probably be easier to read the value of the root from the graph! Instead you should use the result you were shown in Book P1, i.e. if $\mathrm{f}(a) < 0$ and $\mathrm{f}(b) > 0$, or if $\mathrm{f}(a) > 0$ and $\mathrm{f}(b) < 0$ then generally a root of $\mathrm{f}(x) = 0$ lies in the interval $[a, b]$.

12.3 Convergence

The equation $x^3 - 3x - 5 = 0$ in example 1 could be rearranged so that $3x = x^3 - 5$ and hence

$$x = \frac{x^3 - 5}{3}$$

Using the iteration formula

$$x_{n+1} = \frac{x_n^3 - 5}{3}$$

and the starting point $x_0 = 2$, you obtain

$$x_1 = 1$$

$$x_2 = -1.333\,333\,333$$

$$x_3 = -2.456\,790\,123$$

$$x_4 = -6.609\,579\,113$$

$$x_5 = -97.916\,538\,71$$

$$x_6 = -312\,931.4535$$

It should be obvious to you by now that, far from getting close to a root of the equation $x^3 - 3x - 5 = 0$, in fact the sequence of values $x_0, x_1, x_2, x_3, \ldots$ is getting further and further away from a root. The sequence is not convergent. It is a divergent sequence. Successive members of the sequence are increasing negatively without limit.

So not only do you not know when you start using an iteration formula which root of the equation you are likely to end up with, in fact you do not know whether you are going to get to a root at all! This is very frustrating. However, a little help is at hand. Although we will not prove the result, it can be shown that if α is a root of the equation $f(x) = 0$ and if $x = g(x)$ is a rearrangement of this equation so that $x_{n+1} = g(x_n)$ is an iteration formula to be used to find a root of $f(x) = 0$, then:

- **(i) if $|g'(\alpha)| < 1$, the sequence $x_0, x_1, x_2 \ldots$ will converge to a root of $f(x) = 0$. In particular,**
  (a) if $-1 < g'(\alpha) < 0$ then the sequence $x_0, x_1, x_2, \ldots$ will oscillate and converge towards a root of $f(x) = 0$, and
  (b) if $0 < g'(\alpha) < 1$ then the sequence $x_0, x_1, x_2, \ldots$ will converge, without oscillating, towards a root of $f(x) = 0$.
  (ii) if $|g'(\alpha)| \geqslant 1$ then the sequence $x_0, x_1, x_2 \ldots$ will diverge.

Now this creates a bit of a problem, because until you have done the necessary calculations and found a root α of the equation, you cannot calculate $g'(\alpha)$. By the time you have found α, you will not *need* to calculate $g'(\alpha)$ because you will already know that $|g'(\alpha)| < 1$ since if you have found α the iteration formula must have converged! If you have not found α but have found a diverging sequence, then you will not be able to calculate $g'(\alpha)$ but you will know anyway that $|g'(\alpha)|$ must be greater than or equal to one! However, this is not as big a problem as it seems because if you take a value β, close to α, then $g'(\beta)$ will behave in a similar way to $g'(\alpha)$. That is, if β is close to α, then if $|g'(\beta)| < 1$, the sequence will converge and if $|g'(\beta)| \geqslant 0$, the sequence will diverge. So if you take β as your value of x_0 you should be able to test whether your calculations will take you closer to, or further from, a root.

For the equation $x^3 - 3x - 5 = 0$ in example 1, and the rearrangement $x = \sqrt[3]{(3x + 5)}$,

$$g(x) \equiv (3x + 5)^{\frac{1}{3}}$$

So:

$$g'(x) = \tfrac{1}{3}(3x + 5)^{-\frac{2}{3}} \cdot 3 = \frac{1}{(3x + 5)^{\frac{2}{3}}}$$

With $x_0 = 2$ (since this is near a root),

$$g'(2) = \frac{1}{11^{\frac{2}{3}}} \approx 0.202$$

So $|g'(2)| < 1$ and the sequence $x_0, x_1, x_2 \ldots$ will converge to a root as shown in example 1.

For the rearrangement

$$x = \frac{x^3 - 5}{3}, \quad g(x) = \frac{x^3 - 5}{3}$$

So
$$g'(x) = \frac{3x^2}{3} = x^2$$

With $x_0 = 2$,

$$g'(2) = 4$$

So $|g'(2)| > 1$ and the sequence $x_0, x_1, x_2, \ldots$ will diverge as we have already shown.

Example 2
Show that the equation $\ln x - x + 2 = 0$ can be written in the form $x = 2 + \ln x$.

Using the iteration formula $x_{n+1} = 2 + \ln x_n$ and starting with $x_0 = 2$, find, to 3 significant figures, a root of the equation $\ln x - x + 2 = 0$.

$$\ln x - x + 2 = 0$$

$$\Rightarrow \quad -x = -2 - \ln x$$

$$\Rightarrow \quad x = 2 + \ln x$$

If $x_{n+1} = 2 + \ln x_n$ then:

$$x_0 = 2$$
$$x_1 = 2.693\,147\,181$$
$$x_2 = 2.990\,710\,465$$
$$x_3 = 3.095\,510\,973$$
$$x_4 = 3.129\,952\,989$$
$$x_5 = 3.141\,017\,985$$
$$x_6 = 3.144\,546\,946$$
$$x_7 = 3.145\,669\,825$$
$$x_8 = 3.146\,026\,848$$
$$x_9 = 3.146\,140\,339$$

So a root is 3.15 (3 s.f.)

Example 3
Starting with $x_0 = 0$ and using the iteration formula $x_{n+1} = \sin x_n - 0.5$, find, to 4 significant figures, a root of the equation $\sin x - x = 0.5$.

In this question you must first realise that any calculations will only make sense if x is measured in radians.

$$x_0 = 0$$
$$x_1 = -0.5$$
$$x_2 = -0.979\,425\,538$$
$$x_3 = -1.330\,177\,246$$
$$x_4 = -1.471\,190\,631$$
$$x_5 = -1.495\,043\,453$$
$$x_6 = -1.497\,132\,123$$
$$x_7 = -1.497\,288\,019$$
$$x_8 = -1.497\,299\,481$$

So a root is -1.497 (4 s.f.)

Example 4

Show that the equation $x \ln x + x - 3 = 0$ has a root lying in the interval [1,2]. Using a suitable iteration formula, find this root to three significant figures.

Let $f(x) \equiv x \ln x + x - 3$.

Then $\qquad f(1) = 1 \ln 1 + 1 - 3 = -2 < 0$

and $\qquad f(2) = 2 \ln 2 + 2 - 3 = 0.386 > 0.$

So a root lies in the interval [1,2].

Let us first try to find a little more precisely where the root lies.

Take $\qquad x = 1.5$

$$f(1.5) = 1.5 \ln 1.5 + 1.5 - 3 = -0.891 < 0$$

Since $f(1.5) < 0$ and $f(2) > 0$, a root of the equation lies in [1.5,2].

Rearrange the equation

$$x \ln x + x - 3 = 0$$

as

$$x = 3 - x \ln x$$

and so try the iteration formula $x_{n+1} = 3 - x_n \ln x_n$ starting with $x_0 = 2$.

If $\qquad g(x) \equiv 3 - x \ln x$

then $\qquad g'(x) \equiv -x\left(\frac{1}{x}\right) - \ln x$

$$= -1 - \ln x$$

Since the root we are after is close to 2, calculate $g'(2)$.

$$g'(2) = -1 - \ln 2 = -1.693$$

So $|g'(2)| > 1$ and we do not have to waste any more time. The sequence of approximations will diverge.

Rearrange the equation

$$x \ln x + x - 3 = 0$$

as

$$x(\ln x + 1) - 3 = 0$$

$$x(\ln x + 1) = 3$$

$$x = \frac{3}{1 + \ln x}$$

and so try the iteration formula $x_{n+1} = \dfrac{3}{1 + \ln x_n}$

If $$g(x) \equiv \frac{3}{1 + \ln x}$$

then $$g'(x) \equiv \frac{(1 + \ln x) \times 0 - 3 \times \frac{1}{x}}{(1 + \ln x)^2}$$

This time $g'(2) = \dfrac{0 - 1.5}{1.693} = -0.8859$

So $|g'(2)| < 1$ and the iteration formula should converge.

$$x_0 = 2$$
$$x_1 = 1.771\,848\,327$$
$$x_2 = 1.908\,368\,716$$
$$x_3 = 1.822\,324\,789$$
$$x_4 = 1.874\,867\,538$$
$$x_5 = 1.842\,143\,064$$
$$x_6 = 1.862\,278\,771$$
$$x_7 = 1.849\,795\,515$$
$$x_8 = 1.857\,498\,762$$
$$x_9 = 1.852\,731\,52$$
$$x_{10} = 1.855\,676\,553$$
$$x_{11} = 1.853\,855\,219$$
$$x_{12} = 1.854\,980\,846$$
$$x_{13} = 1.854\,284\,891$$
$$x_{14} = 1.854\,715\,076$$

So the root is 1.85 (3 s.f.)

Exercise 12A

1 (a) Show that $x^2 - 3x + 1 = 0$ has one root lying between 0 and 1 and another lying between 2 and 3.

(b) Show that $x^2 - 3x + 1 = 0$ can be rearranged into the form

(i) $x = \dfrac{x^2 + p}{q}$, where p and q are constants.

(ii) $x = r + \dfrac{s}{x}$, where r and s are constants

and state the values of p, q, r and s.

(c) Using the iteration formula

$$x_{n+1} = \frac{x_n^2 + p}{q}$$

together with your values of p and q and starting at $x_0 = 0.5$ find, to 3 decimal places, one root of the equation $x^2 - 3x + 1 = 0$.

(d) Using the iteration formula

$$x_{n+1} = r + \frac{s}{x_n}$$

together with your values of r and s find, to 3 decimal places, the second root of $x^2 - 3x + 1 = 0$.

2 (a) Show that $x^3 = 14$ has a root lying between 2 and 3.

(b) Show further that $x^3 = 14$ can be rearranged into the form

$$x = \frac{p}{x^2} + \frac{x}{2}$$

where p is a constant and state the value of p.

(c) Using the iteration formula

$$x_{n+1} = \frac{p}{x_n^2} + \frac{x_n}{2}$$

together with your value of p and starting with $x_0 = 2.5$, find, to 3 significant figures, a root of $x^3 = 14$.

3 Using the iteration formula

$$x_{n+1} = 2 + \frac{1}{x_n^2}$$

and starting with $x_0 = 2$, find the value, to 4 significant figures, to which the sequence $x_0, x_1, x_2, \ldots$ tends. This sequence leads to one root of an equation. State the equation.

4 Show that the equation $x^3 + 6x^2 - 9x + 2 = 0$ has a root lying between 0 and 0.5. Use the iteration formula

$$x_{n+1} = \frac{6x_n^2 + 2}{9 - x_n^2}$$

with $x_0 = 0$ to find this root to 3 decimal places.

5 Show that $1 + x^2 - x^3 = 0$ has a root lying between 1 and 2. Using the iteration formula

$$x_{n+1} = \sqrt[3]{(1 + x_n^2)}$$

with $x_0 = 1$, find this root to 3 significant figures.

6 Show that $x^2 = \sin x$ has a root lying between 0 and 1. Using the iteration formula

$$x_{n+1} = \frac{\sin x_n}{x_n}$$

find this root to 3 decimal places.

7 By considering the roots of the equation $f'(x) = 0$, or otherwise, prove that the equation $f(x) = 0$, where $f(x) \equiv x^3 + 2x + 4$, has only one real root. Show that this root lies in the interval $-2 < x < -1$.
Use the iteration formula

$$x_{n+1} = -\tfrac{1}{6}(x_n^3 - 4x_n + 4), \quad x_0 = -1$$

to find two further approximations to this root of the equation, giving your final answer to 2 decimal places. [L]

8 Show that the equation $2^x = 8x$ has two roots, one lying between 0 and 1 and the other lying between 5 and 6. Use the iteration formula

$$x_{n+1} = \frac{2^{x_n}}{8}, x_0 = 1$$

to find, to 4 decimal places, the root which lies between 0 and 1.

9 Using the iteration formula

$$x_{n+1} = (22x_n + 50)^{\frac{1}{4}} \quad \text{and} \quad x_0 = 3.5$$

find a root of $x^4 - 22x - 50 = 0$ to 4 significant figures.

10 Show that $x \sin \sqrt{x} = 1$ has a root lying between 1 and 2. Using the iteration formula

$$x_{n+1} = \frac{1}{\sin \sqrt{x_n}}$$

find, to 4 decimal places, the value of this root.

11 Use the iteration formula

$$x_{n+1} = 3^{\frac{1}{x_n}} \quad \text{with} \quad x_0 = 1.5$$

to find the value, to 3 significant figures, to which the sequence $x_0, x_1, x_2, \ldots$ tends. This sequence leads to one root of an equation. State the equation.

SUMMARY OF KEY POINTS

1 In order to find a root of the equation $f(x) = 0$ by iteration, the equation must first be rearranged in the form $x = g(x)$. The iteration formula is then

$$x_{n+1} = g(x_n)$$

2 Each iteration formula with a given starting point can only lead to one root of the equation, at most.

3 Sometimes an iterative procedure with a given starting point will *not* lead to a root of the equation. The sequence $x_0, x_1, x_2, \ldots$ will instead diverge.

4 To decide on a starting point for the iteration, first find an interval in which a root lies, then choose as the starting point either

(i) one end of the interval, or

(ii) the mean value of the interval.

5 If α is a root of $f(x) = 0$ and $x_{n+1} = g(x_n)$ is an iteration formula used to find a root of $f(x) = 0$ then

(i) if $|g'(\alpha)| < 1$ the sequence $x_0, x_1, x_2, \ldots$ will converge to a root

(ii) if $|g'(\alpha)| \geqslant 1$ the sequence $x_0, x_1, x_2, \ldots$ will diverge.

Review exercise 3

1

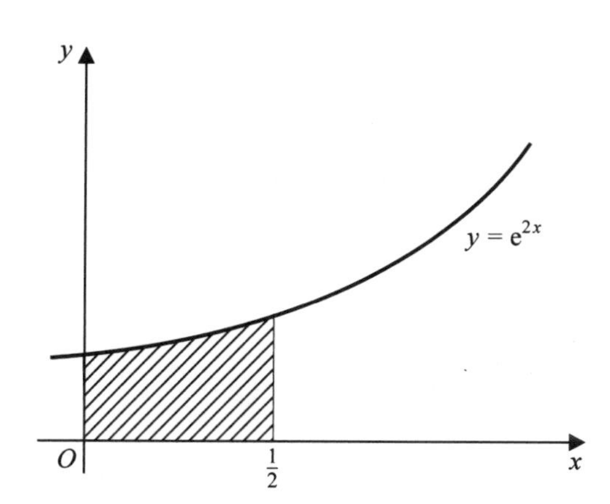

In the figure, the shaded region, R, is bounded by the curve with equation $y = e^{2x}$, the line $x = \frac{1}{2}$ and the coordinate axes. The region R is rotated through $360°$ about the x-axis. Find, in terms of e and π, the volume of the solid generated. [L]

2 Find

(a) $\int x \cos x \, dx$

(b) $\int \cos^2 y \, dy$

Hence find the general solution of the differential equation

$$\frac{dy}{dx} = x \cos x \sec^2 y, \; 0 < y < \frac{\pi}{2}$$

[L]

3 Sketch the curve defined with parameter ϕ by the equations

$$x = a \cos \phi, \quad y = b \sin \phi$$

where a and b are positive constants and $0 \leqslant \phi < 2\pi$.

Given that P is the point at which $\phi = \frac{\pi}{4}$, obtain

(a) an equation of the tangent to the curve at P,

(b) an equation of the normal to the curve at P.

Determine the area of the finite region bounded by the curve.

[L]

4 A and B are two independent events such that $\mathrm{P}(A) = 0.2$ and $\mathrm{P}(B) = 0.15$.

Evaluate the following probabilities.

(a) $\mathrm{P}(A|B)$

(b) $\mathrm{P}(A \cap B)$

(c) $\mathrm{P}(A \cup B)$ [L]

5 Given that the point with coordinates $(2, -1)$ lies on the curve with equation

$$y = \frac{2x+1}{x+k}$$

where k is a constant,

(a) find k.

Hence

(b) write down an equation of each asymptote to the curve

(c) sketch the curve, showing how the curve approaches the asymptotes and the coordinates of the points where the curve intersects the axes. [L]

6 Express as the sum of partial fractions

$$\frac{2}{x(x+1)(x+2)}$$

Hence show that

$$\int_2^4 \frac{2}{x(x+1)(x+2)}\,\mathrm{d}x = 3\ln 3 - 2\ln 5$$ [L]

7 Using the identity $\tan x \equiv \frac{\sin x}{\cos x}$, $-\frac{\pi}{2} < x < \frac{\pi}{2}$, show that

$$\frac{\mathrm{d}}{\mathrm{d}x}(\tan x) = \sec^2 x$$

Use integration by parts to find

$$\int x \sec^2 x\,\mathrm{d}x$$ [L]

8 Sketch the curve defined with parameter t by the equations

$$x = 1 + t, \quad y = 1 - t^2$$

Find the area of the finite region R enclosed by the curve and the x-axis. Find also the volume of the solid generated when R is rotated through 4 right angles about the x-axis. Show that the volume of the solid generated when R is rotated through 2 right angles about the line $x = 1$ is $\frac{\pi}{2}$. [L]

9 Given that $f(x) \equiv 3 + 4x - x^4$, show that the equation $f(x) = 0$ has a root $x = a$, where a is in the interval $1 \leqslant a \leqslant 2$. It may be assumed that if x_n is an approximation to a, then a better approximation is given by x_{n+1}, where

$$x_{n+1} = (3 + 4x_n)^{\frac{1}{4}}$$

Starting with $x_0 = 1.75$, use this result twice to obtain the value of a to 2 decimal places. [L]

10

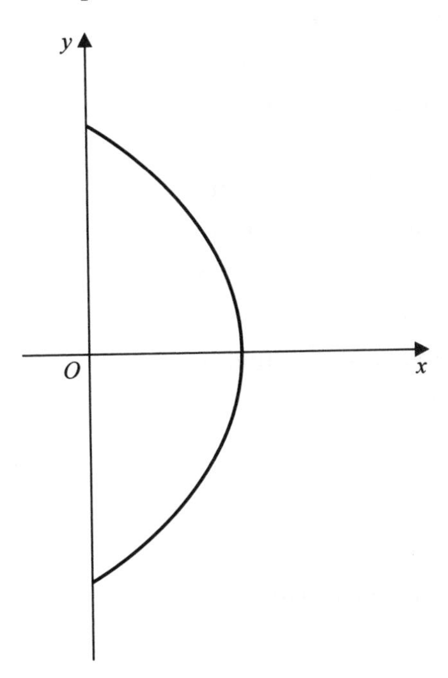

The figure shows a sketch of the curve with parametric equations

$$x = \cos^2 t, \quad y = \tfrac{3}{2}\sin t, \quad -\frac{\pi}{2} \leqslant t \leqslant \frac{\pi}{2}$$

(a) Find the coordinates of the point where the curve crosses the x-axis.

(b) Find $\frac{dx}{dt}$.

(c) Show that the area A of the finite region bounded by the curve and the y-axis is given by $A = 6\int_0^{\frac{\pi}{2}} \sin^2 t \cos t \, dt$.

Using the substitution $u = \sin t$, or otherwise, evaluate A.

[L]

11 Students in a class were given two statistics problems to solve, the second of which was harder than the first. Within the class $\frac{5}{6}$ of the students got the first one correct and $\frac{7}{12}$ got the second one correct. Of those students who got the first problem correct, $\frac{3}{5}$ got the second one correct. One student was chosen at random from the class.

Let A be the event that the student got the first problem correct and B be the event that the student got the second one correct.

(a) Express in words the meaning of $A \cap B$ and of $A \cup B$.

(b) Find $P(A \cap B)$ and $P(A \cup B)$.

(c) Given that the student got the second problem right, find the probability that the first problem was solved correctly.

(d) Given that the student got the second problem wrong, find the probability that the first problem was solved correctly.

(e) Given that the student got the first problem wrong, find the probability that the student also got the second problem wrong. [L]

12

$$f(x) \equiv \frac{x^2}{(3x+2)}$$

(a) Find the coordinates of the points on the curve with equation $y = f(x)$ for which $\frac{dy}{dx} = -1$.

(b) Find an equation of the tangent to the curve at each of these points.

(c) Find the coordinates of the stationary points of the curve with equation $y = f(x)$ and determine their nature.

(d) Write down an equation of the asymptote of the curve which is parallel to the y-axis.

(e) On the same sketch show

(i) the curve, including the coordinates of the stationary points

(ii) the asymptote parallel to the y-axis

(iii) the tangents to the curve with gradient -1. [L]

13 Find the general solution of the differential equation

$$\frac{1}{x}\frac{dy}{dx} + 2y = 8, \quad x > 0$$

Express your answer in the form $y = f(x)$. [L]

14 The finite region bounded by the curve with equation $y = \sin 2x$ and that part of the x-axis for which $0 \leqslant x \leqslant \frac{1}{2}\pi$, is rotated completely about the x-axis. Find, in terms of π, the volume of the solid of revolution generated. [L]

15 A curve C, has parametric equations

$$x = 3t^2 - p, \quad y = 3t^3$$

where p is a non-negative constant and t is a real parameter.

(a) State, in terms of p, the range of values of x.

(b) Express $\frac{dy}{dx}$ in terms of t.

(c) Find the gradient of C at the point $(-p, 0)$.

(d) Sketch C.

When $p = 0$, the tangent to C, at the point where $t = \frac{2\sqrt{2}}{3}$,

meets C again at the point Q.

(e) Find an equation for this tangent.

(f) Verify that at Q, $t = -\frac{\sqrt{2}}{3}$.

(g) Deduce that this tangent is a normal to C at Q. [L]

16 Find the value of each of the constants A, B and C for which

$$\frac{7x^2 - 12x - 1}{(2x+1)(x-1)^2} \equiv \frac{A}{2x+1} + \frac{B}{x-1} + \frac{C}{(x-1)^2}$$

Hence evaluate $\displaystyle\int_2^5 \frac{7x^2 - 12x - 1}{(2x+1)(x-1)^2}\,dx$ [L]

17 The normal at every point on a given curve passes through the point $(2, 3)$. Prove that the equation of the curve satisfies the differential equation

$$(x-2)+\frac{\mathrm{d}y}{\mathrm{d}x}(y-3)=0$$

Hence find a cartesian equation of the family of curves which satisfy the condition above.

Find the equation of the particular curve which passes through the point $(4, 2)$. [L]

18 The tangent at $P(x_n, x_n^2-2)$, where $x_n > 0$, to the curve with equation $y = x^2 - 2$ meets the x-axis at the point $Q(x_{n+1}, 0)$. Show that

$$x_{n+1}=\frac{x_n^2+2}{2x_n}$$

This relationship between x_{n+1} and x_n is used, starting with $x_1 = 2$, to find successive approximations for the positive root of the equation $x^2 - 2 = 0$. Find x_2 and x_3 as fractions and show that $x_4 = \frac{577}{408}$.

Find the error, to 1 significant figure, in using $\frac{577}{408}$ as an approximation to $\sqrt{2}$. [L]

19 (a) The events A, B and C are such that

$$\mathrm{P}(A') = \tfrac{3}{4},\ \mathrm{P}(C) = \tfrac{1}{2},\ \mathrm{P}(A \cup B) = \tfrac{2}{5},\ \mathrm{P}(B \cap C) = \tfrac{1}{10}$$

Given that events A and B are independent, find $\mathrm{P}(B)$ and show that events B and C are also independent.

(b) A bag contains only 9 red balls, 9 blue balls, 12 black balls and 6 white balls. Four balls are taken from the bag at random and without replacement. Find, to three decimal places, the probability that the four balls are

(i) all blue

(ii) all of the same colour

(iii) all of different colours.

Given that all four balls are of the same colour, find the probability that they are all black. [L]

20 The curve C has parametric equations

$$x = at, \quad y = \frac{a}{t}, \quad t \in \mathbb{R}, \ t \neq 0$$

where t is a parameter and a is a positive constant.

(a) Sketch C.

(b) Find $\frac{dy}{dx}$ in terms of t.

The point P on C has parameter $t = 2$.

(c) Show that an equation of the normal to C at P is

$$2y = 8x - 15a$$

This normal meets C again at the point Q.

(d) Find the value of t at Q. [L]

21 Given that $f(x) \equiv x^2 - 6x + 10$, show that $f(x) > 0$ for all real values of x.

Using the same axes sketch the graphs of $y = f(x)$ and $y = \frac{1}{f(x)}$.

[L]

22

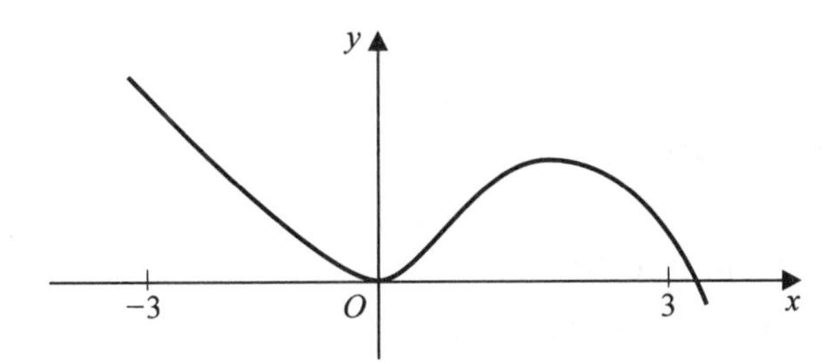

The figure shows a sketch of the curve with equation $y = f(x)$. In separate diagrams show, for $-3 \leqslant x \leqslant 3$, sketches of the curves with equation

(a) $y = f(-x)$

(b) $y = -f(x)$

(c) $y = f(|x|)$

Mark on each sketch the x-coordinate of any point, or points, where a curve touches or crosses the x-axis. [L]

23 (a) $\int \sin\theta \sin 2\theta \, d\theta$

(b) Find the solution of the differential equation

$$(1 + e^{2y}) \frac{dy}{dx} = e^{y} \sin x \sin 2x$$

given that $y = 0$ at $x = \frac{1}{6}\pi$. [L]

24 For married couples the probability that the husband has passed his driving test is $\frac{7}{10}$ and the probability that the wife has passed her driving test is $\frac{1}{2}$. The probability that the husband has passed, given that the wife has passed, is $\frac{14}{15}$. Find the probability that, for a randomly chosen married couple, the driving test will have been passed by

(a) both of them

(b) only one of them

(c) neither of them.

If two married couples are chosen at random, find the probability that only one of the husbands and only one of the wives will have passed the driving test. [L]

25 For the curve with equation $y = \mathrm{f}(x)$ where

$$\mathrm{f}(x) \equiv \frac{2x+1}{x-1}, \quad x \neq 1$$

(a) find the equations of the two asymptotes

(b) find the coordinates of the points at which the curve intersects the coordinate axes.

(c) Sketch the curve with equation $y = \mathrm{f}(x)$ showing clearly the asymptotes and the points where the curve crosses the coordinate axes. [L]

26 Draw the curve $y = \mathrm{e}^{\frac{x}{4}}$ for values of x from 0 to 10.

By drawing a suitable straight line, show that the equation $x = \mathrm{e}^{\frac{x}{4}}$ has two real roots, and estimate their values to one decimal place from your graph.

Use the iteration formula $x_{n+1} = \exp(\frac{1}{4}x_n)$ to evaluate the smaller root to 3 significant figures.

By expressing the equation $x = \mathrm{e}^{\frac{x}{4}}$ in a form involving a logarithm, obtain another iteration formula, and use it to evaluate the larger root to 3 significant figures.

(Note $\exp y \equiv \mathrm{e}^y$.)

[L]

27 (a) Sketch the graph of $y = 8 - 2x^2$, indicating clearly any points of intersection with the coordinate axes.

(b) Find an equation of the tangent to this curve at the point where $x = a$.

(c) Find the value of a for which this tangent is parallel to the line $y = 3x + 6$.

The finite region bounded by the curve and the x-axis is rotated through π radians about the y-axis.

(d) Calculate the volume of the solid formed, leaving your answer in terms of π. [L]

28 A curve is given parametrically by

$$x = \sin t,\ y = \cos^3 t,\ -\pi < t \leqslant \pi$$

Show that

(a) $-1 \leqslant x \leqslant 1$ and $-1 \leqslant y \leqslant 1$

(b) $\dfrac{dy}{dx} = a \sin 2t$, where a is constant, and give the value of a.

Find the value of $\dfrac{dy}{dx}$ when $x = 0$ and show that the curve has points of inflexion where $t = -\dfrac{3\pi}{4}, -\dfrac{\pi}{4}, \dfrac{\pi}{4}$ and $\dfrac{3\pi}{4}$.

Sketch the curve. [L]

29 Express $\dfrac{1}{(3x+2)(x+3)}$ in partial fractions.

Given that $x > 0$, find the general solution of the differential equation

$$(3x+2)(x+3)\frac{dy}{dx} = 7y$$

Given further that $y = 6$ at $x = 4$, express y in terms of x.

[L]

30 Two unbiased dice, faces numbered 1 to 6, are thrown simultaneously. The sum of the two numbers thrown is S.

(a) Find P(S is even).

(b) Show that $P(S < 6) = \frac{5}{18}$.

The two dice are thrown simultaneously three times in succession and their total noted each time.

(c) Show that the probability of exactly two of the three throws resulting in an even total is $\frac{3}{8}$.

Event X is that of the three throws at least one throw results in an odd total and at least one results in an even total.

Event Y is that of the three throws at least two result in an even total.

(d) Find $\mathrm{P}(X)$ and $\mathrm{P}(Y)$ and determine whether or not events X and Y are independent. [L]

31 The function f is defined on the domain $-\mathrm{e} \leqslant x \leqslant \mathrm{e}$ by

$$\mathrm{f}(x) = \sqrt{x}, \qquad 0 \leqslant x \leqslant 1$$
$$\mathrm{f}(x) = 1 - 2\ln x, \; 1 < x \leqslant \mathrm{e}$$
f is an odd function

(a) Sketch the curve $y = \mathrm{f}(x)$, marking the coordinates of the points where it crosses the x-axis.

(b) Calculate the area of the region in the first quadrant enclosed by the curve and the x-axis. [L]

32 Sketch the curve given parametrically by the equations

$$x = 1 + t, \; y = 4 - t^2$$

Write on your sketch the coordinates of any points where the curve crosses the coordinate axes and the coordinates of the stationary point. Show that an equation of the normal to the curve at the point with parameter t is

$$x - 2ty = 2t^3 - 7t + 1$$

The normal to the curve at the point P where $t = 1$ cuts the curve again at the point Q. Determine the coordinates of Q.

[L]

33 Show that the substitution $y = vx$, where v is a function of x, transforms the differential equation

$$xy\frac{\mathrm{d}y}{\mathrm{d}x} = x^2 + y^2$$

into the differential equation

$$vx\frac{\mathrm{d}v}{\mathrm{d}x} = 1$$

Hence solve the original differential equation given that $y = 2$ at $x = 1$. [L]

34 Show that the equation

$$3x - 1 - \cos 2x = 0$$

has only one real root and that this root lies between 0.4 and 0.6. The iterative procedure

$$x_{n+1} = \tfrac{1}{3}(1 + \cos 2x_n)$$

is to be used to find further approximations to the root, starting with the initial value $x = 0.5$. Calculate the root giving your answer to 3 decimal places. [L]

35 (a) By using the substitution $u^2 = x - 1$, or otherwise, find

$$\int \frac{x+1}{\sqrt{(x-1)}}\,\mathrm{d}x$$

(b) Find $\int x \cos 3x \,\mathrm{d}x$

Hence, evaluate $\int_0^{\frac{\pi}{6}} x \cos 3x \,\mathrm{d}x$ [L]

36 The function f is defined by

$$\mathrm{f}\colon x \mapsto -\ln(x-2),\ x \in \mathbb{R},\ x > 2$$

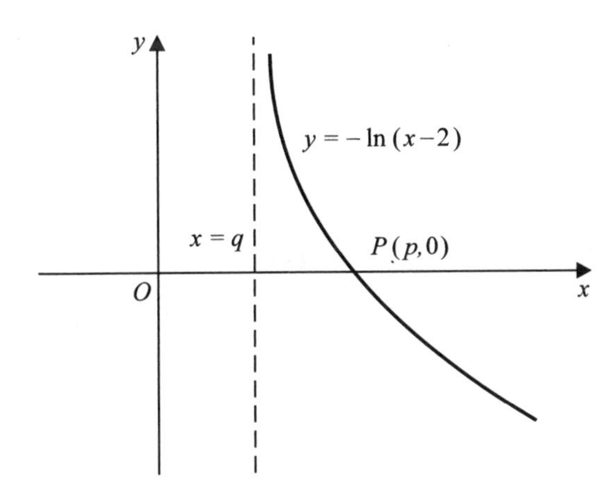

The figure shows a sketch of the curve with equation $y = \mathrm{f}(x)$. The curve crosses the x-axis at the point $P(p, 0)$. The curve has an asymptote, shown by a broken line in the diagram, whose equation is $x = q$.

(a) Write down the value of p and the value of q.

(b) Find the function f^{-1} and state its domain.

(c) Sketch the curve with equation $y = \mathrm{f}^{-1}(x)$ and its asymptote.

Write on your sketch the coordinates of any point where the curve crosses the coordinate axes and the equation of the asymptote. [L]

37 A yellow die and a green die are to be rolled at the same time. The events A, B, C are defined as follows:

A is the event that the number shown on the yellow die is even.

B is the event that the sum of the numbers shown on the two dice is 7.

C is the event that the sum of the numbers shown on the two dice is 8.

(a) Write down (i) $P(A)$ (ii) $P(B)$ (iii) $P(C)$ (iv) $P(A \cap B)$.

(b) Find $P(B|A)$ and $P(A|B)$.

(c) Find $P(C|A)$ and $P(A|C)$.

(d) For each of the pairs of events (A, B), (A, C), state, with reasons, whether the two events are dependent or independent. [L]

38 Express $\dfrac{1}{(1+x)(3+x)}$ in partial fractions.

Hence find the solution of the differential equation

$$\frac{dy}{dx} = \frac{y}{(1+x)(3+x)}, \quad x > -1$$

given that $y = 2$ at $x = 1$.

Express your answer in the form $y = f(x)$. [L]

39

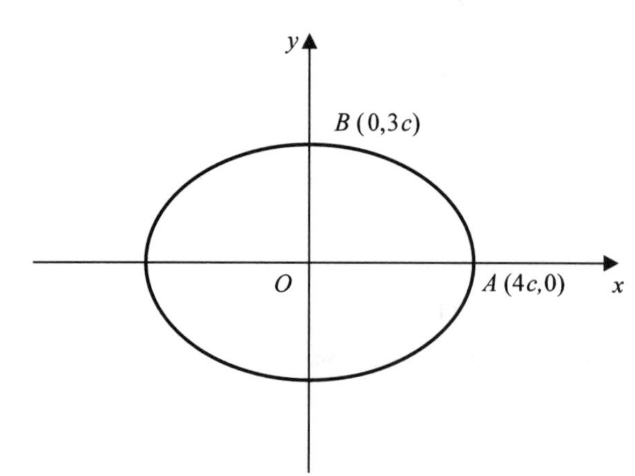

The figure shows a sketch of the curve given parametrically by the equations

$$x = 4c\cos t, \quad y = 3c\sin t, \quad -\pi < t \leqslant \pi$$

where c is a positive constant.

(a) Write down the value of t at the point $A(4c, 0)$ and at the point $B(0, 3c)$.

(b) By considering the integral $\int y \frac{dx}{dt}\,dt$ find, in terms of c, the area of the region enclosed by the curve. [L]

40 In an experiment, two chemicals A and B react yielding another chemical C. The rate of formation of C at time t seconds is k times the product of the concentrations of A and B present at that time, where k is a positive constant. At time t seconds, the concentration of each of A and B is $(p - x)$ units, where x is the concentration of C at that time and p is a positive constant.

(a) Find a differential equation satisfied by x.

(b) Given that $x = 0$ when $t = 0$, obtain x in terms of k, t and p. [L]

41 Sketch the curve given parametrically by

$$x = t^2 - 2, \quad y = 2t, \quad t \in \mathbb{R}$$

indicating on your sketch where (a) $t = 0$ (b) $t > 0$ (c) $t < 0$.

Calculate the area of the finite region enclosed by the curve and the y-axis. [L]

42 Sketch the curve $y = \ln x$.

Show that the equation

$$x + \ln x - 3 = 0$$

has just one real root, and that this root lies between 2 and 2.5.

The sequence defined by the iteration formula:

$$x_{n+1} = 3 - \ln x_n, \quad x_1 = 2$$

is known to converge. Use this sequence to calculate the root of the equation $x + \ln x - 3 = 0$ to two decimal places. [L]

43

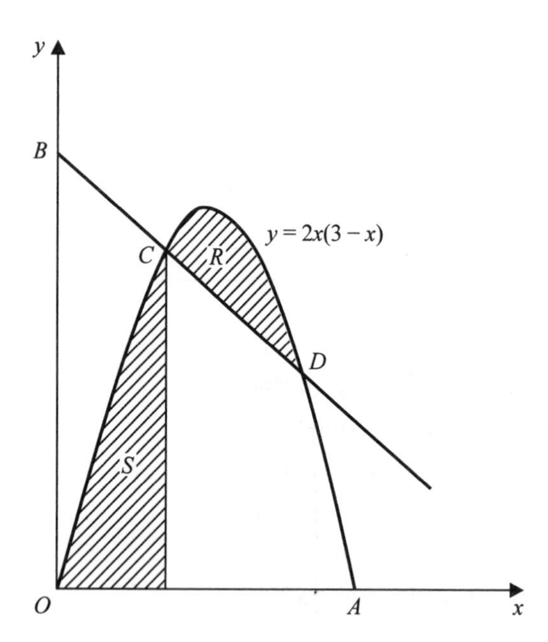

The curve with equation $y = 2x(3 - x)$ crosses the x-axis at O and A.

(a) State the coordinates of the point A.

A straight line, which crosses the y-axis at the point B with coordinates $(0, 5)$, meets the curve at the points C and D, as shown in the figure. The coordinates of the point D are (k, k).

(b) Show that the value of k is $2\frac{1}{2}$.

(c) Find an equation of the straight line passing through B and D.

(d) Show that the x-coordinate of the point C is 1.

(e) Calculate the area of the shaded region R.

(f) Calculate, in terms of π, the volume generated when the shaded region S, bounded by the curve, the x-axis and the line $x = 1$, is rotated through 360° about the x-axis. [L]

44 (a) A student going into the common room reads *Private Eye* with probability 0.75; *Private Eye* but not the *Daily Express* with probability 0.65. The student reads neither with probability 0.20.

(i) Find the probability that the student reads both *Private Eye* and the *Daily Express*.

(ii) Find the probability that the student reads the *Daily Express* but not *Private Eye*.

(iii) If you find the student reading *Private Eye*, what is the probability that the student has (or will have) read the *Daily Express* before leaving the common room?

(b) Let A and B be events such that $\text{P}(A) = \frac{1}{4}$, $\text{P}(B) = \frac{1}{3}$ and $\text{P}(A \cup B) = \frac{5}{12}$.

(i) Find $\text{P}(A|B)$ and $\text{P}(A|B')$.

(ii) Find $\text{P}(A|B)\,\text{P}(B) + \text{P}(A|B')\,\text{P}(B')$.

Comment on your result. [L]

45 Given that $\text{f}(x) \equiv rme^{-x}(1 + x^2)$ find $\text{g}(x)$ such that $\text{f}'(x) \equiv \text{e}^{-x}\text{g}(x)$.

Hence, or otherwise, find the general solution of the differential

equation $\text{e}^x\text{e}^{2y}\dfrac{\text{d}y}{\text{d}x} + (1 - x^2) = 0.$

Express y in terms of x. [L]

46 Sketch the curve given by the parametric equations

$$x = 1 + t,\ y = 2 + \frac{1}{t}\ (t \neq 0)$$

Show that an equation of the normal to the curve at the point where $t = 2$ is $8x - 2y - 19 = 0$.

Find the value of t at the point where this normal meets the curve again. [L]

47 Given that

$$\text{f}(x) \equiv (x - \alpha)(x + \beta),\ \alpha > \beta > 0$$

sketch on separate diagrams the curves with equations

(a) $y = \text{f}(x)$

(b) $y = -\text{f}(x + \alpha)$

On each sketch,

(c) write the coordinates of any points at which the curve meets the coordinate axes

(d) show with a dotted line the axis of symmetry of the curve and state its equation. [L]

48 Given that

$$y = \frac{x}{x+1},\ x \in \mathbb{R},\ x \neq -1$$

show that $\dfrac{\text{d}y}{\text{d}x}$ is always positive.

The function f is defined by

$$f: x \mapsto \frac{x}{x+1}, \quad x \in \mathbb{R}, \; 0 \leqslant x \leqslant 10$$

(a) Determine the range of f.

(b) Find the inverse function f^{-1} and state its domain.

(c) Using the same axes, sketch graphs of f and f^{-1} and state the coordinates of the end points of each graph. [L]

49 (a) By using a suitable substitution, find

$$\int_1^2 x(x-1)^r \, dx, \quad r > 0$$

(b) Sketch the curve C with equation $y^2 = (x-1)^3$, showing clearly the behaviour of the curve near the point with coordinates $(1, 0)$.

(c) Find the area of the finite region R bounded by C and the straight line $x = 2$.

50

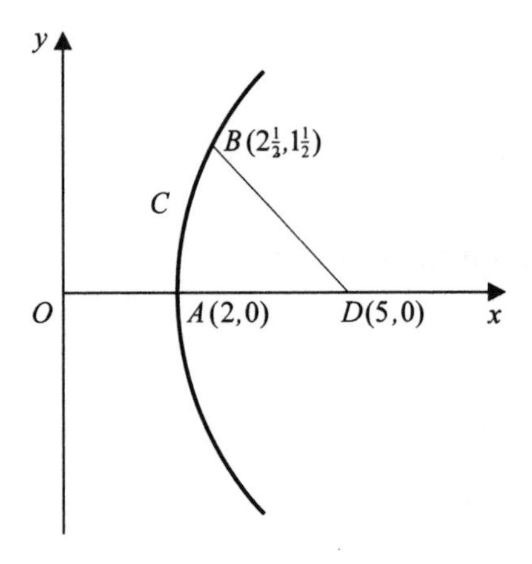

The curve C shown in the figure is given by

$$x = t + \frac{1}{t}, \quad y = t - \frac{1}{t}, \quad t > 0$$

where t is a parameter.

(a) Find $\frac{dy}{dx}$ in terms of t and deduce that the tangent to C at the point $A(2, 0)$ has equation $x - 2 = 0$.

(b) For every point (x, y) on C, show that $x^2 - y^2 = 4$.

The point B has coordinates $(2\frac{1}{2}, 1\frac{1}{2})$ and the point D has coordinates $(5, 0)$. The region R is bounded by the lines AD and

BD and arc AB of the curve C. The region R is rotated through 2π radians about the x-axis to form a solid of revolution.

(c) Find the volume of the solid, leaving your answer in terms of π. [L]

51 Show, graphically or otherwise, that the equation

$$x^3 + 2x - 2 = 0$$

has only one real root. Show also that this root lies between 0.7 and 1. Working to 3 decimal places, obtain approximations to this root by performing two iterations, using the procedure defined by

$$x_{n+1} = \frac{2 - x_n^3}{2}$$

and starting with $x_1 = 0.8$. [L]

52 A curve is given parametrically by the equations

$$x = c(1 + \cos t), \quad y = 2c \sin^2 t, \quad 0 \leqslant t \leqslant \pi$$

where c is a positive constant.

(a) Show that $\dfrac{dy}{dx} = -4 \cos t$.

At the point P on the curve, $\cos t = \frac{3}{4}$.

(b) Show that the normal to the curve at P has equation

$$24y - 8x - 7c = 0$$

(c) Sketch the curve for $0 \leqslant t \leqslant \pi$.

The finite region R is bounded by the curve and the x-axis.

(d) Show that the area of R is given by the definite integral

$$2c^2 \int_0^{\pi} \sin^3 t \, dt$$

(e) Evaluate this integral to find the area of R in terms of c.

[L]

53 Given that $y > \frac{1}{2}$ and $0 < x < \frac{1}{4}\pi$, find the general solution of the differential equation

$$\left(\frac{dy}{dx}\right)\tan 2x = 2y - 1$$

Given further that $y = 1$ at $x = \frac{1}{8}\pi$, find the value of y at $x = \frac{1}{6}\pi$. [L]

54 In a week chosen at random the probabilities that Mr and Mrs Smith are employed are 0.8 and 0.6 respectively and the probability that at least one of them is unemployed is 0.45.

(a) Find the probability that in a week chosen at random they are both unemployed.

Given that Mr Smith is unemployed,

(b) find the probability that Mrs Smith is also unemployed.

The Smiths intend to have 3 children.

Let A be the event that they will have children of both sexes.

Let B be the event that they will have at most 1 girl.

The probability that a new born child is a boy is constant and equal to $p (p \neq 0, p \neq 1)$.

(c) Find $\mathrm{P}(A)$ and $\mathrm{P}(B)$ in terms of p.

Given that the events A and B are independent,

(d) find the value of p. [L]

55 A curve is defined by the equations

$$x = t^2 - 1, \quad y = t^3 - t$$

where t is a parameter.

Sketch the curve for all real values of t.

Find the area of the region enclosed by the loop of the curve. [L]

56 Sketch the curve $y = (2x - 1)^2(x + 1)$, showing the coordinates of

(a) the points where it meets the axes

(b) the turning points

(c) the point of inflexion.

By using your sketch, or otherwise, sketch the graph of

$$y = \frac{1}{(2x - 1)^2(x + 1)}$$

Show clearly the coordinates of any turning points. [L]

57 Evaluate the integrals

(a) $\displaystyle\int_{\frac{1}{3}}^{1} xe^{3x}\,dx$

(b) $\displaystyle\int_{0}^{\frac{\pi}{2}} \frac{\cos x}{1+\sin x}\,dx$

(c) $\displaystyle\int_{0}^{1} \frac{1-x}{(1+x)^2}\,dx$ [L]

58 Express $\dfrac{1}{(3t+1)(t+3)}$ in partial fractions.

Use the substitution $t = \tan x$ to show that

$$\int_{0}^{\frac{\pi}{4}} \frac{1}{3+5\sin 2x}\,dx = \int_{0}^{1} \frac{1}{(3t+1)(t+3)}\,dt$$

Hence show that $\displaystyle\int_{0}^{\frac{\pi}{4}} \frac{1}{3+5\sin 2x}\,dx = \tfrac{1}{8}\ln 3$ [L]

59

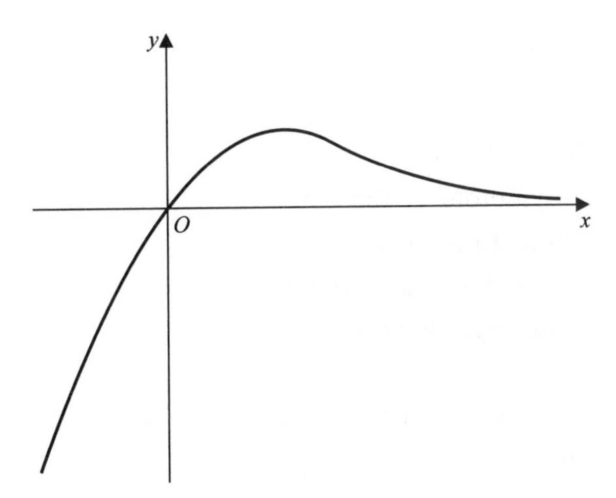

The figure shows the curve with equation

$$y = xe^{-\frac{x}{2}}$$

(a) Determine the coordinates of the turning point.

The region R is bounded by the curve, the x-axis and the line $x = 2$. Find

(b) the area of R

(c) the volume generated when R is rotated through 2π about the x-axis. [L]

60

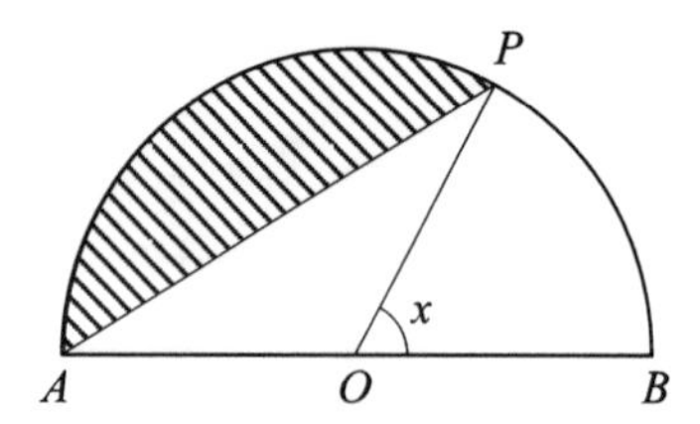

The figure shows a semicircle with O the mid-point of the diameter AB. The point P on the semicircle is such that the area of sector POB is equal to the area of the shaded segment. Angle POB is x radians.

(a) Show that $x = \frac{1}{2}(\pi - \sin x)$.

The iterative method based on the relation $x_{n+1} = \frac{1}{2}(\pi - \sin x_n)$ can be used to evaluate x.

(b) Starting with $x_1 = 1$ perform two iterations to find the values of x_2 and x_3, giving your answers to two decimal places. [L]

61

$$f(x) \equiv \frac{2x^2 + 6}{x - 1}, \; x \in \mathbb{R}, \; x \neq 1$$

(a) Find the set of values of x for which $f(x) > 0$.

(b) Write down the equation of that asymptote of the curve with equation $y = f(x)$ which is parallel to the y-axis.

(c) Find the coordinates of the maximum and minimum points of the curve with equation $y = f(x)$, distinguishing between these points.

(d) Hence sketch the curve with equation $y = f(x)$, showing the results you have found in (a), (b) and (c). [L]

62 Express $\dfrac{x}{(1+x)(1+2x)}$ in the form $\dfrac{A}{1+x} + \dfrac{B}{1+2x}$, where

A and B are constants.

Given that $x > 0$, find the general solution of the differential equation

$$(1+x)(1+2x)\frac{\mathrm{d}y}{\mathrm{d}x} = xe^{2y}$$

[L]

63 (a) Of the households in Edinburgh, 35% have a freezer and 60% have a colour TV set. Given that 25% of the households have both a freezer and a colour TV set, calculate the probability that a household has either a freezer or a colour TV set but not both.
State, with your reasons, whether the events of having a freezer and of having a colour TV set are or are not independent.

(b) State in words the meaning of the symbol $\mathrm{P}(B|A)$, where A and B are two events.

A shop stocks tinned cat food of two makes A and B, and two sizes, large and small. Of the stock, 70% is of brand A, 30% is of brand B. Of the tins of brand A, 30% are small size whilst of the tins of brand B, 40% are small size. Using a tree diagram, or otherwise, find the probability that

(i) a tin chosen at random from the stock will be of small size,

(ii) a small tin chosen at random from the stock will be of brand A. [L]

64 Evaluate the integrals

(a) $\displaystyle\int_0^{\frac{3}{4}} x\sqrt{(1+x^2)}\,\mathrm{d}x$

(b) $\displaystyle\int_0^{\frac{\pi}{4}} \tan^2 x\,\mathrm{d}x$

(c) $\displaystyle\int_0^{\pi} x\sin x\,\mathrm{d}x$

(d) $\displaystyle\int_3^4 \frac{1}{x^2-3x+2}\,\mathrm{d}x$

Express your answer to (d) as a natural logarithm. [L]

65 The finite region R is bounded by the curve $y = \mathrm{e}^{2x}$, the coordinate axes and the line $x = 1$. The region R is rotated through 4 right angles about the x-axis. Find the volume generated, giving your answer to 2 significant figures. [L]

66 Sketch the curve given by the parametric equations

$$x = t^2 - 1, \quad y = 2t + 2, \quad t \in \mathbb{R}$$

The normal to this curve at the point P given by $t = 2$ meets the curve again at Q. Find the value of t at Q. [L]

67

$$f(x) \equiv \frac{3x}{(x+2)(x-1)}$$

Express f(x) in partial fractions.

Hence evaluate $\int_{-1}^{0} f(x)\,dx$ [L]

68 (a) The independent probabilities that Amanda, Betty and Carol shampoo their hair on any particular evening are $\frac{1}{4}$, $\frac{1}{5}$ and $\frac{1}{6}$ respectively. Find the probability that

(i) only Amanda out of the three girls will shampoo her hair tomorrow

(ii) just two out of the three girls will shampoo their hair tomorrow.

(b) There are three boxes labelled P, Q and R containing coloured discs.

Box P contains one white disc, three blue discs and two green discs.

Box Q contains one white disc and two blue discs.

Box R contains one white disc, one blue disc and three green discs.

A box is to be selected at random and from that box a disc is to be selected at random. Construct a tree diagram showing all possible outcomes and their associated probabilities. Hence find the probability that a white disc will be the one chosen. [L]

69 

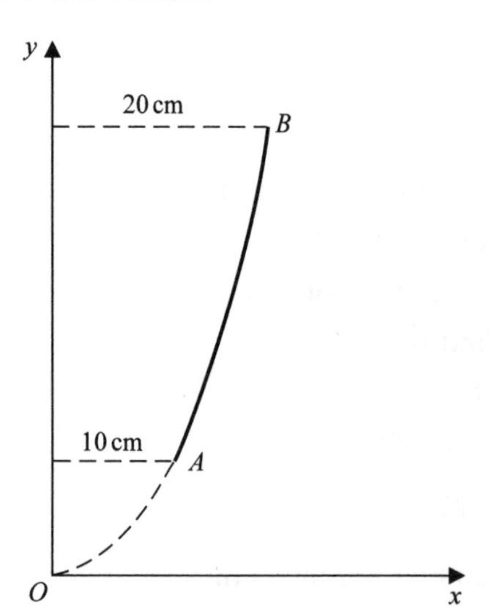

In the figure AB is a sketch of part of the curve with equation $10y = x^2$. The curved surface of an open bowl with a flat circular base is traced out by the complete revolution of the arc AB about the y-axis. The radius of the base is 10 cm and the radius of the top rim of the bowl is 20 cm.

(a) Calculate the capacity of this bowl, in litres to 1 decimal place.

The point C lies on the arc AB. The curved surface of another bowl is traced out by rotating the arc AC through a complete revolution about the y-axis.

The capacity of this bowl is 10 litres.

(b) Calculate the depth of this bowl, in cm to 1 decimal place.

[L]

70 (a) Express $\dfrac{1}{x(x+3)}$ in partial fractions.

Hence show that

$$\int_1^3 \frac{1}{x(x+3)}\,dx = \tfrac{1}{3}\ln 2$$

(b) Evaluate

$$\int_0^{\frac{\pi}{4}} x\cos 2x\,dx$$

(c) By using the substitution $u^2 = a^2 - x^2$, or otherwise, evaluate

$$\int_0^a x(a^2 - x^2)^{\frac{1}{2}}\,dx$$

[L]

71 Using the same axes sketch the curves

$$y = \frac{1}{x-1}, \quad y = \frac{x}{x+3}$$

giving the equations of the asymptotes. Hence, or otherwise, find the set of values of x for which

$$\frac{1}{x-1} > \frac{x}{x+3}$$

[L]

72 (a) Express $\dfrac{x}{(x+1)(x+2)}$ in partial fractions.

Evaluate

$$\int_0^2 \frac{x}{(x+1)(x+2)}\,dx$$

(b) Using the substitution $u = \cos x$, or otherwise, evaluate

$$\int_0^{\frac{\pi}{2}} \sin^3 x(\cos x)^{\frac{1}{2}}\,dx$$

[L]

73 (a) The events A and B are such that

$$P(A|B) = \tfrac{7}{10},\ P(B|A) = \tfrac{7}{15} \text{ and } P(A \cup B) = \tfrac{3}{5}$$

Find the values of

(i) $P(A \cap B)$

(ii) $P(A' \cap B)$

(b) A hand of four cards is to be drawn without replacement and at random from a pack of fifty two playing cards. Giving your answer in each case to three significant figures, find the probabilities that this hand will contain

(i) four cards of the same suit

(ii) either two aces and two kings *OR* two aces and two queens. [L]

74 Find the coordinates of the points of intersection of the curves

$$y = \frac{x}{x+3} \text{ and } y = \frac{x}{x^2+1}$$

Sketch the curves on the same diagram, showing any asymptotes or turning points.
Show that the area of the finite region in the first quadrant enclosed by the two curves is

$$\tfrac{7}{2}\ln 5 - 3\ln 3 - 2$$

[L]

75 Solve the differential equation

$$\frac{dy}{dx} = xe^{x+3y}$$

given that $y = 0$ at $x = 1$. [L]

76 A curve is defined by the equations

$$x = at^2, \quad y = at^3$$

where a is a positive constant and t is a parameter. Sketch the curve and note on your sketch the set of values of t corresponding to each branch of the curve. [L]

77 It is given that

$$f(x) \equiv \frac{3x+7}{(x+1)(x+2)(x+3)}$$

(a) Express $f(x)$ as the sum of three partial fractions.

(b) Find $\int f(x)\,dx$.

(c) Hence show that the area of the finite region bounded by the curve with equation $y = f(x)$ and the lines with equations $y = 0$, $x = 0$ and $x = 1$ is $\ln 2$.

(d) Given that k is a positive constant, show that

$$\frac{d}{dx}\left[\ln\left(\frac{e^x}{e^x+k}\right)\right] = \frac{k}{e^x+k}$$

(e) Find the general solution of the differential equation

$$\frac{dy}{dx} = yf(e^x)$$

You may leave your answer in an unsimplified form.

[L]

78 The function g is defined by

$$g\colon x \mapsto \frac{2x+5}{x-3}, \quad x \in \mathbb{R}, \quad x \neq 3$$

Sketch the graph of the function g. Find an expression for $g^{-1}(x)$, specifying its domain. [L]

79 The function f is defined by

$$f\colon x \mapsto \frac{2ax}{x+a}, \quad x \in \mathbb{R}_0^+$$

where a is a positive constant.
Show that $f'(x) > 0$, and find the range of f.
Sketch the curve $y = f(x)$, for $x \in \mathbb{R}_0^+$, and state an equation of the asymptote of this curve.

Find, in terms of a, the area of the finite region bounded by the curve $y = \mathrm{f}(x)$, the x-axis and the line $x = 2a$. [L]

80 The bag P contains ten balls of which four are white and six are red. The bag Q contains eight balls of which five are white and three are red. Two balls are to be drawn at random from the bag P and placed in Q. One ball is then to be drawn at random from the ten balls in Q and placed in P so that at the end of these stages there will be nine balls in each bag.

The event A is '*The two balls drawn from P are of the same colour*'.

The event B is '*The ball drawn from Q is white*'.

By using a tree diagram, or otherwise,

(a) find the value of $\mathrm{P}(A)$

(b) show that $\mathrm{P}(B) = \frac{29}{50}$

(c) show that $\mathrm{P}(A \cap B) = \frac{13}{50}$.

Hence deduce the value of $\mathrm{P}(A \cup B)$.

Find the conditional probability that all three balls are red, given that they are of the same colour. [L]

81
$$\mathrm{f}(x) \equiv \frac{x^2 + 6x + 7}{(x+2)(x+3)}, \quad x \in \mathbb{R}$$

Given that $\mathrm{f}(x) \equiv A + \frac{B}{x+2} + \frac{C}{x+3}$,

(a) find the values of the constants A, B and C.

(b) Show that $\int_0^2 \mathrm{f}(x)\,\mathrm{d}x = 2 + \ln\left(\frac{25}{18}\right)$ [L]

82 Express $\frac{x}{(x+1)(x+2)}$ in partial fractions.

Solve the differential equation

$$(x+1)(x+2)\frac{\mathrm{d}y}{\mathrm{d}x} = x(y+1)$$

for $x > -1$, given that $y = \frac{1}{2}$ at $x = 1$. Express your answer in the form $y = \mathrm{f}(x)$. [L]

83 The curve C has equation

$$xy - 4y - 2x + 7 = 0$$

(a) Show that this equation may be written as

$$y = \frac{2x - 7}{x - 4}$$

(b) Express the equation in the form $x = f(y)$.

(c) Write down the equations of the asymptotes to C.

(d) Sketch the curve C, indicating clearly the asymptotes and the coordinates of any points of intersection with the coordinate axes. [L]

84 Express

$$\frac{2}{(1 + x)(1 + 3x)}$$

in partial fractions.

Hence, or otherwise, solve the differential equation

$$\frac{dy}{dx} = \frac{2(y + 2)}{(1 + x)(1 + 3x)}$$

given that $y = -1$ at $x = 0$. [L]

85 (a) A pupil either walks to school or travels there by bus. The probability that she will arrive early or on time is $\frac{19}{28}$. The probability that she will arrive late, given that she walked to school, is twice the probability that she will arrive late, given that she travelled by bus. Whenever she travels to school by bus, the probability that she will arrive early or on time is $\frac{3}{4}$. By using a tree diagram, or otherwise, find the probability that, on a randomly chosen day

(i) she walks to school

(ii) she will have travelled by bus, given that she arrives late

(b) Events A and B are such that $P(B|A) = 0.2$, $P(A' \cap B) = 0.3$ and $P(A \cup B) = 0.8$. By using a Venn diagram, or otherwise,

(i) find the values of $P(B)$ and $P(A|B)$.

Event C is such that $P(C \cap B') = 0.03$. Given that B and C are independent,

(ii) find the values of $P(C)$ and $P(B \cap C)$. [L]

86

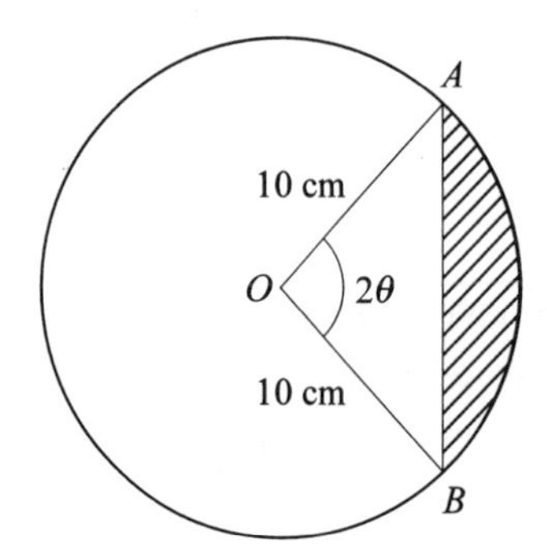

In the figure, O is the centre of a circle, radius 10 cm, and the points A and B are situated on the circumference so that $\angle AOB = 2\theta$ radians. The area of the shaded segment is $44\,\text{cm}^2$. Show that

$$2\theta - \sin 2\theta - 0.88 = 0$$

Show further that a root of this equation lies between 0.9 and 1. By taking 0.9 as a first approximation to this root, use the iterative procedure

$$\theta_{n+1} = \tfrac{1}{2}(\sin 2\theta_n + 0.88)$$

to find this root, giving your answer to 3 decimal places. [L]

87 Given that $e^{2x+y}\dfrac{dy}{dx} = x$ and that $y = 0$ at $x = 0$, express e^y in terms of x. [L]

88 The points W, X, Y and Z are marked in order on a horizontal line with the point W furthest to the left.Four beads, one red, one green, one orange and one blue are to be placed at random at W, X, Y and Z, one bead at each point.

The event A is 'the red bead is to the left of both the green bead and the blue bead'.

The event B is 'the orange bead is either at X or at Y'.

The event C is 'the blue bead is next to both the red bead and the green bead'.

(a) Show that $P(A) = \frac{1}{3}$, and find the values of $P(B)$ and $P(C)$.

(b) Evaluate $P(B \cup C)$.

(c) Evaluate $P(A \cap B)$ and $P(A \cap C)$ and hence determine whether the events A and B, or A and C, or both of these pairs of events are independent. [L]

89

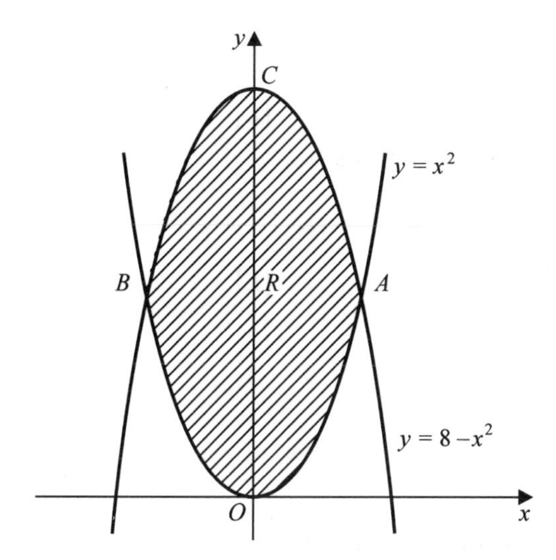

The figure shows the shaded region R which is bounded by the curves with equations $y = x^2$ and $y = 8 - x^2$. Find

(a) the coordinates of the points A, B and C

(b) the area of R

(c) the volume generated when R is rotated through π radians about **the y-axis**. [L]

90 A curve is defined in terms of the parameter t by the equations

$$x = t^3, \quad y = 3t^2$$

The points P and Q have parameters $t = 2$ and $t = 3$ respectively on this curve and O is the origin. Show that the gradient of the line OQ is equal to the gradient of the tangent to the curve at P.

Obtain an equation of the normal to the curve at Q.

Calculate the area of the finite region bounded by the curve and the line OQ. [L]

91 Using the same axes, sketch the curves

$$y = \frac{1}{x} \quad \text{and} \quad y = \frac{x}{x + 2}$$

State the equations of any asymptotes, the coordinates of any points of intersection with the axes and the coordinates of any points of intersection of the two curves.

Hence, or otherwise, find the set of values of x for which

$$\frac{1}{x} > \frac{x}{x+2}$$

[L]

92 Express $f(x) \equiv \dfrac{5x^2 + 8x + 3}{(x-2)(x^2 + 3x + 3)}$ in partial fractions.

Evaluate, to 2 decimal places, $\displaystyle\int_0^1 f(x)\,dx$. [L]

93 Find the general solution of the differential equation

$$e^x \frac{dy}{dx} + y^2 = xy^2$$

Sketch, for positive values of x, the integral curve for which $y = e$ at $x = 1$, showing all asymptotes and turning points.

[L]

94

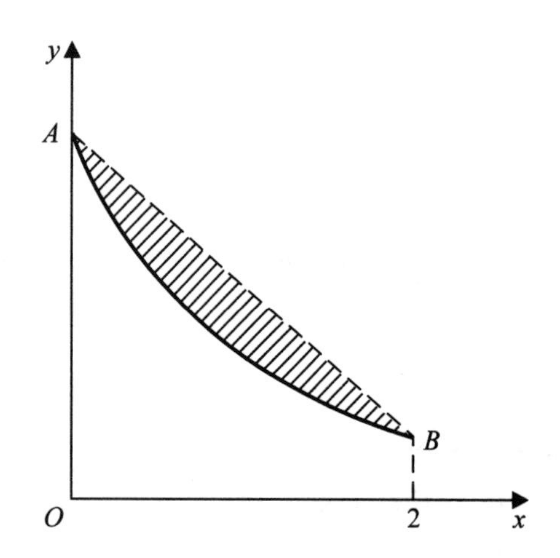

The figure shows a straight line joining the points A and B, together with the sketch-graph of a certain function. The function is known to be

either (i) $f: x \mapsto x^2 - 3.6x + 4, \quad 0 \leqslant x \leqslant 2$

or (ii) $g: x \mapsto \dfrac{4}{2x+1}, \quad 0 \leqslant x \leqslant 2$

(a) Show that $f(0) = g(0)$ and that $f(2) = g(2)$.

(b) Calculate the area of the shaded region, giving your answer to two decimal places,

(i) for the function f

(ii) for the function g. [L]

95 (a) Using the substitution $u^2 = x - 1$, or otherwise, show that

$$\int_2^5 \frac{1}{1 + \sqrt{(x - 1)}}\, dx = 2 - \ln(\tfrac{9}{4})$$

(b) Express

$$\frac{x^2 + x + 1}{(x + 1)(x^2 + 1)}$$

in partial fractions and hence show that

$$\int_0^1 \frac{x^2 + x + 1}{(x + 1)(x^2 + 1)}\, dx = \tfrac{3}{4}\ln 2 + \tfrac{\pi}{8}$$

[L]

96 For the curve whose equation is

$$y = (x^2 - 36)(3 - 2x)$$

find the set of values of x for which $\frac{dy}{dx} > 0$.

Sketch the curve and mark on your sketch

(a) the coordinates of the turning points,

(b) the coordinates of those points at which the curve crosses the coordinates axes. [L]

97 Show that the normal at the point $P(\frac{\pi}{4}, 1)$ to the curve with equation $y = \tan x$ cuts the x-axis at the point $Q[(\frac{\pi}{4} + 2), 0]$. The finite region A is bounded by the arc of the curve with equation $y = \tan x$ from the origin O to the point P, the line PQ and the line OQ. Find the area of A.
The region A is rotated through four right angles about the x-axis to generate the solid of revolution S. Find the volume of S. [L]

98

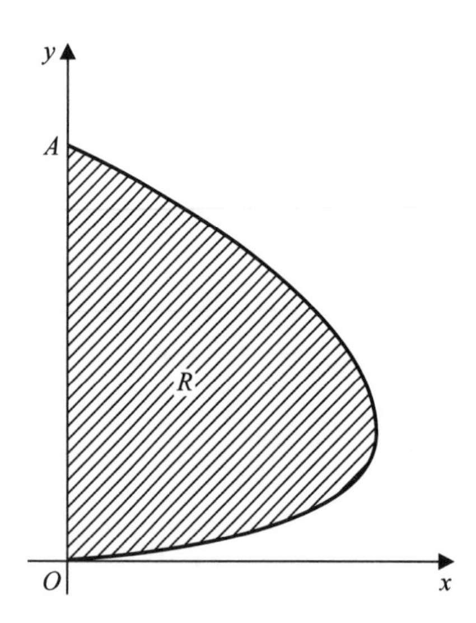

The figure shows part of the curve with equation

$$x^2 = y(4 - y)^2$$

(a) Find the coordinates of the point A where the curve meets the y-axis.

The finite region R shaded in the figure is bounded by the curve with equation $x^2 = y(4 - y)^2$ and the line OA, where O is the origin.

(b) Calculate the area of R.

The finite region R is rotated through 2π about the y-axis.

(c) Calculate the volume of the solid formed, leaving your answer in terms of π. [L]

99 The carburettor for a particular motor car is manufactured at one of three factories X, Y, Z and then delivered to the main assembly line. Factory X supplies 45% of the total number of carburettors to the line, Factory Y 30% and Factory Z 25%. Of the carburettors manufactured at Factory X, 2% are faulty and the corresponding percentages for factories Y and Z are 4% and 3% respectively.

Let X, Y and Z represent the events that a carburettor chosen at random from the assembly line was manufactured at Factory X, Y or Z respectively and let F denote the event that this carburettor is faulty.

(a) Express in words the meaning of $Y \cap F'$ and of $Y \cup Z$.

(b) Calculate $P(X \cap F)$, $P(Y \cap F)$ and $P(Z \cap F)$.

(c) Sketch a Venn diagram to illustrate the events X, Y, Z and F. Include in your diagram the probabilities corresponding to the different regions within the diagram.

(d) Find the probability that a carburettor, selected at random from the main assembly line, is faulty.

(e) Given that a carburettor is found to be faulty, find, to 3 decimal places, the probability that it was manufactured at Factory X.

(f) Given that a carburettor is not faulty, find, to 3 decimal places, the probability that it was manufactured at Factory X. [L]

Examination style paper P2

Answer all questions **Time 90 minutes**

1. Find the values of x for which

$$2 \ln 2x - 6 \ln 2 = \ln(x - 3).$$

(6 marks)

2. The line p_1 through the point $C(-1, 5)$ is perpendicular to the line p_2 which passes through the points $A(6, 1)$ and $B(-8, -6)$.
   (a) Find an equation for p_1 and hence determine the coordinates of the point M where the lines p_1 and p_2 meet.
   (b) Find the ratio $AM : AB$.

(7 marks)

3. (a) Find the binomial expansion of $(1 + kx)^{11}$ in ascending powers of x up to and including the term in x^3, where k is a negative constant.
   Given that the coefficient of the term in x^2 is $\frac{55}{9}$, find
   (b) the value of k, and hence
   (c) the coefficient of x^3.

(8 marks)

4. Giving your answers in radians to 2 decimal places, solve the equation $\sin x - 3 \cos 2x + 2 = 0$ for $0 < x \leqslant 2\pi$.

(8 marks)

5. (a) Show that the equation $x = \ln(x + 4)$ has a root between 1 and 2.
   (b) Use the iteration formula $x_{n+1} = \ln(x_n + 4)$, $x_0 = 2$ to find this root correct to 5 decimal places. Show all your intermediate results clearly.

(9 marks)

6. The journey times, to the nearest minute, of 170 passenger trains running between two stations were recorded and the results are given in the following table:

Time (minutes)	16	17	18	19	20	21	22
Frequency	5	19	29	55	37	21	4

By drawing a cumulative frequency polygon, estimate

(a) the median journey time,

(b) the probability that for two trains taken at random, one will have a journey time greater than 20 minutes and the other a journey time less than 20 minutes.

(9 marks)

7. (a) Given that $3y = 3\tan x + \tan^3 x$, show that

$$\frac{dy}{dx} = \sec^4 x$$

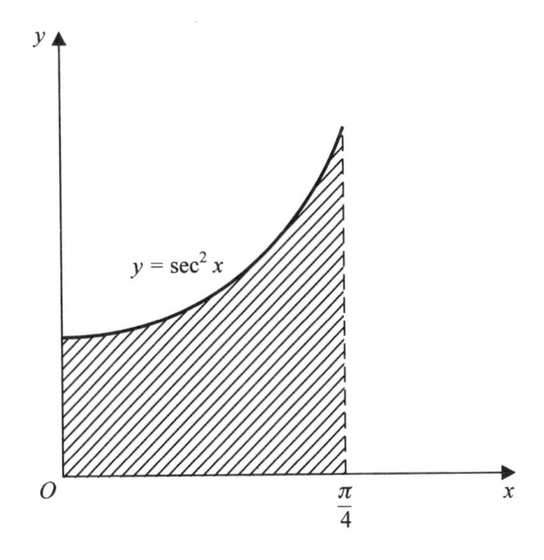

The shaded region is bounded by the x-axis, the y-axis, the line $x = \frac{\pi}{4}$ and the curve with equation $y = \sec^2 x$. The region is rotated through 2π about the x-axis to form a solid of revolution.

(b) By integration, determine the volume of this solid, giving your answer in terms of π.

(12 marks)

8. At time t, the rate of increase in the concentration x of microorganisms in controlled surroundings is equal to k times the concentration, where k is a positive constant.

(a) Write down a differential equation in x, t and k.

(b) Given that $x = c$ at time $t = 0$, find x in terms of c, k and t.

(c) Sketch the graph of x against t for $t \geqslant 0$.

(d) Given further that $k = 10^{-2}$ and that t is measured in hours, find the time taken for the concentration to increase to $\frac{3c}{2}$ from when it was c.

(12 marks)

9.
$$f(x) \equiv 6x^3 + Ax^2 + x - 2$$

where A is a constant.

Given that $x + 2$ is a factor of $f(x)$,

(a) find the value of A.

Using this value of A,

(b) express $\dfrac{1}{f(x)}$ in partial fractions,

(c) find the value of $\dfrac{d^2}{dx^2}\left(\dfrac{1}{f(x)}\right)$ at $x = 0$.

(13 marks)

10. The curve C is given by

$$x = 2t - 1, \quad y = 4 - t^2,$$

where t is a parameter.

The line with equation $x + 2y = 3$ meets C at the points P and Q.

(a) Determine the coordinates of P and Q.

The tangent at the point T to the curve C is parallel to PQ.

(b) Find the coordinates of T.

The finite region R is bounded by C and the line PQ.

(c) Draw a sketch to show the curve C, the line PQ and the region R.

(d) Using integration, find the area of R.

(16 marks)

Answers

The University of London Examination and Assessment Council accepts no responsibility whatsoever for the accuracy or method of working in the answers given for examination questions.

Exercise 1A

1 (i) b (ii) b (iii) a (iv) c (v) c

2 $A = 1, B = 3$ **3** $A = 2, B = -1$

4 $A = B = 3$ **5** $A = 5, B = -4$

6 $A = 10, B = -2$ **7** $A = 1, B = 2, C = 3$

8 $A = C = 1, B = -2$

9 $A = 1, B = 2, C = -3$

10 $A = 2, B = 4, C = -3$

11 $A = 10, B = 3, C = -2$

12 $A = 1, B = 0, C = -1$

13 $A = 2, B = 12, C = -3$

14 $A = 1, B = 0, C = -1$

15 $A = \frac{1}{3}, B = -\frac{1}{3}, C = 1$

16 $A = 3, B = 1, C = -5$

17 $A = 2, B = -3, C = -4$

18 $A = 3, B = 2, C = 0$

19 $A = 4, B = 2, C = -1$

20 $A = 5, B = -2, C = -1$

21 $A = 1, B = -2, C = -2$

22 $A = 4, B = -\frac{3}{2}, C = 16$; Minimum value $= 16$

23 $A = 3, B = 3, C = -32$; Minimum value $= -32$

24 $A = 20, B = 2, C = 1$; Maximum value $= 20$

25 $A = 25, B = 5$

Exercise 1B

1 $2x^2 + x - 1$ **2** $x^2 + 2x + 1$

3 $2x^2 - x + 7$ **4** $-3x^3 + 2x^2 - 7x + 5$

5 $6x^2 - 7x + 5$ **6** $2x^3 + 4x^2 - 2x$

7 $2x^2 + x - 2$ remainder 2

8 $5x^3 - 2x^2 + 3x - 1$ remainder 6

9 $-3x^3 + 2x^2 + 11x - 20$ remainder -10

10 $4x^3 - 2x^2 + x - 6$ remainder 4

11 $3x^3 - 2x^2 + x + 1$ **12** $-x^2 + 6x + 5$

13 $-2x^2 - 3x + 1$ **14** $x^3 + 3x^2 + 5x - 7$

15 $2x^3 + 4x^2 - 7x + 6$

16 $3x^2 - 4x + 6$ remainder $2x - 1$

17 $x^3 - x^2 + 2x + 6$ remainder $-2x + 7$

18 $-2x^3 + x^2 + 6$ remainder $-3x - 2$

19 $2x^2 - 4x + 3$ remainder $-8x - 4$

20 $3x^3 - 2x + 1$ remainder $2x - 2$

Exercise 1C

1 $\dfrac{4}{3x^3(x+2)}$ **2** $\dfrac{6}{(x-4)(x+1)}$

3 $\dfrac{3(1-x)x}{2(7+x)(3-2x)}$

4 $\dfrac{x+3}{2(x+2)}$ **5** $\dfrac{2x^2(x-2)(x-5)}{2x+1}$

6 $\dfrac{10x}{3(x-5)(3x-2)}$ **7** $\dfrac{x(2-x)(2+x)}{3(x+4)(2x-3)}$

8 $\dfrac{20x^2}{7(x+1)(3x-2)}$ **9** $\dfrac{(x+3)(2x+3)}{5x^2(x-3)}$

10 $\dfrac{(x+2)(2x-1)}{(x-2)(x+3)}$ **11** $\dfrac{x+2}{4(x-2)}$

12 $\dfrac{5x+5}{(2x+1)(x+3)}$ **13** $\dfrac{7x-14}{(4x-5)(x-3)}$

14 $\dfrac{11+42x-8x^2}{2(4x+3)(2x-1)}$ **15** $\dfrac{9}{2x-3}$

16 $\dfrac{3x^2+16x+7}{(x+1)(x+3)(2x-1)}$

17 $\dfrac{-3x^2+6x+2}{(x-1)(2x+1)(x+4)}$

18 $\dfrac{-13x^2-38x-12}{(5-2x)(6+x)(1-x)}$

19 $\dfrac{9x^2+28x-2}{10(x-2)(x+2)(x+3)}$

20 $\dfrac{28-59x-14x^2}{(3-x)(7+x)(1-2x)}$

Exercise 1D

1 $\dfrac{1}{x+2}+\dfrac{1}{x+3}$ **2** $\dfrac{1}{x-1}+\dfrac{1}{x+3}$

3 $\dfrac{3}{x+4}-\dfrac{2}{x+3}$ **4** $\dfrac{5}{x+2}-\dfrac{4}{x+3}$

5 $\dfrac{1}{x-1}+\dfrac{2}{2x-1}-\dfrac{1}{x+3}$

6 $\dfrac{x+1}{x^2+4}+\dfrac{2}{x-2}$ **7** $\dfrac{1}{x-3}-\dfrac{2}{x^2+3}$

8 $\dfrac{2}{x^2+5}-\dfrac{2}{2x+3}$ **9** $\dfrac{1}{5+2x^2}-\dfrac{3}{x+3}$

10 $\dfrac{2}{2x+1}-\dfrac{2x+1}{x^2+2x+7}$

11 $\dfrac{2}{x-5}+\dfrac{3}{(x-5)^2}$

12 $\dfrac{1}{x+3}-\dfrac{2}{(x+3)^2}+\dfrac{4}{(x+3)^2}$

13 $\dfrac{4}{(x-1)^2}-\dfrac{3}{x+2}$

14 $\dfrac{4}{(x-1)^2}-\dfrac{2}{x-1}-\dfrac{3}{x+2}$

15 $\dfrac{1}{2x+1}-\dfrac{1}{2x+3}+\dfrac{3}{(2x+3)^2}$

16 $1+\dfrac{1}{x-1}$ **17** $x+1\dfrac{1}{x-1}$

18 $1+\dfrac{1}{x-1}-\dfrac{1}{x+1}$ **19** $1-\dfrac{2}{x}+\dfrac{3}{x-1}$

20 $x+\dfrac{1}{2(x-1)}+\dfrac{1}{2(x+1)}$

21 $2+\dfrac{3}{x+1}+\dfrac{1}{x-2}$

22 $2-\dfrac{1}{1-x}-\dfrac{3}{1+2x}$

23 $2x+\dfrac{1}{x+2}-\dfrac{1}{x+3}$

24 $x+1\dfrac{3}{x-2}-\dfrac{1}{x+2}$

25 $-x+\dfrac{1}{x}-\dfrac{2}{x^2}+\dfrac{1}{x+1}$

26 $\dfrac{2}{2x-3}-\dfrac{3}{3x+2}$

27 $\dfrac{2}{2x-1}+\dfrac{1}{x+2}-\dfrac{3}{(x+2)^2}$

28 $\dfrac{1}{x^2+1}+\dfrac{1}{x+1}+\dfrac{2}{(x+1)^2}$

29 $2x+\dfrac{2}{2x+3}-\dfrac{1}{x^2}+\dfrac{1}{x}$

30 $\dfrac{2}{2x-1}-\dfrac{1}{3x+1}$

31 $\dfrac{3}{x-1}-\dfrac{1}{(x-1)^2}+\dfrac{3-2x}{x^2+2}$

32 $\dfrac{11}{7(x-2)}-\dfrac{1}{x+2}+\dfrac{13}{7(2x+3)}$

33 $1-\dfrac{1}{x}-\dfrac{x}{x^2+x+1}$

34 $\frac{3}{x^2} - \frac{1}{x} + \frac{2}{x+1}$

35 $\frac{1}{x+1} + \frac{-x+2}{x^2-x+1}$

Exercise 1E

1 $(x+2)(x+3)(x-2)$
2 $(x-1)(x-2)(x+4)$
3 $(2x-1)(x+3)(x-2)$
4 $(x-2)(2x+5)(3x-2)$
5 $(x-1)(3x-1)(x+7)$
6 21 7 448 8 -154 9 $-6\frac{3}{4}$
10 $-2\frac{1}{2}$ 11 2 12 7 13 3 14 3
15 2 16 $a=1, b=-4$ 17 $a=2, b=1$
18 $a=-4$ $(x+1)(3x-1)(x-2)$
19 $a=9, x=1, -\frac{3}{2}, -4$
20 $a=2, b=-11, x=-2, \frac{1}{2}, 3$
21 $a=1, b=-6$ $(x^2+1)(x-2)(x+3)$
22 $p=-3, q=-11$ $(2x-1)(x-3)(x+2)$
23 $k=-6, p(x) \equiv (x-3)^2+2>0$ for all real x
24 $A=5, B=-4; x=1, -\frac{1}{2}, -3$

Exercise 2A

1 (i) quantitative and discrete
(ii) quantitative and continuous
(iii) qualitative
(iv) quantitative and discrete
(v) quantitative and continuous

2 Set up a checkpoint on a busy road and then stop and scrutinise every nth commercial vehicle.

3 There are always some non-respondents to a postal questionnaire.

4 (a) See text section 2.2
(b) (i) Not all of those sampled would be local residents living in Avon, and also most sampled would be in employment; this omits sections of population, making the sample biased.
(ii) This will exclude people who did their shopping during the week or who do not shop in this precinct anyway. There could be also many tourists and visitors.

5 (a) During several periods over a day collect bulbs from the production line, number them all and then use a table of random digits to collect a sample of the required size.
(b) After selecting a sample at random from the total employment register for the firm, use the random sample to fill in a questionnaire.
(c) Record the total numbers absent over each day in several weeks chosen at random over a year.

Exercise 2B

6

x	1–15	16–20	21–30	31–40
f	27	32	37	14

9

Length	$0 \leqslant x \leqslant 20$	$20 < x \leqslant 30$	$30 < x \leqslant 40$	$40 < x \leqslant 70$
f	4	6	4	6

10

Time	$0 < t \leqslant 20$	$20 < t \leqslant 40$	$40 < t \leqslant 60$	$60 < t \leqslant 80$
f	25	135	65	25

Exercise 2C

1 4, 2 2 40, 20 3 13, 2 4 5.2, 2.75
5 19.6, 2.94 6 3.94, 1.62 7 4, 14.5, 4
8 (a) 4 (b) 7 (c) 2
9 (a) 5.16 (b) 6 (c) 10

(d) 4 (e) 6.01 (f) 2.45

10 5.99 g, 0.08 g, 0.04 g **11** 6, 5.27, 2.29

12 30.1 lb, 11.34 lb

13 (a) 18 500 miles (b) 16 000 miles (c) 7500 miles

14

Mark	$m \leqslant 20$	$20 < m \leqslant 30$	$30 < m \leqslant 40$	$40 < m \leqslant 60$	$60 < m \leqslant 100$
f	12	20	43	74	51

Mean 50.0, SD 20.9

15 47.8, 26.6 **16** 267.5 min, 10.7 min

17 (a) 270.5 min

20 (a) $\mu_x = 5, \mu_y = 6$ (b) $\sigma_x^2 = 25, \sigma_y^2 = 4$; $\mu_{x+y} = 5.6, \sigma_{x+y}^2 = 12.64$

Exercise 3A

1 3 **2** 3 **3** 5 **4** 1296 **5** 16 **6** 3
7 3 **8** $\frac{1}{3}$ **9** $\frac{1}{5}$ **10** 2187 **11** 4
12 4 **13** 1 **14** −2 **15** 3 **16** $\frac{1}{3}$
17 3 **18** −3 **19** $\frac{1}{3}$ **20** $\frac{1}{3}$ **21** 1.58
22 1.92 **23** 0.613 **24** 0.774
25 1.09 **26** $\log_3 14$ **27** $\lg 3$ **28** $\ln 3$
29 $\ln 1280$ **30** $\log_a \frac{9}{8}$ **31** $\log_a \frac{4}{27}$
32 $\log_a 441 - 2$ **33** $\log_a 10$
34 $5 + \log_a 6$ **35** $-\log_a 12$
36 $\log_a x + \log_a y - \log_a z$
37 $2\log_a x + \log_a y - 3\log_a z$
38 $\log_a x + 2\log_a y + 3\log_a z$
39 $\frac{1}{2}\log_a x + \log_a y + \frac{1}{2}\log_a z$
40 $\log_a x + \log_a y - \frac{3}{2}\log_a z$

Exercise 3B

1 2.81 **2** 2.68 **3** 1.73 **4** 1.37
5 2.58 **6** 1.5 **7** 1.85 **8** 0.112
9 1.71 **10** −13.8 **11** 6.05 **12** −3.89
13 1 or 1.58 **14** −0.631 or 1.26
15 −0.792 **16** 1 **17** 1.58
18 1 or 1.79 **19** ±0.693
20 0.431 or 0.683

Exercise 3C

1 (a) 69.1°C (b) 60.2°C (c) 44.4°C
2 (a) 5.0 (b) 16.2 (c) 26.4
3 (a) 49.7 million (b) 67.1 million (c) 122.4 million
4 (a) 16.0 (b) 28.7 (c) 33.6
5 $P = 36.4e^{0.01t}$ (a) 38.7 million (b) 46.4 million (c) 48.4 million
6 (a) 40.8°C (b) 34.8°C (c) 23.9°C
7 $k = 0.112$ (a) $0.46\,m_0$ (b) $0.33\,m_0$ (c) $0.15\,m_0$
8 $k = 0.0263$ (a) 8.5 (b) 19.4
9 (a) $0.66\,m_0$ (b) 21

Exercise 4A

1 $a = 1.7, b = -1.5$
2 $c = 450, k = -0.58$
3 a = 27, b = 1.6 **4** $f = 10$
5 $a = 0.5, b = 12$
6 $a = 2, b = 34$ **7** $a = 140, b = 0.36$
8 $p = -10, q = 84$ **9** $a = 2.0, x = 3.3$
10 $a = 19.4, k = 0.5$
11 $m = 1.4, c = 0.74, y = 5.5x^{1.4}$
12 $a = 2.0, b = 1.8$

Review exercise 1

1 (a) −1; $(x+1)(x-2)(2x+1)$ (b) 18
2 (b) $k = 7, V = 0.05$
3 $\dfrac{3x-7}{x(x-1)(x-2)}$
4 (a) 29 (b) (i) 8.54 (ii) 11.40
5 $k = -13, p = -35, q = 27$ **6** 5
7 $A = -4, B = 4; x = 1, \pm 2$
9 $A = 6, B = -2, C = 5$
10 (b) $n = 1.7, k = 0.79$

11 2.32

12 (a) median 46.8, SIQR 12.4

13 $k = 0.1088$, 1943 people

14 (a) 0.778 15 (b) 1.397 94 (c) 0.176 09

15 (a) 2^{2+p+2q} (b) $p + 6q$

16 $\frac{2}{x+2} + \frac{1}{x-1}$, $\ln 2$

17 $a = 3, b = 1$; remainder 15

18 (a) 0.301 03 (b) −0.602 06

19 0.631 **20** $a = 90, k = -0.3$; −280

21 (a) $A = 1, B = -1, C = 2$

23 $(x-2)(2x-1)(2x+3)$

24 $\frac{1}{x+2} + \frac{21-9x}{4x^2+3}$

25 $\frac{7-11x}{2(x+1)(x-1)(x-2)}$

26 (a) 138 (c) 90

27 $A = \frac{1}{3}, B = -\frac{1}{3}, C = \frac{2}{3}$

28 (a) 2 (b) 3.5, 2, 5
(c) mean 3.98, SD 2.41

29 (a) $4p$ (b) $-2p$ (c) $1 + p$

30 0, 0.526

31 $\frac{1}{8}$

32 $\frac{19x^2+x-31}{(x-3)(2x+5)(3x+2)}$

33 (a) 0.1 (b) 1820

34 1.5

35 $\frac{1}{x} - \frac{2}{x+1} + \frac{1}{x+2}$

36 $c = 6$, $x = -2, \frac{1}{2}, 3$

37 (a) $2p$ (b) 210

39 5.5, 2.87 (a) 2.87 (b) 8.62
(c) 0.287

40 $k = 2, n = 0.5$

41 $a = 0.90$, $k = 0.039$

42 $a = 8000$, $b = 2000$, $c = \frac{1}{10}$; 13.9 days

43 (a)(i) $r = 4, s = -5$
(ii) $p = 2, q = 11$
(b)(ii) $(x+1)(2x+5)(x-4)$
(iii) $-1, -2\frac{1}{2}, 4$

44 (a) 2 to < 4 (b) mean 4.6, SD 3.21
(c) median 3 years 10 months,
IQR 4 years 11 months

45 (a) −38; $(x+4)(3x-2)(x-3)$ (b) −5

46 $\frac{4}{x^2-4}$

47 $a = 1.9$, $b = 1.8$

48 $A = B = \frac{2}{9}$, $C = \frac{2}{3}$

49 $A = 79$, $k = 0.023$; 21 minutes

50 $p = 7.4$, $q = 0.5$

51 (a) $\frac{2}{x+1} - \frac{1}{x+2} - \frac{1}{x+3}$
(b) $\ln\frac{k(x+1)^2}{(x+2)(x+3)}$

52 $1 + \frac{3}{x+2} + \frac{2}{x+1}$

53 median 36, IQR 15; 24%

54 $(x-2)(3x-1)(x+1)$; $x = \ln 2$ or $-\ln 3$

55 $\frac{17-13x}{3(x+1)(x-1)(x-2)}$

56 (b) $x = \frac{\sqrt{3}}{4}, y = \frac{\sqrt{3}}{2}$

57 $A = 2, B = -3$

58 $a = 6.0, b = 4.5, u = 1.8$

59 $\frac{6}{(x-2)(x+1)}$

60 median 3.2, IQR 0.5

61 1.24×10^{-4}

62 (b) 1.46 (c) 0.37, 0.63

63 $\frac{1}{3}\left(\frac{1}{x}-\frac{1}{x+3}\right)$

64 (b) $(x+5)(2x-1)(x+1)$
(c) $\pm 1, -5$

65 12

66 3.4

67 $p = 6$, $q = \frac{1}{2}$ or $p = 10$, $q = 8\frac{1}{2}$; $\frac{3}{2}$ or $\frac{5}{2}$

68 (a) $e^{-\frac{1}{2}}$ (b) 6.56 (c) 1 (d) 51.31
(e) 6.91

70 (a) $\frac{1}{3}, 5$ (c) -1, 1.46

71 (a) mean 54, SD 90.3

72 32°C

73 $\frac{1}{1-3x}+\frac{2x}{1+6x^2}$

74 $\frac{3x+2}{(x+1)(x+2)(x+3)}$

75 $a = \frac{1}{2}, b = 12$

76 (a) 1.26 (c) $x = 27, y = 3$

77 (a) (i) 12
(ii) $(x+3)(5x-7)(x-4)$ (b) 4

78 LCM $= x(x-2)(x-3)$; $\frac{6}{x(x-3)}$

79 (a) 8 (b) -2

80 $A = 200, B = 1.5$

81 mode 50, median 51, mean 50.61

82 $\frac{5}{x(x+2)}$

83 $p = -28, q = -16$

84 (a) $\frac{x+y}{2}$ (b) $2x - y$ (c) $2 + x + y$

85 mean 906.72, SD 2.64

86 (b) $n = 0.50, k = 55$

87 $\frac{13x+31}{5(x-3)(x-1)(x+2)}$

88 $\frac{x+1}{x^2+1}-\frac{3}{3x+2}$

89 (b) $(3x+2)(x+2)(x-1)$
(c) $90°, 222°, 318°$

90 (a) $\frac{1}{4}$ (b) 1

91 $\frac{3x+1}{(x+2)(2x+3)}$

92 10, 5.75 (a) 5.5, 2.87 (b) 19, 5.75

93 $a = 2, b = -5$; 90

94 For A: mean 1000.02, SD 0.68
For B: mean 999.32, SD 0.42

95 mean 40.5, SD 17.3

96 (a) (i) 9 (ii) 12 (iii) $9\frac{1}{9}$
(b) (i) $2\log_a p + \log_a q$
(ii) $\log_a p = 4, \log_a q = 1$

97 (a) $\frac{1-x}{2(x+1)(x+2)(x+5)}$ (b) $-\frac{1}{15}$

98 mean 1.697, SD 0.276

99 (a) (i) xy (ii) x^2 (b) (ii) 3

100 $N_0 = 100$, $K = 0.005$

Exercise 5A

1 (a) $1 + 6x + 9x^2$ (b) $1 - 10x + 25x^2$
(c) $1 - 2x^2 + x^4$ (d) $1 + 4x^3 + 4x^6$
(e) $9 - 24x + 22x^2 - 8x^3 + x^4$

2 (a) $1 + 6y + 12y^2 + 8y^3$
(b) $1 - 3x^3 + 3x^6 - x^9$
(c) $1 + 9x^{-1} + 27x^{-2} + 27x^{-3}$

3 (a) $1 + 4x + 6x^2 + 4x^3 + x^4$
(b) $1 + 5x + 10x^2 + 10x^3 + 5x^4 + x^5$

4 (a) $1 - 12x + 54x^2 - 108x^3 + 81x^4$
(b) $1 + 10x + 40x^2 + 80x^3 + 80x^4 + 32x^5$

Exercise 5B

1 $1 + 15x + 75x^2 + 125x^3$; $2 + 150x^2$

2 $1 - x + \frac{1}{4}x^2$, $1 - \frac{3}{2}x + \frac{3}{4}x^2 - \frac{1}{8}x^3$,
$1 - 2x + \frac{3}{2}x^2 - \frac{1}{2}x^3 + \frac{1}{16}x^4$

3 $8 - 36x + 54x^2 - 27x^3$,
$27 + 54x + 36x^2 + 8x^3$,
$-30 - 216x + 90x^2 - 97x^3$

4 (a) $1 - 12x + 60x^2 - 160x^3$
(b) $128 - 448x + 672x^2 - 560x^3$

5 $-524\,288y^9 - 1\,179\,648y^7 - 258\,048y^5$

6 0.850 763 0

7 (a) 1.049 070 08 (b) 0.953 041 92

8 $37 - 252x - 36x^2 - 91x^3$

9 $625x^4 - 1000x^3 + 600x^2$

10 $m = 6,\ A = 2,\ B = 240$

11 (a) 56 (b) 98

12 $z^9 - 9z^7 + 36z^5 - 84z^3 + 126z - 126z^{-1}$
$+84z^{-3} - 36z^5 + 9z^{-7} - z^{-9}$

13 (a) 364 (b) 1080 (c) −80
(d) 86 016

14 −20 **15** $6\frac{19}{35}$

Exercise 5C

1 $1 - 23x + 253x^2$ **2** $A = 15,\ B = 105,$
$C = 455$

3 $78\,125x^7 - 328\,125x^6 + 590\,625x^5$

4 $162 + 432x^2 + 32x^4$

5 (a) $1 - 40y + 700y^2 - 7000y^3$
(b) $1 - 44y + 860y^2 - 9800y^3$

6 $1 + 10x + 45x^2 + 120x^3$, 0.980 179 04

7 $40x^8 + 1920x^4 + 16\,128$

8 $a = 1351,\ b = -780$

9 6726 **10** (a) 27 (b) 3 247 695

11 (a) 1.012 066 22
(b) 998 501 049 500 000

12 $n = 22,\ k = 2$, 12 320

13 $1 + 7ax + 21a^2x^2 + 35a^3x^3$; $a = 2,\ b = 5$;
434; 1484

15 $k = 2,\ n = 7$; 280, 560

16 $18\,564y^6$, 5.543×10^{10} (4 s.f.)

17 $1 + 6y + 15y^2 + 20y^3 + 15y^4 + 6y^5 + y^6$,
1.062 15

18 $a = 2,\ b = \frac{1}{2}$; $5x^3 + \frac{5}{8}x^4 + \frac{x}{32}$

19 $x^{80} + 40x^{78} + 780x^{76} + 9880x^{74}$

20 $k = \frac{1}{4},\ n = 17$; $\frac{85}{8}x^3$, $\frac{595}{64}x^4$

Exercise 6A

1 (a) 1.235 (b) 1.466 (c) 1.122
(d) 1.086 (e) −1.664 (f) −1.662
(g) −3.420 (h) 1.287 (i) −0.5774
(j) −1.026

2 (a) −0.02921 (b) −1.034 (c) 1.341
(d) −0.08784 (e) −1.043 (f) 1.041
(g) 1.293 (h) −1.516 (i) 1.171
(j) 1.075

3 (a) 56.5, 303.5 (b) 207.5, 332.5
(c) 140.9, 320.9 (d) 153.9, 206.1
(e) 152.1, 207.9 (f) 81.6, 261.6
(g) 51.7, 308.3 (h) 33.4, 146.6
(i) 39.5, 219.5 (j) 61.1, 118.9

4 (a) 0.907, 5.376 (b) 0.663, 2.478
(c) 0.948, 4.090 (d) 2.114, 4.169
(e) 0.519, 2.623 (f) 2.660, 5.802
(g) 0.714, 5.569 (h) 3.745, 5.680
(i) 1.883, 5.025 (j) 2.080, 4.204

Exercise 6B

1 (a) $\dfrac{\sqrt{3}+1}{2\sqrt{2}}$ (b) $\dfrac{\sqrt{3}-1}{2\sqrt{2}}$

(c) $\dfrac{1+\sqrt{3}}{1-\sqrt{3}}$ (d) $\dfrac{\sqrt{3}-1}{2\sqrt{2}}$

(e) $\dfrac{1-\sqrt{3}}{2\sqrt{2}}$ (f) $\dfrac{\sqrt{3}-1}{\sqrt{3}+1}$

(g) $\dfrac{\sqrt{3}-1}{2\sqrt{2}}$ (h) $-\dfrac{\sqrt{3}+1}{2\sqrt{2}}$

2 (a) 1 (b) $\frac{1}{2}$ (c) $-\sqrt{3}$ (d) $\frac{1}{2}$
(e) $\frac{1}{2}$ (f) $\frac{1}{\sqrt{3}}$

3 (a) $\sin 4\theta$ (b) $\cos 3\theta$ (c) $\tan 2\theta$
(d) $3\sin 5\theta$ (e) $4\cos 2\theta$ (f) $2\cos 3\theta$

4 (a) $\frac{63}{65}$ (b) $\frac{56}{65}$ (c) $-\frac{63}{16}$ (d) $-\frac{56}{33}$

5 (a) $\frac{63}{65}$ (b) $-\frac{65}{56}$ (c) $\frac{63}{16}$ (d) $\frac{56}{33}$

6 (a) $\frac{36}{325}$ (b) $-\frac{325}{204}$ (c) $\frac{253}{325}$
(d) $-\frac{325}{323}$ (e) $-\frac{204}{253}$ (f) $-\frac{323}{36}$

7 (a) (i) 1 when $\theta = 20°$
(ii) -1 when $\theta = 200°$
(b) (i) 1 when $\theta = 110°$
(ii) -1 when $\theta = 290°$
(c) (i) 1 when $\theta = 65°$
(ii) -1 when $\theta = 245°$

8 (a) -1 when $\theta = 230°$
(b) -1 when $\theta = 223°$
(c) -3 when $\theta = 75°$

Exercise 6D

1 (a) $\frac{1}{2}$ (b) $\frac{1}{\sqrt{2}}$ (c) -1 (d) $\frac{\sqrt{3}}{2}$
(e) $1 - \frac{\sqrt{3}}{2}$ (f) $-2\sqrt{2}$ (g) $2\sqrt{3}$
(h) $\frac{1}{2} + \frac{\sqrt{3}}{4}$

2 (a) $\frac{24}{25}$ (b) $-\frac{7}{25}$ (c) $-\frac{24}{7}$

3 (a) $\frac{120}{169}$ (b) $-\frac{119}{169}$ (c) $-\frac{120}{119}$

4 (a) $-\frac{336}{625}$ (b) $\frac{625}{527}$ (c) $-\frac{527}{336}$

5 (a) $\frac{25}{7}$ (b) $-\frac{25}{24}$ (c) $-\frac{24}{7}$

6 (a) $\frac{1}{\sqrt{2}}$ (b) $\frac{1}{\sqrt{2}}$ (c) 1

7 $\frac{1}{3}$ or -3 **8** $\frac{1}{7}$ or -7

9 $\sin 3\theta \equiv 3\sin\theta - 4\sin^3\theta$; $\frac{47}{128}$

10 $\cos 3\theta = 4\cos^3\theta - 3\cos\theta$; $-\frac{11}{16}$

11 (a) $(x-2)^2 = 4y^2(1-y^2)$
(b) $y = 3(2x^2 - 4x + 1)$
(c) $x(1-y^2) = 2y$
(d) $x - 3 = \frac{2}{y^2} - 1$
(e) $x^2y = 2$

Exercise 6E

1 8.6°, 141.4° **2** 68.5°, 225.5°
3 170.7°, 350.7° **4** 169.6°, 349.6°
5 60°, 240° **6** 126.2°, 306.2°
7 85.9°, 265.9°
8 22.5°, 112.5°, 202.5°, 292.5°
9 0, 60°, 180°, 300°, 360°
10 0, 120°, 180°, 240°, 360°
11 21.5°, 158.5°
12 0, 60°, 120°, 180°, 240°, 300°
13 19.5°, 160.5°, 270°
14 33.7°, 63.4°, 213.7°, 243.4°
15 33.7°, 153.4°, 213.7°, 333.4°
16 30°, 90°, 150°, 270°
17 120°, 240°
18 15°, 75°, 195°, 255°
19 0, 180°, 360°, 60°, 300°, 120°, 240°
20 0, 180°, 360°, 45°, 135°, 225°, 315°
21 45°, 225°, 26.6°, 206.6°
22 51.3°, 128.7°
23 105°, 165°, 285°, 345°
24 30°, 150°, 180°
25 22.5°, 112.5°, 202.5°, 292.5°
26 45°, 225°, 121°, 301°
27 14.2°, 74.2°, 134.2, 194.2, 254.2, 314.2
28 0, 180°, 360°, 138.6°, 221.4°

Exercise 7A

1 $AC - 4.5$ cm, $BC = 5.7$ cm, $C = 91°$
2 $C = 67.4°$, $AB = 12.2$ cm, $BC = 9.11$ cm
3 $A = 69.5°$, $AB = 22.5$ cm, $BC = 24.0$ cm
4 $B = 79.3°$, $AB = 17.2$ cm, $AC = 17.9$ cm
5 $B = 50°$ $AB = 6.7$ cm, $BC = 5.7$ cm
6 $B = 56.2°$, $AB = 7.6$ cm, $BC = 5.9$ cm
7 $C = 33.3°$, $B = 99.7°$, $AC = 11.3$ cm
8 $A = 35.3°$, $B = 80.7°$, $AC = 21.2$ cm
9 $A = 33.3°$, $C = 73.7°$, $AB = 12.7$ cm

10 $A = 30.2°$, $C = 96.6°$, $AB = 11.7$ cm
11 $A = 56.3°$, $B = 55.0°$, $BC = 15.5$ cm
12 $A = 34.6°$, $B = 73.1°$, $AC = 21.4$ cm
13 $B = 44.8°$, $C = 99.0°$, $AB = 19.1$ cm *or* $B = 135.2°$, $C = 8.6°$, $AB = 2.9$ cm
14 $B = 75.6°$, $A = 52.1°$, $BC = 15.6$ cm *or* $B = 104.4°$, $A = 23.3°$, $BC = 7.8$ cm
15 $C = 30°$, $A = 132.6°$, $BC = 36.2$ cm *or* $C = 150°$, $A = 12.6°$, $BC = 10.7$ cm
16 $B = 59°$, $A = 73.7°$, $BC = 20.4$ cm *or* $B = 121°$, $A = 11.7°$, $BC = 4.3$ cm
17 $C = 23.6°$, $A = 25.4°$, $BC = 4.6$ cm
18 $B = 82°$, $C = 66.8°$, $AB = 14.0$ cm *or* $B = 98°$, $C = 50.8°$, $AB = 11.8$ cm
19 $B = 24.4°$, $C = 139.9°$, $AB = 50$ cm *or* $B = 155.6°$, $C = 8.7°$, $AB = 11.7$ cm
20 $A = 58.3°$, $C = 84.0°$, $AB = 14.1$ cm *or* $A = 121.7°$, $C = 20.6°$, $AB = 5.0$ cm

Exercise 7B

1 (a) 4.39 cm (b) 11.2 cm (c) 6.42 cm (d) 7.46 cm (e) 8.00 cm (f) 9.82 cm (g) 16.1 cm (h) 13.9 cm (i) 19.5 cm (j) 14.4 cm
2 (a) 44.0° (b) 60° (c) 109.5° (d) 82.8° (e) 141.4° (f) 57.1° (g) 73.4° (h) 137° (i) 15.9° (j) 22.6°
3 (a) $AB = 2.45$ cm, $A = 60.2°$, $B = 76.8°$, area 2.39 cm^2
(b) $BC = 2.87$ cm $B = 88.1°$ $C = 56.9°$, area 5.74 cm^2
(c) $AB = 18.8$ cm, $A = 22.1°$ $B = 17.9°$, area 31.8 cm^2
(d) $AC = 12.6$ cm, $A = 21.9°$, $C = 48.1°$, area 23.5 cm^2
(e) $AC = 4.70$ cm, $A = 33.2°$, $C = 106.8°$, area 9.00 cm^2

Exercise 7C

1 4.3 km, 257°
2 21.8 nautical miles
3 2.51 km
4 554 m
5 2.62 nautical miles, 341°
6 6.40 km
7 224°
8 5.46 km, 280°
9 13 km, 055°
10 6.21 km

Exercise 7D

1 (a) 16.2° (b) 59.0°
2 (a) 28.1° (b) 33.7°
3 (a) 10.1° (b) 11.3° (c) 61.7°
4 (a) 19.8° (b) 28.5°
5 (a) 54.7° (b) 63.4°
6 (a) 63.7° (b) 70.7°
7 (a) 27.6° (b) 26.1°
8 (a) 9.43 cm, 9.90 cm
(b) (i) 53.9° (ii) 17.6°
9 (a) 20 m (b) 29 m (c) 46.4° (d) 60.3°
10 (b) $VM = 5.83$ cm, $AC = 7.21$ cm $VC = 6.16$ cm (c) 71.6°, 61.9°
11 (a) 10 cm (b) 8 cm (c) 62.2° (d) 67.2°
12 (a) 13.3 m (b) 55.7° (c) 71.2°
13 (a) 54.7° (b) 70.5°

Exercise 7E

1 $\sqrt{3}$ 2 $2\sqrt{11}$ 3 $\sqrt{34}$ 4 $\sqrt{14}$
5 $\sqrt{195}$ 6 $t = 1 \pm \sqrt{39}$ 7 $\frac{1}{2}$ 8 4.9°
9 $-1 \pm \sqrt{7}$

10 $\dfrac{-7 \pm \sqrt{21}}{2}$

11 $82.0°$

13 (a) $11t^2 + 22t + 43$ (b) -1 (c) 32

14 (a) $9t^2 - 10t + 35$ (b) $\frac{5}{9}$ (c) $\frac{290}{9}$

15 (a) $3t^2 - 24t + 86$ (b) 4 (c) 38

16 $73.1°$ **17** $63.2°$

18 (a) $B(2, 11, 1)$, $C(6, 11, 1)$, $D(6, 7, 1)$, $H(6, 7, 5)$ (b) 6.93 (c) $35.3°$

19 (a) $B(5, 5, -2)$, $E(5, -3, 3)$, $F(5, 5, 3)$, $H(9, -3, 3)$ (b) $51.3°$

20 $73.3°$

Exercise 8A

1 $2(x+2)$ **2** $7(x-5)^6$ **3** $-4(3-x)^3$

4 $6(2x-1)^2$ **5** $12(3x-7)^3$

6 $-2(x-4)^{-3}$ **7** $3(3-x)^{-4}$

8 $-(x-7)^{-2}$ **9** $3(5-x)^{-2}$

10 $48x(6x^2+1)^3$ **11** $(x-2)^{-1}$ **12** $2xe^{x^2}$

13 $-e^{3-x}$ **14** $-2x(6-x^2)^{-1}$

15 $\dfrac{-(2x-7)}{(x^2-7x+2)^2}$

16 $\dfrac{-72x^2}{(3-4x^3)^2}$

17 $\dfrac{2}{\sqrt{(5+4x)}}$ **18** $-2x^{-\frac{1}{2}}(1-x^{\frac{1}{2}})^3$

19 $-2x^{\frac{1}{3}}(x^{\frac{2}{3}}+5)^{-4}$ **20** $(2x-3)e^{x^2-3x}$

21 $-4x^{-2}(1+\frac{1}{x})^3$

22 $-3(2x+x^{-2})(x^2+\frac{1}{x})^{-4}$

23 $-x(1-x^2)^{-\frac{1}{2}}$

24 $\dfrac{-4x}{(1+x^2)^3}$ **25** $\dfrac{6x-4}{3x^2-4x+3}$

26 $(\frac{1}{2}x^{-\frac{1}{2}} - \frac{1}{2}x^{-\frac{3}{2}})(x^{\frac{1}{2}} + x^{-\frac{1}{2}})^{-1}$

27 $\dfrac{e^x}{e^x+5}$ **28** 1

29 $\dfrac{-54x^2}{(6x^3-5)^4}$ **30** $\dfrac{-1}{x(\ln x)^2}$

31 $-\frac{1}{108}$ **32** $\frac{7}{5}$ **34** $\frac{28}{27}$ **35** $-\frac{4}{243}$

Exercise 8B

1 $x^2(5x-3)(x-1)$ **2** $2x^3(7x+2)(2x+1)^2$

3 $xe^x(2+x)$ **4** $2\ln x + \dfrac{(2x-1)}{x}$

5 $2e^{2x}\ln x + \dfrac{e^{2x}}{x}$ **6** $\ln(3x-1) + \dfrac{3(x+1)}{3x-1}$

7 $(12x-1)(2x-1)^4$

8 $2(2x-3)^2(x^2+1)(7x^2-6x+3)$

9 $(1+9x^2)(x^2+1)^3$

10 $\dfrac{4x^2-3x-2}{\sqrt{(x^2-1)}}$ **11** $\frac{5}{2}x^{\frac{3}{2}} - 3x^{\frac{1}{2}} + \frac{1}{2}x^{-\frac{1}{2}}$

12 $\frac{2}{3}(x-2)^{-\frac{1}{3}}(x+2)^{\frac{3}{4}} + \frac{3}{4}(x+2)^{-\frac{1}{4}}(x-2)^{\frac{2}{3}}$

13 $\dfrac{20x^2-12x-2}{\sqrt{(4x-3)}}$

14 $\frac{1}{2}x^{-\frac{1}{2}}\ln(x^2-1) + \dfrac{2x^{\frac{3}{2}}}{x^2-1}$

15 $(x-1)^2e^{x^2-2x}[2x^2-4x+5]$

16 $\dfrac{-1}{(x-1)^2}$ **17** $-1 + \dfrac{1}{(1-x)^2}$

18 $\dfrac{6x^2-x^4}{(2-x^2)^2}$ **19** $\dfrac{e^x(2x-1)}{(2x+1)^2}$

20 $-3x^{-4}\ln(1-x^2) - \dfrac{2x^{-2}}{1-x^2}$

21 $\dfrac{6x^2}{(x^3+1)^2}$ **22** $\dfrac{x-2}{(x-1)^{\frac{3}{2}}}$

23 $\dfrac{2x-2-2x^2}{(x^2-1)^2}$

24 $\dfrac{1}{2(x^4+1)\sqrt{x}} - \dfrac{4x^{\frac{7}{2}}}{(x^4+1)^2}$ **25** $\dfrac{6e^x}{(3e^x+1)^2}$

26 $\dfrac{2x(2x+1)e^{x^2}\ln(2x+1) - 2e^{x^2}}{(2x+1)[\ln(2x+1)]^2}$

27 $\dfrac{2x(x^2-1) - 2x(x^2+1)\ln(x^2+1)}{(x^2+1)(x^2-1)^2}$

28 $\dfrac{2x^5+4x^3-2x}{(x^2+1)^2}$

29 $\dfrac{6x^2(4x^3-1)^{-\frac{1}{2}}\ln(2x-3)-2(2x-3)^{-1}(4x^3-1)^{\frac{1}{2}}}{[\ln(2x-3)]^2}$

30 $\dfrac{2x(x^2+1)^{-1}(e^x+1)-e^x\ln(x^2+1)}{(e^x+1)^2}$

31 ± 1

32 $\dfrac{x}{(1+x^2)^{\frac{1}{2}}}$; $0, \frac{3}{3}$ **33** (0,0), (3,0), $(\frac{9}{5}, 8.40)$

34 $\frac{1}{48}$ **35** $6+\ln 4$

Exercise 8C

1 $2\,\text{cm}^2\,\text{s}^{-1}$ **2** $0.0244\,\text{m}\,\text{s}^{-1}$

3 $112\,\text{cm}^2\,\text{s}^{-1}$

4 (a) $2.88\,\text{cm}^3\,\text{s}^{-1}$ (b) $2.88\,\text{cm}^2\,\text{s}^{-1}$

5 (a) $0.004\,77\,\text{cm}\,\text{s}^{-1}$ (b) $2.4\,\text{cm}^2\,\text{s}^{-1}$

6 7.78 **7** (a) e

(b) $A(1,0)$, $B(e, e^{-1})$, $C(e^{\frac{3}{2}}, \frac{3}{2}e^{-\frac{3}{2}})$

9 stationary values at $x=-\frac{1}{2}, x=1$;

$\max y=-2, \min y=-8$

10 $0.005\,15\,\text{cm}\,\text{s}^{-1}$

11 $\dfrac{1}{x-2}+\dfrac{1}{x+2}, -\dfrac{5}{18}, 0$ **12** $-\dfrac{6}{125}$

13 $\dfrac{e}{4}, \dfrac{e}{4}$

14 (a) $4x-y-5=0$ (b) $6x+y+3=0$

(c) $3x+2y+4=0$

15 (a) $x+3y-22=0$

(b) $x-5y-28=0$

(c) $3x-2y-4=0$

16 (b) $y=1\frac{1}{2}$ or -5 (c) (0,2), (7,1), 25 units ^{2}

17 $4y-3x=7$, $3y+4x=24$

18 (a) $A(-2,0), B(0,\sqrt{2})$

(b) $2y-x=3$

(c) $\dfrac{3\sqrt{5}}{2}$ (d) $1+\sqrt{2}$

20 $y=2x-e$, $2y+x=3e$, $\dfrac{5e}{2}, \dfrac{5e^2}{4}$

22 $3y+4x=10$, $3y+10x=95$

23 $(\frac{1}{2}, -\frac{1}{12})$ minimum, $(-\frac{1}{2}, \frac{1}{12})$ maximum

24 1,-6, 9

25 (a) tangent: $y=4x-1$,

normal: $x+4y=4\frac{1}{2}$ (b) $(9, -\frac{9}{8})$

26 (a) $(3, \frac{1}{12})$ (b) $(6, \frac{2}{27})$

27 (b) (1, 1) maximum, $(-1,-1)$ minimum

28 (a) $(e^{-\frac{1}{2}}, -\frac{1}{2}e^{-1})$ (b) $y=3ex-2e^2$

29 $\dfrac{2}{x+1}-\dfrac{1}{x^2+1}, 0$

30 $A=2, B=1, C=-1, 2, -3\frac{7}{8}$

Exercise 8D

1 $3\cos 3x$ **2** $\frac{1}{2}\cos\frac{1}{2}x$ **3** $-4\sin 4x$

4 $-\frac{2}{3}\sin\frac{2}{3}x$ **5** $2\sec^2 2x$ **6** $\frac{1}{4}\sec^2\frac{x}{4}$

7 $5\sec 5x\tan 5x$ **8** $-\frac{1}{2}\operatorname{cosec}\frac{x}{2}\cot\frac{x}{2}$

9 $-6\operatorname{cosec}^2 6x$

10 $\frac{1}{2}\sec\frac{x}{2}\tan\frac{x}{2}$ **11** $-\frac{3}{2}\operatorname{cosec}^2\frac{3x}{2}$

12 $-\frac{2}{3}\operatorname{cosec}\frac{2x}{3}\cot\frac{2x}{3}$ **13** $2\sin x\cos x$

14 $3\sin^2 x\cos x$

15 $\dfrac{\cos x}{2\sqrt{(\sin x)}}$ **16** $-4\cos^3 x\sin x$

17 $-5\cos^4 x\sin x$ **18** $-\frac{1}{3}\sin x(\cos x)^{\frac{-2}{3}}$

19 $2\tan x\sec^2 x$

20 $\dfrac{\sec^2 x}{2\sqrt{(\tan x)}}$ **21** $\dfrac{-\sec^2 x}{\tan^2 x}$ **22** $\dfrac{-2\cos x}{\sin^3 x}$

23 $\dfrac{16\sin x}{\cos^5 x}$ **24** $-2\operatorname{cosec}^2 x\cot x$

25 $4\sin 2x\cos 2x$ **26** $-6\cos 3x\sin 3x$

27 $6\tan^2 2x\sec^2 2x$ **28** $4\sec^2 2x\tan 2x$

29 $3\cos(3x+5)$

30 $-4\cos(2x-4)\sin(2x-4)$

31 $-4\tan(1-2x)\sec^2(1-2x)$
32 $-6\cot^3 x - \operatorname{cosec}^2 3x$
33 $-8\operatorname{cosec}^2 4x \cot 4x$ **34** 0
35 $2(\cot x - \operatorname{cosec} x)(-\operatorname{cosec}^2 x + \operatorname{cosec} x \cot x)$
36 $-2\cos 2x$
37 $2\sin x\cos^2 x - \sin^3 x$
38 $\cos^4 x - 3\sin^2 x\cos^2 x$
39 $\sec^3 x + \tan^2 x\sec x$
40 $\sec x\tan x\operatorname{cosec} x - \sec x\operatorname{cosec} x\cot x$
41 $\sin^2 x + 2x\sin x\cos x$
42 $2x\cos x - x^2\sin x$
43 $2x\tan 2x + 2x^2\sec^2 2x$
44 $\dfrac{\sin x - x\cos x}{\sin^2 x}$
45 $\dfrac{x\cos x - \sin x}{x^2}$
46 $\dfrac{2x\cos x + x^2\sin x}{\cos^2 x}$
47 $\dfrac{-3(x\sin 3x + \cos 3x)}{x^4}$
48 $2\sec^2 2x\sec 3x + 3\tan 2x\tan 3x\sec 3x$
49 $2\sin x\cos^4 x - 3\cos^2 x\sin^3 x$
50 $e^x(\sin x + \cos x)$
51 $\dfrac{2e^{2x}(\cos x + \sin x)}{\cos^3 x}$
52 $2\sin x\cos x\ln x + \dfrac{\sin^2 x}{x}$
53 $\dfrac{e^{2x}(3\sin x + \cos x)}{(\sin x + \cos x)^2}$
54 $\dfrac{2\cos 3x + 6(2x-5)\sin 3x\ln(2x-5)}{(2x-5)\cos^3 3x}$
57 $\frac{3}{7}$ **58** $y - 0.6 = 0.8(x - 0.644)$ is tangent; $y - 0.6 = -1.25(x - 0.644)$ is normal
59 2 **60** $\dfrac{-4\sin 2x}{(1-\cos 2x)^2}$
63 $\frac{\pi}{2} - 1$ **64** 1.00 **65** $y - 1 = -\frac{1}{2}x$
66 $\frac{2\sqrt{3}}{27}\pi a^3$
67 maximum at (1.11, 4.09); minimum at (4.25, −2190)

Exercise 8E

1 $\dfrac{1}{y}$ **2** $-\dfrac{x}{y}$ **3** $\dfrac{-y}{2x}$ **4** $\dfrac{-(x+y)}{x+3y}$
5 $\cot x\cot y$ **6** $-\dfrac{y}{x}$ **7** -1
8 $\dfrac{-y\ln y}{x\ln x}$ **9** $y(e^{-x} - \ln y)$
10 $\dfrac{-y}{x+2y}$ **11** $\dfrac{1}{3t}$ **12** $\dfrac{3t}{2}$ **13** $-\dfrac{2}{t^3}$
14 $-\dfrac{b}{a}\cot t$ **15** $\operatorname{cosec} t$
16 $\dfrac{\sin t + \cos t}{\cos t - \sin t}$ **17** $\dfrac{\sin t + t\cos t}{\cos t - t\sin t}$
18 $\dfrac{\cos\theta}{1+\sin\theta}$ **19** $1 + e^{-2u}$ **20** $\cot 2\theta$
21 $y - 4 = 4\ln 2(x-2)$, $y - 32 = 32\ln 2\,(x-5)$; $\dfrac{(38\ln 2 - 7)}{7\ln 2}$
22 $4y - 9x = 34$
23 tangent: $2y = 3 - 5x$; normal: $5y = 2x - 7, x^5 = y^2$
24 $3y = 12x + 4$ **25** (2, 3)
26 $y = 2x - 3$
27 $2t - t^2, 6y + 8x = 11$ **28** $-\frac{40}{33}$
29 (a) $10^x\ln 10$ (b) $2x(2^{x^2})\ln 2$ (c) $-5^{-x}\ln 5$
30 $2y + 2x - \pi = 0$

Review exercise 2

1 (a) $e^{-x}(\cos x - \sin x)$ (b) $2\tan x$

2 $2.88\,\text{cm}^2\,\text{s}^{-1}$

3 (i) (a) $x^3 + 3x + 3x^{-1} + x^{-3}$
(b) $x^5 + 5x^3 + 10x + 10x^{-1} + 5x^{-3} + x^{-5}$, 123 (ii) $A = 45$, $B = 945$, $C = 12\,285$

5 (a) 9.71 cm (b) 7.34 cm
(c) $18.1\,\text{cm}^2$

6 normal: $y + tx = 2t + t^3$,
tangent: $ty = x + t^2$

7 $(1, 1)$, $(3\frac{1}{3}, -3\frac{1}{3})$

8 $x^4 - 4x^5 + 6x^6 - 4x^7 + x^8$

9 (a) 4.51 cm (b) 37° (c) 33°

10 $e^{2x}(2\cos x - \sin x)$ (b) $\dfrac{2(x+1)}{(2x+1)^2}$

11 0.6, 0.8

12 (a) 8 (b) $\frac{7}{64}$

13 $1.5\,\text{cm}^2\,\text{s}^{-1}$

14 $\frac{3}{4}$, 12.5

15 (a) 59.1 m (b) 117 m (c) 090°

16 (b) 3, (c) $1 + \frac{7}{3}x + \frac{7}{3}x^2$

17 $-\frac{5\pi}{3}, -\frac{\pi}{3}, \pi$

18 (a) 17 cm (b) 18.4 cm (c) 22.4°
(d) 79.7 cm (e) 4.33 cm

19 (a) $(1, -2)$ (b) $3\sqrt{5}$ **20** $\sqrt{33}$

21 (a) $4\cos 4x$ (b) $x^{-3}e^{3x}(3x - 2)$

22 (a) $6e^{2t} - 1$ (b) $3e^{3t} - 2$
(c) $\dfrac{3e^{3t} - 2}{6e^{2t} - 1}$ (d) $\ln 4$

23 (a) 74.4° (b) 10.6 (c) $69.5\,\text{cm}^2$
(d) 8.34 cm

24 (a) 71.2° (b) 52.8 cm (c) 74.6°
(d) $519\,\text{cm}^2$

25 142 506, 142 506

26 (a) 15.6°, 195.6°
(b) $\frac{\pi}{6}, \frac{5\pi}{6}, \frac{7\pi}{6}, \frac{11\pi}{6}$ (c) 68.0 **27** 9, −19

28 (b) $x + y + 2 = 0$

29 $A = 44$, $B = 924$, $C = 12\,320$

30 (a) $x^{-2}(2x\cos 2x - \sin 2x)$
(b) $xe^{3x}(3x + 2)$ (c) $-6x\sin(3x^2)$

31 (a) 60° (b) 37.2° **32** $0.22\,\text{m}\,\text{s}^{-1}$

33 (a) $-\frac{1}{t^2}$ (c) $-\frac{1}{8}$

34 (i) (a) 23.1, 276.9 (b) 116.6, 296.6
(ii) (a) 2.26 (b) 69.4°

35 $\dfrac{E^2(r - R)}{(r + R)^3}$, $R = r$

36 (b) $\frac{1}{3}$ (c) 7

37 (a) 20.9 cm (b) $361\,\text{cm}^2$

38 (a) 125 km (b) 098° (c) $2380\,\text{km}^2$

39 $\frac{\pi}{3} \leqslant \theta \leqslant \frac{2\pi}{3}$ or $\frac{4\pi}{3} \leqslant \theta \leqslant \frac{5\pi}{3}$

40 (a) (i) $\dfrac{2}{x}$
(ii) $2x\sin 3x + 3x^2\cos 3x$ (b) $\frac{1}{7}$

41 $e^{2x}(2\cos x - \sin x)$; (b) $y = 2x + 1$

43 $\frac{4}{5}p\,\text{cm}^2\,\text{s}^{-1}$

44 (b) ± 8

45 (a) (i) $\dfrac{-(x^3 + 4)}{2x^3(1 + x^3)^{\frac{1}{2}}}$
(ii) $\dfrac{2\cos x - 3\sin x + 1}{(2 + \cos x)(3 - \sin x)}$

46 (a) 8 cm (b) 34° (c) $130\,\text{cm}^2$
(d) 15.3 cm

47 −4.61, 2.61

48 (a) $-\frac{63}{65}, \frac{56}{63}$ (b) $\frac{\pi}{6}, \frac{5\pi}{6}$

49 $a^6 + 6a^5b + 15a^4b^2 + 20a^3b^3 + 15a^2b^4 + 6ab^5 + b^6$, 0.982 134 46, 7.29×10^{-16}, 18

50 (a) $n(1 + y^2)$, $2n^2y(1 + y^2)$
(b) 0 and $4e^{-2}$

51 (a) 12.1 cm (b) 14 cm (c) 52°
(d) $159\,\text{cm}^3$

52 (i) (a) $x^2 + 2 + \frac{1}{x^2}$
(b) $x^4 + 4x^2 + 6 + \frac{4}{x^2} + \frac{1}{x^4}$, $p^4 - 4p^2 + 2$
(ii) $p = \frac{5}{3}, q = \frac{1}{3}$ or $p = \frac{-20}{7}, q = \frac{-4}{7}$

53 (a) 148.2, 211.8 (b) 323.1 (c) 19.5, 160.5, 228.6, 311.4

56 (i) (a) $\dfrac{2\sec x(\sec x + \tan x)}{\sec x - \tan x}$

(b) $e^{x\sin x}(\sin x + x\cos x)$ (ii) $\frac{1}{16}l^2$

58 (b) 3.79, −0.79

59 $1.25\,\text{cm s}^{-1}$

60 (a) $+\frac{1}{8}$ (b) $\sin\theta = \frac{3\sqrt{7}}{8}$

61 (a) 10.51 cm (b) 008° (c) 8 min (d) 1.46 km

62 (a) 9.59 cm (b) 40.8° (c) 11.5 cm (d) 96.4°

63 $1 + 8ax + 28a^2x^2$, $a = 1, b = -8$

64 (i) (a) 67.5, 157.5, 247.5, 337.5 (b) 63.4, 90, 243.4, 270 (c) 270 (ii) (a) 46.7° (b) 29.0°

65 (a) 0, 48.6°, 131.4°, 180°, 270°, 360° (b) $\frac{-5\pi}{3}, \frac{-\pi}{3}, \pi$

67 (a) $y = n(\sin 4nx - \sin 2nx)$

69 (a) 87.5 m, 153 m, (b) $12\ 900\,\text{m}^2$

70 (a) $k = \frac{3}{2}, p = 63, q = 189$ (b) 126

71 (a) $6\,\text{cm}^3$ (b) 4.24 cm (c) 62.1° (d) $9.60\,\text{cm}^2$ (e) 1.9 cm

72 (b) (i) $\frac{\pi}{4}, \frac{\pi}{2}, \frac{3\pi}{4}, \frac{5\pi}{4}, \frac{3\pi}{2}, \frac{7\pi}{4}$ (ii) $\frac{\pi}{2}$ (c) 33.6

73 3

74 (a) (i) 75,255 (ii) 60,135,300,315 (ii) 90,270 (c) $(x-1)^2 + 4y^2 = 1$

75 (a) (i) $e^{2x}(2\cos\pi x - \pi\sin\pi x)$

(ii) $\dfrac{2x}{1+x^2} - \dfrac{2}{1+2x}$ (iii) $\dfrac{4x}{\sqrt{(1+4x^2)}}$

76 164° **77** (a) −2 (b) 80

78 (a) (i) $\frac{1}{2}$ (ii) $\frac{1}{2}$ (iii) $\frac{1}{\sqrt{3}}$ (b) 23.8, 203.8

79 (a) $\frac{3}{5}$ (b) 64

80 4.3 km, 257°

81 (26, 9)

82 (a) $-\frac{1}{4}$ (b) 4

83 $1 - 14x + 84x^2, a = -15, b = 98$

84 (a) 16.6 cm (b) $123\,\text{cm}^2$ (c) 20.4 cm

85 (a) (i) 300, 420 (ii) 325, 505 (b) (i) $3a$ (ii) 19.5°, (iii) 35°

86 $3y = x + 14$, $y - \dfrac{10}{\sqrt{6}} = \frac{1}{3}(x - \sqrt{6})$,

$y + \dfrac{10}{\sqrt{6}} = \frac{1}{3}(x + \sqrt{6})$

87 (a) 2 at $\theta = 45$ (b) (i) 53.1° (ii) 62.1°

88 (a) 3 (b) 540

89 (a) 53° (c) 60° (d) 44°

90 $0.032\,\text{cm s}^{-1}$ **91** 1

92 (a) (i) $\frac{5}{2}(5x+2)^{-\frac{1}{2}}$ (ii) $3\sin^2 x\cos x$ (b) (i) 0.0007 (ii) 0

93 (a) $\frac{\pi}{3}, \frac{4\pi}{3}$ (b) (ii) −0.464, 2.68, −0.322, 2.820

94 (b) $0.016\,\text{cm s}^{-1}$ (c) $0.203\,\text{cm}^2\,\text{s}^{-1}$

95 $\frac{5\pi}{4}, \frac{7\pi}{4}, y - x + 4 = 0, y + x + 2 = 0$

96 (c) $0, \frac{\pi}{18}, \frac{5\pi}{18}, \frac{13\pi}{18}, \frac{17\pi}{18}, \pi$

97 (a) 20.74° (b) 252 m (c) 32.09° (d) 63.7 m

98 (a) (i) $\dfrac{2(1-x^2)}{(1+x^2)^2}$ (ii) $\dfrac{\sin x\cos x}{1+\sin^2 x}$

(b) $\dfrac{2+y-2x}{2y-x-1}, \frac{4}{3}, -\frac{1}{3}$

Exercise 9A

[The constant of integration is omitted in indefinite integration.]

1 $\frac{1}{4}\sin 4x$ **2** $-\frac{1}{3}\cos 3x$ **3** $-2\cos\frac{1}{2}x$

4 $\frac{2}{3}\sin\frac{3}{2}x$ **5** $\tan x$ **6** $-\frac{1}{3}\cot 3x$

7 $\frac{1}{2}\tan 2x$ **8** $\frac{1}{3}e^{3x-2}$ **9** $\frac{1}{2}\ln|2x-5|$

10 $\frac{1}{16}(4x-3)^4$ **11** $\frac{1}{5}\sin(5x+4)$

12 $\frac{1}{4}\cos(3-4x)$ **13** $-\frac{1}{6}(3-2x)^3$

14 $\dfrac{1}{2(3-2x)}$ **15** $\frac{1}{2}e^{2x} - 2x - \frac{1}{2}e^{-2x}$

16 $-2\cot\frac{1}{2}x$ **17** $\frac{1}{3}\sec 3x$

18 $-\frac{1}{2}\text{cosec } 2x$ **19** $\tan x - \frac{1}{3}x^3$

20 $x^2 - \frac{1}{2}\cos 2x$ **21** $\frac{1}{2}$ **22** $1\frac{1}{2}$ **23** 0

24 1 **25** $1 + \frac{1}{8}\pi^2$ **26** $2 - 2\sqrt{3}$

27 $\frac{1}{3}\ln\frac{5}{2}$ **28** 10 **29** $\frac{1}{2}(e - e^{-3})$ **30** 2

Exercise 9B

[The constant of integration is omitted in indefinite integration.]

1 $\frac{1}{2}x - \frac{1}{4}\sin 2x$ **2** $-\cot x - x$

3 $\frac{1}{2}\tan 2x - x$ **4** $\dfrac{3x}{2} + 2\sin x + \frac{1}{4}\sin 2x$

5 $3x + 4\cos x - \sin 2x$

6 $\dfrac{5x}{2} + \frac{1}{4}\sin 2x + \tan x$ **7** $\frac{1}{4}\ln\left|\dfrac{x-2}{x+2}\right|$

8 $\ln|x-3| - 2\ln|x-2|$

9 $\ln|x-1| - 2\ln|2x+1|$

10 $\frac{1}{2}\ln|2x+1| - \frac{1}{3}\ln|3x+1|$

11 $\ln\left|\dfrac{2x+1}{3x+1}\right|$ **12** $\ln\left|\dfrac{3+2x}{3-2x}\right|$

13 $A = 1, B = \frac{1}{2}, C = -\frac{1}{2}$;

$x + \frac{1}{2}\ln\left|\dfrac{x-1}{x+1}\right|$

14 $\frac{\pi}{3} + \frac{1}{4}$ **15** 0

16 $\frac{\sqrt{3}}{4} + \frac{1}{3}$ **17** (a) $\frac{\pi}{2} - 1$ (b) $\frac{\pi}{2} + 1$

Exercise 9C

[The constant of integration is omitted in indefinite integration]

1 $\frac{1}{4}\sin^4 x$ **2** $\frac{1}{3}\tan^3 x$ **3** $\frac{1}{8}(x^2+1)^4$

4 $\frac{1}{6}(x^4-1)^{\frac{3}{2}}$ **5** $(x^2-1)^{\frac{1}{2}}$ **6** $\frac{1}{2}\sec^2 x$

7 $\frac{1}{2}e^{x^2}$ **8** $\frac{1}{3}(\ln|x|)^3$

9 $x + 1 - 2\ln|x+1| - \dfrac{1}{x+1}$

10 $\frac{2}{3}(x-2)(x+1)^{\frac{1}{2}}$ **11** $6\frac{2}{3}$ **13** $-\frac{1}{12}$

14 (a) $e - 1$ (b) $\frac{1}{4}\sqrt{3}$ (c) $\ln\frac{5}{4}$

15 (a) $\frac{1}{2}\ln 2$ (b) $\frac{1}{2}\ln 2$

Exercise 9D

[The constant of integration is omitted in indefinite integration.]

1 $-e^{-x}(x+1)$ **2** $\frac{1}{3}xe^{3x} - \frac{1}{9}e^{3x}$

3 $-x\cos x + \sin x$

4 $\dfrac{x^2}{2}\ln|x| - \dfrac{x^2}{4}$

5 $x\ln|x-1| - x - \ln|x-1|$

6 $\frac{1}{3}x\sin 3x + \frac{1}{9}\cos 3x$

7 $\dfrac{(5x+1)(x-1))^5}{30}$ **8** $\frac{2}{15}(3x+2)(x-1)^{\frac{3}{2}}$

9 $e^x(x^2 - 2x + 2)$

10 $x^2\sin x + 2x\cos x - 2\sin x$

11 $-e^{-x}(x^2 + 2x + 2)$

12 $\dfrac{x^4}{16}(4\ln x - 1)$ **13** π

14 $\dfrac{\pi}{\sqrt{2}} + \dfrac{4}{\sqrt{2}} - 4$ **15** $\frac{2}{9}e^3 + \frac{1}{9}$ **16** $-\frac{1}{20}$

17 8.4 **18** $\frac{1}{9}(1 - 4e^{-3})$ **19** $e - 2$

20 $\frac{1}{2}(e^{\frac{\pi}{2}} + 1)$

Exercise 9E

[The constant of integration is omitted in indefinite integration.]

1 $\frac{1}{6}(4x-5)^{\frac{3}{2}}$ **3** $\frac{1}{4}\ln|4x+5|$

3 $x - 2\ln|x| - \dfrac{1}{x}$ **4** $\frac{1}{2}\sin^2 x$

5 $\frac{1}{3}\ln|\sec 3x|$ **6** $-\frac{1}{3}x\cos 3x + \frac{1}{9}\sin 3x$

7 $2x^{\frac{1}{2}} + \frac{2}{3}x^{\frac{3}{2}}$ **8** $x - \ln|x+1|$

9 $-\frac{1}{5}\cos^5 x$ **10** $3x\ln x - 3x$
11 $3\ln|x-1| - 2\ln|x|$
12 $-\frac{1}{2}(1+\tan x)^{-2}$ **13** $\frac{1}{2}x - \frac{1}{8}\sin 4x$
14 $\frac{1}{2}x^2 + 2x + 4\ln|x-2|$
15 $\frac{5}{2}x + \frac{3}{4}\sin 2x - \cos 2x$
16 $2e^{\frac{1}{2}x}(x^2 - 4x + 8)$ **17** $\frac{1}{4}\ln\left|\frac{x-2}{x+2}\right|$
18 $\frac{1}{18}\ln|9x^2+1|$ **19** $x + 2x^{-1} - \frac{1}{3}x^{-3}$
20 $\frac{1}{3}(2-3x)^{-1}$ **21** $-\frac{1}{5}\ln|4-5x|$
22 $\frac{1}{3}\ln|\sin 3x|$ **23** $-\frac{1}{2}\operatorname{cosec} 2x$
24 $-\frac{1}{3}\cot 3x - x$ **25** $\frac{1}{5}x\sin 5x + \frac{1}{25}\cos 5x$
26 $\frac{2}{3}(x+2)\sqrt{(x-1)}$
27 $-e^{-x}(x^2+2x+2)$
28 $-\frac{2}{3}\cos^3 x + \cos x$ **29** $-\frac{2}{3}\cos^3 x$
30 $\frac{1}{2}\sec 2x$ **31** $x + \ln|1+x^2|$
32 $\ln\left|\frac{x-4}{x-2}\right|$ **33** $\frac{1}{x} + \ln\left|\frac{x-1}{x}\right|$
34 $x - \frac{1}{2}\cot 2x$ **35** $x + 8\ln|x-4|$
36 $\frac{1}{2}\ln\left|\frac{x^2-1}{x^2}\right|$ **37** $\frac{1}{3}\ln|x^3+1|$
38 $\frac{1}{2}e^{2x} + \frac{1}{3}x^3 + 2xe^x - 2e^x$
39 $\frac{1}{16}x^4(4\ln x - 1)$ **40** $\frac{1}{2}e^{x^2}(x^2-1)$
41 $\frac{1}{3}\cos^3 x - \cos x$
42 $\sin x - \frac{1}{3}\sin^3 x, -\cos x + \frac{2}{3}\cos^3 x - \frac{1}{5}\cos^5 x$
43 $\frac{1}{3}\tan^3 x - \tan x + x$
44 $\frac{1}{3}\tan^3 x + \tan x, x + \cot x - \frac{1}{3}\cot^3 x$
45 (a) $-\frac{1}{10}\cos 10x - \frac{1}{2}\cos 2x$
(b) $-\frac{1}{3}\cos\frac{3x}{2} - \cos\frac{x}{2}$
46 $\frac{5}{2}\ln 3 - \ln 2$
48 $\frac{1}{3}\left[8 - \frac{8}{3\sqrt{3}}\right]$ **49** $\ln 2 + \frac{1}{2}\ln 13 - \frac{1}{2}\ln 20$
50 $\frac{1}{2}\ln\frac{5}{2}$

Exercise 9F

1 $\frac{\sqrt{3}-1}{2}$ **2** 1 **3** $2e^3$
4 $5\ln 5 - 2\ln 2 - 3$ **5** $\frac{\pi}{6}$ **6** $2 - \frac{\pi}{4}$
7 0.02 **8** $\frac{1}{2}(e^4 - e)$ **9** $\ln\frac{4}{3}$ **10** $\frac{26}{27}\sqrt{3}$
11 8π **12** 168π **13** $\frac{1}{2}\pi^2$ **14** $\pi\ln\frac{5}{2}$
15 $\frac{\pi}{4}e^2(3e^2-1)$ **16** $347\frac{1}{15}\pi$
17 $\frac{\pi}{4}(4-\pi)$ **18** $\pi[3(\ln 3)^2 - 6\ln 3 + 4]$
19 $\frac{\pi}{4}[15 + 8\ln 4]$ **20** $\frac{64}{15}\pi$
24 $60, \frac{16\,256\pi}{7}$
25 (a) $\sqrt{2}-1$ (b) $\frac{\pi}{4}(\pi-2)$
26 (a) $12\frac{2}{3}$ (b) $32\frac{1}{2}\pi$
27 (a) 27 (b) $\frac{1773\pi}{5}$
28 $16\ln\frac{16}{3}, \frac{208}{3}\pi$ **29** (a) $\frac{3\pi}{2}$ (b) 12π

Exercise 9G

1 $y = \frac{1}{2}e^{2x-1} + C$ **2** $2x + e^{1-2y} = C$
3 $4y = 2x + \sin 2x + C$ **4** $\tan y = x + C$
5 $2\ln y = x^2 + C$ **6** $e^{-y} + e^x = C$
7 $\sin y = x\ln x - x + C$
8 $y + 2 = C(x+1)$
9 $\ln y = \ln x + \frac{x^2}{2} + C$
10 $y^2 = 2\operatorname{cosec} x + C$ **11** $6y = 2x^3 + 3x^2$
12 $3y = \sin^3 x - 1$ **13** $\ln(3y+1) = 3x - 3$
14 $2e^{-y} + x^2 - 3 = 0$ **15** $2\sin y = \sec x$
16 $2\tan y = x - \sin x\cos x$
17 $\sin^2 y = \frac{11}{4} - \frac{2}{x}$
18 $y^2 = 2\tan x(\tan x + 1) + 5$
19 $y\cos y - \sin y = \frac{\pi}{2} - x\sin x - \cos x$
20 $\frac{y-1}{y+1} = \frac{2\sin^2 x}{3}$

22 (a) 19 minutes (b) 53.6°

23 26.6 days

24 $\ln y = \ln x + 1 - x^2$

25 $y = x\ln x - x + e + 1$

Exercise 10A

1 (a) (2, 4) (b) $(4\frac{1}{2}, 4\frac{1}{2})$ (c) $(2, 2\frac{1}{2})$
(d) $(1, 1\frac{1}{2})$ (e) $(-4, -4\frac{1}{2})$
(f) $(\frac{1}{2}, -5)$ (g) $(1\frac{1}{2}, 1\frac{1}{2})$
(h) $(-\frac{1}{2}, -4\frac{1}{2})$ (i) $(\frac{1}{2}, -6\frac{1}{2})$
(j) $(2\frac{3}{8}, \frac{3}{4})$

2 (a) $(8, -1)$ (b) (0, 6) (c) $(5, -1)$
(d) (13, 14) (e) (8, 10)
(f) $(-14, 13)$ (g) $(-9, -9)$
(h) $(-5, -5)$ (i) (7, 12)
(j) $(-5, -6)$

3 (a) $y = 2x - 2\frac{1}{2}$ (b) $y - 3x + 8 = 0$
(c) $11y + x + 15 = 0$ (d) $y - x = 1$
(e) $7x - 3y = 28$ (f) $2x - 3y = 6\frac{1}{2}$
(g) $4y + 2x = 5$ (h) $13y + x = 8$
(i) $4x + y = 16\frac{1}{2}$ (j) $3y + 7x + 4 = 0$

4 $(3\frac{1}{2}, -2), 6\frac{1}{2}$

6 (0,0),(-1, 7), (6,8)

Exercise 11A

1 $A' = \{4, 7, 8, 10, 12\}, B' = \{5, 7, 9, 11\}$
$A \cup B = \{4, 5, 6, 8, 9, 10, 11, 12\}$
$A' \cap B' = \{7\}$

2 (a) 17, 12 (b) 2

3 (a) 50 (b) 47 (c) 23 (d) 74

4 $P' = \{1, 3, 7, 10, 12\}, P \cap Q = \{2, 5, 6, 11\}$,
$P' \cap Q = \{1, 3, 7, 10, 12\}$

5 $A \cap B = \{x : -2 \leqslant x \leqslant 3\}$,
$A \cup B = \{x : -4 \leqslant x \leqslant 5\}$

6 $P \cap Q = \{x : 0 \leqslant x \leqslant 6\}$,
$P \cup Q = \{x : -4 \leqslant x \leqslant 8\}$

7 21

Exercise 11B

1 (a) $\frac{1}{6}$ (b) $\frac{1}{6}$ (c) $\frac{1}{3}$

2 (a) $\frac{1}{13}$ (b) $\frac{3}{13}$ (c) $\frac{7}{13}$

3 (a) $\frac{7}{13}$ (b) $\frac{1}{26}$

4 (a) $\frac{17}{22}$ (b) $\frac{10}{11}$ (c) $\frac{8}{11}$ (d) $\frac{21}{22}$

5 $HH, HT, TH, TT; \frac{1}{2}; \frac{5}{9}$

6 0.42, 0.58

Exercise 11C

1 0.74, 0.26 **2** 0.8 **3** 0.98, 0.417

4 0.15, 0.25 **5** (a) $\frac{1}{10}$ (b) $\frac{7}{10}$ (c) $\frac{1}{6}$

6 (a) $\frac{1}{3}$ (b) $\frac{1}{4}$ (c) $\frac{1}{6}$ **7** $\frac{7}{15}$ **8** $\frac{8}{9}$

9 $\frac{9}{14}, \frac{2}{5}, \frac{1}{3}$ **10** $\frac{3}{5}, \frac{3}{5}, \frac{4}{5}, \frac{2}{3}, \frac{2}{3}$

11 (a) 0.52 (b) 0.4

12 (a) 0.4 (b) 0.6

13 $\frac{24}{40}, \frac{19}{40}, \frac{26}{40}, \frac{17}{19}$; 0.018

14 (a) $\frac{7}{22}$ (b) $\frac{31}{66}$ **15** (a) $\frac{1}{4}$ (b) $\frac{1}{24}$

16 0.5 **17** (a) $\frac{19}{45}$ (b) $\frac{27}{52}$

18 $\frac{13}{60}$ **19** 0.2

20 (a) $\frac{1}{221}$ (b) $\frac{13}{17}$ (c) $\frac{32}{221}$

Exercise 12A

1 (b) (i) $p = 1, q = 3$ (ii) $r = 3, s = -1$
(c) 0.382 (d) 2.618

2 (b) 7 (c) 2.410

3 2.206, $x^3 - 2x^2 - 1 = 0$ **4** 0.275

5 1.47 **6** 0.877 **7** −0.38 **8** 0.1375

9 3.332 **10** 1.1411

11 1.83, $x = 3^{\frac{1}{x}}$

Review exercise 3

1 $\frac{\pi}{4}(e^2 - 1)$

2 (a) $x\sin x + \cos x + C$
(b) $\frac{y}{2} + \frac{1}{4}\sin 2y + C$;

$\frac{1}{4}\sin 2y + \frac{1}{2}y = x\sin x + \cos x + C$

3 (a) $ay\sqrt{2} + bx\sqrt{2} = 2ab$
(b) $by\sqrt{2} - ax\sqrt{2} = b^2 - a^2$; πab

4 (a) 0.2 (b) 0.03 (c) 0.32

5 (a) -7 (b) $x = 7, y = 2$

6 $\dfrac{1}{x} - \dfrac{2}{x+1} + \dfrac{1}{x+2}$

7 $x\tan x + \ln\cos x + C$

8 $\frac{4}{3}, \frac{16\pi}{15}$ **9** 1.78

10 (a) (1, 0) (b) $-\sin 2t$ (c) 2

11 (b) $\frac{1}{2}, \frac{11}{12}$ (c) $\frac{6}{7}$ (d) $\frac{4}{5}$ (e) $\frac{1}{2}$

12 (a) $(-\frac{1}{3}, \frac{1}{9}), (-1, -1)$
(b) $9x + 9y + 2 = 0, x + y + 2 = 0$
(c) (0, 0) minimum, $(-\frac{4}{3}, -\frac{8}{9})$ maximum
(d) $x = -\frac{2}{3}$

13 $y = 4 - ke^{-x^2}$ **14** $\frac{1}{4}\pi^2$

15 (a) $x \geqslant -p$
(b) $\dfrac{3t}{2}$ (c) 0 (e) $y = x\sqrt{2} - \dfrac{8\sqrt{2}}{9}$

16 $A = 3, B = 2, C = -2$; $\frac{3}{2}\ln\frac{11}{5} + 2\ln 4 - \frac{3}{2}$

17 $(y-3)^2 + (x-2)^2 = C$,
$(y-3)^2 + (x-2)^2 = 5$

18 $\frac{3}{2}, \frac{17}{2}, 2 \times 10^{-6}$

19 (a) $\frac{1}{5}$ (b) (i) 0.002 (ii) 0.013
(iii) 0.099; 0.650

20 (b) $-\dfrac{1}{t^2}$ (d) $-\frac{1}{8}$

23 (a) $\frac{2}{3}\sin^3\theta + C$
(b) $e^y - e^{-y} = \frac{2}{3}\sin^3 x - \frac{1}{12}$

24 (a) $\frac{7}{15}$ (b) $\frac{4}{15}$ (c) $\frac{4}{15}; \frac{7}{10}$

25 (a) $x = 1, y = 2$ (b) $(-\frac{1}{2}, 0), (0, -1)$

26 1.4, 8.6, 1.43, 8.61

27 (b) $y + 4ax = 2a^2 + 8$ (c) $-\frac{3}{4}$
(d) 16π

28 (b) $a = -\frac{3}{2}, \dfrac{dy}{dx} = 0$

29 $\dfrac{3}{7(3x+2)} - \dfrac{1}{7(x+3)}$;

$y = \dfrac{C(3x+2)}{x+3}$; $y = \dfrac{3(3x+2)}{x+3}$

30 (a) $\frac{1}{2}$ (d) $\frac{3}{4}, \frac{1}{2}$

31 (b) $2e^{\frac{1}{2}} - \frac{7}{3}$

32 $(-\frac{1}{2}, \frac{7}{4})$

33 $y^2 = 2x^2(\ln x + 2)$

34 0.509

35 (a) $\frac{2}{3}(x-1)^{\frac{3}{2}} + 4(x-1)^{\frac{1}{2}} + C$
(b) $\dfrac{x}{3}\sin 3x + \frac{1}{9}\cos 3x + C$; $\frac{\pi}{18} - \frac{1}{9}$

36 (a) $p = 3, q = 2$,
(b) $f^{-1} : x \mapsto e^{-x} + 2, x \in \mathbb{R}$

37 (a) (i) $\frac{1}{2}$ (ii) $\frac{1}{6}$ (iii) $\frac{5}{36}$ (iv) $\frac{1}{12}$
(b) $\frac{1}{6}, \frac{1}{2}$ (c) $\frac{1}{6}, \frac{3}{5}$

38 $\dfrac{1}{2(1+x)} = \dfrac{1}{2(3+x)}$;

$y = \sqrt{\left[\dfrac{8(1+x)}{3+x}\right]}$

39 (a) $0, \frac{\pi}{2}$ (b) $12\pi c^2$

40 (a) $\dfrac{dx}{dt} = k(p-x)^2$
(b) $x = \dfrac{p^2kt}{pkt+1}$

41 $\dfrac{16\sqrt{2}}{3}$ **42** 2.21

43 (a) (3, 0) (c) $x + y = 5$ (e) 1.125
(f) $\dfrac{34\pi}{5}$

44 (a) (i) 0.1 (ii) 0.05 (iii) 0.13

(b) (i) $\frac{1}{2}, \frac{1}{8}$ (ii) $\frac{1}{4}$

45 $g(x) \equiv 2x - 1 - x^2$;

$y = \frac{1}{2}\ln[2e^{-x}(1 + x^2) + C]$

46 $-\frac{1}{8}$

48 (a) $[0, \frac{10}{11}]$

(b) $f^{-1} : x \mapsto \dfrac{x}{1 - x}, 0 \leqslant x \leqslant \frac{10}{11}$

49 (a) $\dfrac{2r + 3}{(r + 1)(r + 2)}$ (c) $\frac{4}{5}$

50 (a) $\dfrac{t^2 + 1}{t^2 - 1}$ (c) $\dfrac{29\pi}{12}$

51 0.744, 0.794

52 (e) $\dfrac{8c^2}{3}$

53 $\dfrac{1}{2}\ln | 2y - 1 | = \frac{1}{2}\ln | \sin 2x | + C$;

$y = \frac{1}{2}(1 + \sqrt{\frac{3}{2}})$

54 (a) 0.15 (b) 0.75

(c) $3p(1 - p), p^2(3 - 2p)$ (d) $\frac{1}{2}$

55 $\frac{8}{15}$

57 (a) $\frac{2}{9}e^3$ (b) $\ln 2$ (c) $1 - \ln 2$

58 $\dfrac{3}{8(3t + 1)} - \dfrac{1}{8(t + 3)}$

59 (a) $(2, 2e^{-1})$ (b) $4 - 8e^{-1}$

(c) $\pi(2 - 10e^{-2})$

60 (b) $x_2 = 1.15$, $x_3 = 1.11$

61 (a) $x > 1$ (b) $x = 1$

(c) (3, 12) minimum, (−1, −4) maximum

62 $\dfrac{1}{1 + x} - \dfrac{1}{1 + 2x}$,

$-\frac{1}{2}e^{-2y} = C + \ln\dfrac{1 + x}{\sqrt{(1 + 2x)}}$

63 (a) $\frac{9}{20}$ (b) (i) $\frac{33}{100}$ (ii) $\frac{7}{11}$

64 (a) $\frac{61}{192}$ (b) $1 - \frac{\pi}{4}$ (c) π (d) $\ln\frac{4}{3}$

65 42

66 −3

67 $\dfrac{2}{x + 2} + \dfrac{1}{x - 1}$; $\ln 2$

68 (a) (i) $\frac{1}{6}$ (ii) $\frac{1}{10}$ (b) $\frac{7}{30}$

69 (a) 23.6 litres (b) 17.1 cm

70 (a) $\dfrac{1}{3x} - \dfrac{1}{3(x + 3)}$ (b) $\frac{\pi}{8} - \frac{1}{4}$

(c) $\dfrac{a^3}{3}$

71 $1 < x < 3$

72 (a) $\dfrac{2}{x + 2} - \dfrac{1}{x + 1}$; $\ln\frac{4}{3}$ (b) $\frac{8}{21}$

73 (a) (i) $\frac{7}{30}$ (ii) $\frac{1}{10}$ (b) (i) 0.0106

(ii) 0.000 266

74 $(0, 0), (-1, -\frac{1}{2}), (2, \frac{2}{5})$

75 $-\frac{1}{3}e^{-3y} = xe^x - e^x - \frac{1}{3}$

77 (a) $\dfrac{2}{x + 1} - \dfrac{1}{x + 2} - \dfrac{1}{x + 3}$

(b) $\ln\left[\dfrac{k(x + 1)^2}{(x + 2)(x + 3)}\right]$

(e) $\ln y + A = 2\ln\left(\dfrac{e^x}{e^x + 1}\right) - \dfrac{1}{2}$

$\ln\left(\dfrac{e^x}{e^x + 2}\right) - \frac{1}{3}\left(\ln\dfrac{e^x}{e^x + 3}\right)$

78 $\dfrac{5 + 3x}{x - 2}$, $x \in \mathbb{R}, x \neq 2$

79 range $(0, 2a]$, asymptote $y = 2a$;

$2a^2[2 + \ln\frac{1}{3}]$

80 (a) $\frac{7}{15}; \frac{59}{75}; \frac{25}{39}$

81 (a) $A = 1,\ B = -1,\ C = 2$

82 $\dfrac{2}{x+2} - \dfrac{1}{x+1}; y = \dfrac{x^2+x+1}{3(x+1)}$

83 (b) $x = \dfrac{4y-7}{y-2}$ (c) $x = 4, y = 2$

84 $\dfrac{3}{1+3x} - \dfrac{1}{1+x}; y+2 = \dfrac{1+3x}{1+x}$

85 (i) $\frac{2}{7}$ (ii) $\frac{5}{9}$ (b) (i) 0.4, 0.25
(ii) 0.05, 0.02

86 0.922

87 $e^y = \frac{1}{4}(5 - 2xe^{-2x} - e^{-2x})$

88 (a) $P(B) = \frac{1}{2}, P(c) = \frac{1}{6}$ (b) $\frac{2}{3}$
(c) $P(A \cap B) = \frac{1}{6}, P(A \cap C) = \frac{1}{12}$

89 (a) (2, 4), $(-2, 4), (0, 8)$ (b) $\frac{64}{3}$
(c) 48π

90 $2y + 3x = 135; 72.9$

91 $0 < x < 2$

92 $\dfrac{3}{x-2} + \dfrac{2x+3}{x^2+3x+3}; -1.23$

93 $\dfrac{1}{y} = xe^{-x} + C$

94 (b) (i) 1.33 (ii) 1.58

95 (b) $\dfrac{1}{2}\left[\dfrac{1}{x+1} + \dfrac{1+x}{x^2+1}\right]$

96 $-3 < x < 4$

97 $\dfrac{1}{2}\ln 2 + 1; \dfrac{5\pi}{3} - \dfrac{\pi^2}{4}$

98 (a) (0, 4) (b) $\dfrac{128}{15}$ (c) $\dfrac{64\pi}{3}$

99 (b) 0.009, 0.012, 0.0075 (d) 0.0285
(e) 0.316 (f) 0.454

Examination style paper P2

1 4 or 12

2 (a) $2x + y - 3 = 0; (2, -1)$ (b) 2:7

3 (a) $1 + 11kx + 5.5k^2x^2 + 16.5k^3x^3$
(b) $-\frac{1}{3}$ (c) $-\frac{55}{9}$

4 0.34, 2.80, 3.67, 5.76

5 (b) 1.749 03

6 (a) 18.6 min (b) 0.125

7 (b) $\dfrac{4\pi}{3}$

8 (a) $\dfrac{dx}{dt} = kx$ (b) $x = ce^{kt}$

(d) 40.5 hours

9 (a) 13

(b) $\frac{1}{105}\left[\dfrac{5}{x+2} + \dfrac{27}{3x-1} - \dfrac{28}{2x+1}\right]$

(c) -6.75

10 (a) $P(-3, 3), Q(3, 0)$ (b) $(0, \frac{15}{4})$ (d) 9

List of symbols and notation

The following symbols and notation are used in the London modular mathematics examinations:

$\{ \quad \}$	the set of
$\mathrm{n}(A)$	the number of elements in the set A
$\{x : \quad \}$	the set of all x such that
$\in$	is an element of
$\notin$	is not an element of
$\varnothing$	the empty (null) set
$\mathscr{E}$	the universal set
$\cup$	union
$\cap$	intersection
$\subset$	is a subset of
A'	the complement of the set A
PQ	operation Q followed by operation P
$\mathrm{f} : A \to B$	f is a function under which each element of set A has an image in set B
$\mathrm{f} : x \mapsto y$	f is a function under which x is mapped to y
$\mathrm{f}(x)$	the image of x under the function f
f^{-1}	the inverse relation of the function f
fg	the function f of the function g
○—○—○	open interval on the number line
●—●—●	closed interval on the number line
$\mathbb{N}$	the set of positive integers and zero, $\{0, 1, 2, 3, \ldots\}$
$\mathbb{Z}$	the set of integers, $\{0, \pm 1, \pm 2, \pm 3, \ldots\}$
$\mathbb{Z}^+$	the set of positive integers, $\{1, 2, 3, \ldots\}$
$\mathbb{Q}$	the set of rational numbers
$\mathbb{Q}^+$	the set of positive rational numbers, $\{x : x \in \mathbb{Q}, x > 0\}$
$\mathbb{R}$	the set of real numbers
$\mathbb{R}^+$	the set of positive real numbers, $\{x : x \in \mathbb{R}, x > 0\}$
$\mathbb{R}_0^+$	the set of positive real numbers and zero, $\{x : x \in \mathbb{R}, x \geqslant 0\}$
$\mathbb{C}$	the set of complex numbers
$\sqrt{\ }$	the positive square root
$[a, b]$	the interval $\{x : a \leqslant x \leqslant b\}$
$(a, b]$	the interval $\{x : a < x \leqslant b\}$
(a, b)	the interval $\{x : a < x < b\}$

$\|x\|$	the modulus of $x = \begin{cases} x \text{ for } x \geqslant 0 \\ -x \text{ for } x < 0 \end{cases}, x \in \mathbb{R}$
$\approx$	is approximately equal to
A^{-1}	the inverse of the non-singular matrix A
A^T	the transpose of the matrix A
det A	the determinant of the square matrix A
$\sum_{r=1}^{n} f(r)$	$f(1) + f(2) + \ldots + f(n)$
$\prod_{r=1}^{n} f(r)$	$f(1)f(2)\ldots f(n)$
$\binom{n}{r}$	the binomial coefficient $\begin{cases} \dfrac{n!}{r!(n-r)!} \text{ for } n \in \mathbb{Z}^+ \\ \dfrac{n(n-1)\ldots(n-r+1)}{r!} \text{ for } n \in \mathbb{Q} \end{cases}$
$\exp x$	e^x
$\ln x$	the natural logarithm of $x, \log_e x$
$\lg x$	the common logarithm of $x, \log_{10} x$
arcsin	the inverse function of sin with range $[-\pi/2, \pi/2]$
arccos	the inverse function of cos with range $[0, \pi]$
arctan	the inverse function of tan with range $(-\pi/2, \pi/2)$
arsinh	the inverse function of sinh with range $\mathbb{R}$
arcosh	the inverse function of cosh with range $\mathbb{R}_0^+$
artanh	the inverse function of tanh with range $\mathbb{R}$
$f'(x), f''(x), f'''(x)$	the first, second and third derivatives of $f(x)$ with respect to x
$f^{(r)}(x)$	the rth derivative of $f(x)$ with respect to x
$\dot{x}, \ddot{x}, \ldots$	the first, second, . . . derivatives of x with respect to t
z	a complex number, $z = x + iy = r(\cos\theta + i\sin\theta) = re^{i\theta}$
Re z	the real part of z, Re $z = x = r\cos\theta$
Im z	the imaginary part of z, Im $z = y = r\sin\theta$
z^*	the conjugate of $z, z^* = x - iy = r(\cos\theta - i\sin\theta) = re^{-i\theta}$
$\|z\|$	the modulus of $z, \|z\| = \sqrt{(x^2 + y^2)} = r$
arg z	the principal value of the argument of z, arg $z = \theta$, where $\left.\begin{matrix}\sin\theta = y/r \\ \cos\theta = x/r\end{matrix}\right\} -\pi < \theta \leqslant \pi$
a	the vector **a**
$\overrightarrow{AB}$	the vector represented in magnitude and direction by the directed line segment AB
â	a unit vector in the direction of **a**
i,j,k	unit vectors in the directions of the cartesian coordinate axes
$\|\mathbf{a}\|$	the magnitude of **a**
$\|\overrightarrow{AB}\|$	the magnitude of $\overrightarrow{AB}$
a.b	the scalar product of **a** and **b**
a × **b**	the vector product of **a** and **b**

A'	the complement of the event A
$\mathrm{P}(A)$	probability of the event A
$\mathrm{P}(A\|B)$	probability of the event A conditional on the event B
$\mathrm{E}(X)$	the mean (expectation, expected value) of the random variable X
$X, Y, R,$ etc.	random variables
$x, y, r,$ etc.	values of the random variables X, Y, R, etc.
$x_1, x_2 \ldots$	observations
$f_1, f_2, \ldots$	frequencies with which the observations $x_1, x_2, \ldots$ occur
$\mathrm{p}(x)$	probability function $\mathrm{P}(X = x)$ of the discrete random variable X
$p_1, p_2, \ldots$	probabilities of the values $x_1, x_2, \ldots$ of the discrete random variable X
$\mathrm{f}(x), \mathrm{g}(x), \ldots$	the value of the probability density function of a continuous random variable X
$\mathrm{F}(x), \mathrm{G}(x), \ldots$	the value of the (cumulative) distribution function $\mathrm{P}(X \leqslant x)$ of a continuous random variable X
$\mathrm{Var}(X)$	variance of the random variable X
$\mathrm{B}(n, p)$	binomial distribution with parameters n and p
$\mathrm{N}(\mu, \sigma^2)$	normal distribution with mean μ and variance σ^2
μ	population mean
σ^2	population variance
σ	population standard deviation
$\bar{x}$	sample mean
s^2	unbiased estimate of population variance from a sample, $$s^2 = \frac{1}{n-1}\sum(x - \bar{x})^2$$
ϕ	probability density function of the standardised normal variable with distribution N(0, 1)
Φ	corresponding cumulative distribution function
α, β	regression coefficients
ρ	product-moment correlation coefficient for a population
r	product-moment correlation coefficient for a sample
$\sim p$	not p
$p \Rightarrow q$	p implies q (if p then q)
$p \Leftrightarrow q$	p implies and is implied by q (p is equivalent to q)

Index